by John T. Moore, Ed.D

Wiley Publishing, Inc.

Chemistry For Dummies®

Published by
Wiley Publishing, Inc.
909 Third Avenue
New York, NY 10022
www.wiley.com

Published simultaneously in Canada

Library of Congress Control Number: 2002106030

ISBN: 0-7645-5430-1

Manufactured in the United States of America

10 9 8 7 6 5 4 3

1B/RV/RR/QS/IN

Chemistry For Dummies®

...For Dummies™ Bestselling Book Series

Cheat Sheet

PERIODIC TABLE OF THE ELEMENTS

1 H Hydrogen 1.00797																	2 He Helium 4.0026
3 Li Lithium 6.939	4 Be Beryllium 9.0122											5 B Boron 10.811	6 C Carbon 12.01115	7 N Nitrogen 14.0067	8 O Oxygen 15.9994	9 F Flourine 18.9984	10 Ne Neon 20.183
11 Na Sodium 22.9898	12 Mg Magnesium 24.312											13 Al Aluminum 26.9815	14 Si Silicon 28.086	15 P Phosphorus 30.9738	16 S Sulfur 32.064	17 Cl Chlorine 35.453	18 Ar Argon 39.948
19 K Potassium 39.102	20 Ca Calcium 40.08	21 Sc Scandium 44.956	22 Ti Titanium 47.90	23 V Vanadium 50.942	24 Cr Chromium 51.996	25 Mn Manganese 54.9380	26 Fe Iron 55.847	27 Co Cobalt 58.9332	28 Ni Nickel 58.71	29 Cu Copper 63.546	30 Zn Zinc 65.37	31 Ga Gallium 69.72	32 Ge Germanium 72.59	33 As Arsenic 74.9216	34 Se Selenium 78.96	35 Br Bromine 79.904	36 Kr Krypton 83.80
37 Rb Rubidium 85.47	38 Sr Strontium 87.62	39 Y Yttrium 88.905	40 Zr Zirconium 91.22	41 Nb Niobium 92.906	42 Mo Molybdenum 95.94	43 Tc Technetium (99)	44 Ru Ruthenium 101.07	45 Rh Rhodium 102.905	46 Pd Palladium 106.4	47 Ag Silver 107.868	48 Cd Cadmium 112.40	49 In Indium 114.82	50 Sn Tin 118.69	51 Sb Antimony 121.75	52 Te Tellurium 127.60	53 I Iodine 126.9044	54 Xe Xenon 131.30
55 Cs Cesium 132.905	56 Ba Barium 137.34	57 La Lanthanum 138.91	72 Hf Hafnium 179.49	73 Ta Tantalum 180.948	74 W Tungsten 183.85	75 Re Rhenium 186.2	76 Os Osmium 190.2	77 Ir Iridium 192.2	78 Pt Platinum 195.09	79 Au Gold 196.967	80 Hg Mercury 200.59	81 Tl Thallium 204.37	82 Pb Lead 207.19	83 Bi Bismuth 208.980	84 Po Polonium (210)	85 At Astatine (210)	86 Rn Radon (222)
87 Fr Francium (223)	88 Ra Radium (226)	89 Ac Actinium (227)	104 Rf Rutherfordium (261)	105 Db Dubnium (262)	106 Sg Seaborgium (266)	107 Bh Bohrium (264)	108 Hs Hassium (269)	109 Mt Meitnerium (268)	110 Uun Ununnilium (269)	111 Uuu Unununium (272)	112 Uub Ununbium (277)	113 Uut §	114 Uuq Ununquadium (285)	115 Uup §	116 Uuh Ununhexium (289)	117 Uus §	118 Uuo Ununoctium (293)

Lanthanide Series	58 Ce Cerium 140.12	59 Pr Praseodymium 140.907	60 Nd Neodymium 144.24	61 Pm Promethium (145)	62 Sm Samarium 150.35	63 Eu Europium 151.96	64 Gd Gadolinium 157.25	65 Tb Terbium 158.924	66 Dy Dysprosium 162.50	67 Ho Holmium 164.930	68 Er Erbium 167.26	69 Tm Thulium 168.934	70 Yb Ytterbium 173.04	71 Lu Lutetium 174.97
Actinide Series	90 Th Thorium 232.038	91 Pa Protactinium (231)	92 U Uranium 238.03	93 Np Neptunium (237)	94 Pu Plutonium (242)	95 Am Americium (243)	96 Cm Curium (247)	97 Bk Berkelium (247)	98 Cf Californium (251)	99 Es Einsteinium (254)	100 Fm Fermium (257)	101 Md Mendelevium (258)	102 No Nobelium (259)	103 Lr Lawrencium (260)

§ Note: Elements 113, 115, and 117 are not known at this time, but are included in the table to show their expected positions.

Chemistry For Dummies®

Bonding

- In bonding, atoms lose, gain, or share electrons in order to have the same number of electrons as the nearest noble gas.
- Metal + nonmetal = ionic bond
- Nonmetal + nonmetal = covalent bond
- Electron filling pattern: 1s, 2s, 3s, 3p, 4s, 3d, 4p, 5s, 4d, 5p, 6s, 4f, 5d, 6p, 7s, 5f

Isotope Representation

$${}^{A}_{Z}X$$

X = element symbol, Z= atomic number (# of protons), A= mass number (# of protons + # of neutrons)

Useful Conversions And Metric Prefixes

Temperature conversions:

- $^{\circ}F = \frac{9}{5}(^{\circ}C) + 32$
- $^{\circ}C = 5/9(^{\circ}F - 32)$
- $K = ^{\circ}C + 273$

Metric/English conversions:

- 2.54 cm = 1 in
- 454 g = 1 lb
- 0.946L = 1 qt

Pressure conversion: 1atm = 760 mmHg = 760 torr

Common metric prefixes:

- *milli-* = 0.001
- *centi-* = $\frac{1}{100}$
- *kilo-* = 1000

Mole Concept

1 mole = 6.022×10^{23} particles/mol = formula weight expressed in grams

Solution Concentrations

weight/weight (w/w) % = (grams solute/grams solution) × 100

molarity (M) = moles solute/liters solution

parts-per-million (ppm) = grams solute/1,000,000 grams solution = mg/l

Acids and Bases

An *acid* is an H^+ donor, and a *base* is an H^+ acceptor.

$pH = -\log[H^+]$; $[H^+] = 10^{-pH}$

pH = 7 is *neutral;* pH less than 7 is *acidic;* pH greater than 7 is *basic.*

Redox

Oxidation is loss of electrons; *reduction* is gain of electrons.

Gas Laws

Combined Gas Law:
$(P_1V_1)/T_1 = (P_2V_2)/T_2$ (T must be in kelvins)

Ideal Gas Law:
$PV = nRT$ (where R = 0.0821 l·atm/K·mol)

Item 5430-1.

For more information about Wiley Publishing, call 1-800-762-2974.

For Dummies: Bestselling Book Series for Beginners

About the Author

John T. Moore, Ed.D grew up in the foothills of western North Carolina. He attended the University of North Carolina-Asheville where he received his bachelor's degree in chemistry. He earned his Master's degree in chemistry from Furman University in Greenville, South Carolina. After a stint in the United States Army, he decided to try his hand at teaching. In 1971, he joined the chemistry faculty of Stephen F. Austin State University in Nacogdoches, Texas, where he still teaches chemistry. In 1985, he started back to school part-time and in 1991 received his Doctorate in Education from Texas A&M University.

John's area of specialty is chemical education. He has developed several courses for students planning on teaching chemistry at the high school level. In the early 1990s, he shifted his emphasis to training elementary education majors and in-service elementary teachers in hands-on chemical activities. He has received four Eisenhower grants for professional development of elementary teachers and for the last five years has been the co-editor (along with one of his former students) of the "Chemistry for Kids" feature of *The Journal of Chemical Education.*

Although teaching has always been foremost in his heart, John found time to work part-time for almost five years in the medical laboratory of the local hospital and has been a consultant for a textbook publisher. He is active in a number of local, state, and national organizations, such as the Nacogdoches Kiwanis Club and the American Chemical Society.

John lives in the Piney Woods of East Texas with his wife Robin and their three dogs and cat. He enjoys brewing his own beer and mead. And he loves to cook. In fact, he and his wife have recently bought a gourmet food & kitchen shop called *The Cottage.* ("I was spending so much there it was cheaper to just go ahead and buy the store.") His two boys, Jason and Matt, remain in the mountains of North Carolina.

Dedication

This book is dedicated to those children, past, present, and future, who will grow to love chemistry, just as I have done. You may never make a living as a chemist, but I hope that you will remember the thrill of your experiments and will pass that enjoyment on to your children. This book is also dedicated to my wife Robin, who took time out of her busy campaign schedule to encourage me and have faith in me during those times when I didn't have much faith in myself. This time you were the wind beneath my wings. And it's dedicated to my close friends who helped keep me grounded in reality, especially Sue Mary, who always had just the right quote from a Jimmy Buffett song to lift me up, and Jan, whose gift of a tie-dyed lab coat kept me from taking myself too seriously. And finally, this book is dedicated to my sons, Matthew and Jason, and my wonderful daughter-in-law, Sara. I love you guys.

Author's Acknowledgments

I would not have had the opportunity to write this book without the encouragement of my agent, Grace Freedson. She took the time to answer my constant e-mails and teach me a little about the publishing business. I owe many thanks to the staff at Wiley, especially acquisitions editor Greg Tubach, project editor Tim Gallan, copy editor Greg Pearson, and technical reviewer Bill Cummings, for their comments and help with this project. Special thanks also to the MMSEC elementary teachers of Nacogdoches ISD, especially Jan, Derinda, and Sondra. You made me a better teacher, and you showed your support and concern for me as I was writing this book. Special thanks also to Andi and The Cottage Girls, Kim, Jonell, Stephanie, Amanda, and Laura, for taking such good care of the shop while I was involved in this project. Thanks to my colleagues who kept asking me how it was going and especially Rich Langley, who was always there to point out my procrastination. And let me offer many thanks to all my students over the past thirty years, especially the ones who became teachers. I've learned from you and I hope that you've learned from me.

Publisher's Acknowledgments

We're proud of this book; please send us your comments through our Dummies online registration form located at www.dummies.com/register/.

Some of the people who helped bring this book to market include the following:

Acquisitions, Editorial, and Media Development

Senior Project Editor: Tim Gallan

Acquisitions Editors: Greg Tubach, Kathy Cox

Copy Editors: Greg Pearson, Sandy Blackthorn

Technical Editor: Bill Cummings

Editorial Manager: Christine Meloy Beck

Editorial Assistant: Melissa Bennett

Cover Photos: © Chris Salvo/ Getty Imasges/FPG

Cartoons: Rich Tennant, www.the5thwave.com

Production

Project Coordinator: Erin Smith

Layout and Graphics: Melanie DesJardins, Carrie Foster, Joyce Haughey, LeAndra Johnson, Barry Offringa, Laurie Petrone, Heather Pope, Jacque Schneider, Betty Schulte, Erin Zeltner

Illustrators: Kelly Hardesty, Rashell Smith, Kathie Schutte

Proofreaders: Laura Albert, John Bitter, John Tyler Connoley, Andy Hollandbeck, Arielle Carole Mennelle

Indexer: Sherry Massey

Publishing and Editorial for Consumer Dummies

Diane Graves Steele, Vice President and Publisher, Consumer Dummies

Joyce Pepple, Acquisitions Director, Consumer Dummies

Kristin A. Cocks, Product Development Director, Consumer Dummies

Michael Spring, Vice President and Publisher, Travel

Brice Gosnell, Publishing Director, Travel

Suzanne Jannetta, Editorial Director, Travel

Publishing for Technology Dummies

Andy Cummings, Vice President and Publisher, Dummies Technology/General User

Composition Services

Gerry Fahey, Vice President of Production Services

Debbie Stailey, Director of Composition Services

Contents at a Glance

Table of Contents

Introduction

You've passed the first hurdle in understanding a little about chemistry: You've picked up this book. I imagine that a large number of people looked at the title, saw the word *chemistry,* and bypassed it like it was covered in germs.

I don't know how many times I've been on vacation and struck up a conversation with someone, and the dreaded question is asked: "What do you do?"

"I'm a teacher," I reply.

"Really? And what do you teach?"

I steel myself, grit my teeth, and say in my most pleasant voice, "Chemistry."

I see The Expression, followed by, "Oh, I never took chemistry. It was too hard." Or "You must be smart to teach chemistry." Or "Goodbye!"

I think a lot of people feel this way because they think that chemistry is too abstract, too mathematical, too removed from their real lives. But in one way or another, all of us do chemistry.

Remember as a child making that baking soda and vinegar volcano? That's chemistry. Do you cook or clean or use fingernail polish remover? All of that is chemistry. I never had a chemistry set as a child, but I always loved science. My high school chemistry teacher was a great biology teacher but really didn't know much chemistry. But when I took my first chemistry course in college, the labs hooked me. I enjoyed seeing the colors of the solids coming out of solutions. I enjoyed *synthesis,* making new compounds. The idea of making something nobody else had ever made before fascinated me. I wanted to work for a chemical company, doing research, but then I discovered my second love: teaching.

Chemistry is sometimes called the central science (mostly by chemists) because to have a good understanding of biology or geology or even physics, you must have a good understanding of chemistry. Ours is a chemical world, and I hope that you enjoy discovering the chemical nature of it — and that afterward, you won't find the word *chemistry* so frightening.

About This Book

My goal with this book is not to make you into a chemistry major. My goal is simply to give you a basic understanding of some chemical topics that commonly appear in high school or college introductory chemistry courses. If you're taking a course, use this book as a reference in conjunction with your notes and textbook.

Simply watching people play tennis, no matter how intently you watch them, will not make you a tennis star. You need to practice. And the same is true with chemistry. It's *not* a spectator sport. If you're taking a chemistry course, then you need to practice and work on problems. I show you how to work certain types of problems — gas laws, for example — but use your textbook for practice problems. It's work, yes, but it really can be fun.

How to Use This Book

I've arranged this book's content in a logical (at least to me) progression of topics. But this doesn't mean you have to start at the beginning and read to the end of the book. I've made each chapter self-contained, so feel free to skip around. Sometimes, though, you'll get a better understanding if you do a quick scan of a background section as you're reading. To help you find appropriate background sections, I've placed "see Chapter XX for more information" cross-references here and there throughout the book.

Because I'm a firm believer in concrete examples, I've also included lots of illustrations and figures with the text. They really help in the understanding of chemistry topics. And to help you with the math, I've broken up problems into steps so that it's easy to follow exactly what I'm doing.

I've kept the material to the bare bones, but I've included a few sidebars. They're interesting reading (at least to me) but not really necessary for understanding the topic at hand, so feel free to skip them. This is *your* book; use it any way you want.

Assumptions (And You Know What They Say about Assumptions!)

I really don't know why you bought this book (or will buy it — in fact, if you're still in the bookstore and *haven't* bought it yet, buy two and give one as a gift), but I assume that you're taking (or retaking) a chemistry course or preparing to take a chemistry course. I also assume that you feel relatively comfortable with arithmetic and know enough algebra to solve for a single unknown in an equation. And I assume that you have a scientific calculator capable of doing exponents and logarithms.

And if you're buying this book just for the thrill of finding out about something different — with no plan of ever taking a chemistry course — I applaud you and hope that you enjoy this adventure.

How This Book Is Organized

I've organized the topics in a logical progression — basically the same way I organize my courses for non-science and elementary education majors. I've included a couple chapters on environmental chemistry — air and water pollution — because those topics appear so often in the news. And I've included some material in appendixes that I think might help you — especially Appendix C on the unit conversion method of working problems.

Following is an overview of each part of the book.

Part I: Basic Concepts of Chemistry

In this part, I introduce you to the really basic concepts of chemistry. I define chemistry and show you where it fits among the other sciences (in the center, naturally). I show you the chemical world around you and explain why chemistry should be important to you. I also show you the three states of matter and talk about going from one state to another — and the energy changes that occur.

Besides covering the macroscopic world of things like melting ice, I cover the microscopic world of atoms. I explain the particles that make up the atom — protons, neutrons, and electrons — and show you where they're located in the atom.

I discuss how to use the Periodic Table, an indispensable tool for chemists. And I introduce you to the atomic nucleus, which includes discussions about radioactivity, carbon-14 dating, fission and fusion nuclear reactors, and even cold fusion. You'll be absolutely *glowing* after reading this stuff.

Part II: Blessed Be the Bonds That Tie

In this part, you get into some really good stuff: bonding. I show you how table salt is made in Chapter 6, which covers ionic bonding, and I show you the covalent bonding of water in Chapter 7. I explain how to name some ionic compounds and how to draw Lewis structural formulas of some covalent ones. I even show you what some of the molecules look like. (Rest assured that I define all these techno-buzzwords on the spot, too.)

I also talk about chemical reactions in this part. I give some examples of the different kinds of chemical reactions you may encounter and show you how to balance them. (You really didn't think I could resist that, did you?) I cover factors that affect the speed of reactions and why chemists rarely get as much product formed as expected. And I discuss electron transfer in the redox reactions involved in electroplating and flashlight batteries. I hope that you'll see the light in this part!

Part III: The Mole: The Chemist's Best Friend

In this part, I introduce the mole concept. Odd name, yes. But the mole is central to your understanding of chemical calculations. It enables you to figure the amount of reactants needed in chemical reactions and the amount of product formed. I also talk about solutions and how to calculate their concentrations. And I explain why I leave the antifreeze in my radiator during the summer and why I add rock salt to the ice when I'm making ice cream.

In addition, I give you the sour and bitter details about acids, bases, pHs, and antacids. And I present the properties of gases. In fact, in the gas chapter, you'll see so many gas laws (Boyle's Law, Charles' Law, Gay-Lussac's Law, the Combined Gas Law, the Ideal Gas Law, Avogadro's Law, and more) that you might feel like a lawyer when you're done.

Part IV: Chemistry in Everyday Life: Benefits and Problems

In this part, I show you the chemistry of carbon, called *organic chemistry.* I spend some time talking about hydrocarbons because they're so important in our society as a source of energy, and I introduce you to some organic functional groups. In Chapter 15, I show you a practical application of organic chemistry — the refining of petroleum into gasoline. In Chapter 16, I show you how that same petroleum can be used in the synthesis of polymers. I cover some of the different types of polymers, how they're made, and how they're used.

In this part, I also show you a familiar chemistry lab — the home — and tell you about cleaners, detergents, antiperspirants, cosmetics, hair-care products, and medicines. And I discuss some of the problems society faces due to the industrial nature of our world: air and water pollution. I hope that you don't get lost in the smog!

Part V: The Part of Tens

In this part, I introduce you to ten great serendipitous chemical discoveries, ten great chemistry nerds (nerds rule!), and ten useful chemistry Internet sites. I started to put in my ten favorite chemistry songs, but I could only think of nine. Bummer.

I also include some appendixes that can give you help when dealing with mathematical problems. I cover scientific units, how to handle really big or small numbers, a handy unit conversion method, and how to report answers using what are called *significant figures*.

Icons Used in This Book

If you've read other *For Dummies* books, you'll recognize the icons used in this book, but here's the quickie lowdown for those of you who aren't familiar with them:

This icon gives you a tip on the quickest, easiest way to perform a task or conquer a concept. This icon highlights stuff that's good to know and stuff that'll save you time and/or frustration.

The Remember icon is a memory jog for those really important things you shouldn't forget.

I use this icon when safety in doing a particular activity, especially mixing chemicals, is described.

I don't use this icon very much because I've kept the content pretty basic. But in those cases where I've expanded on a topic beyond the basics, I warn you with this icon. You can safely skip this material, but you may want to look at it if you're interested in a more in-depth description.

Where to Go from Here

That's really up to you and your prior knowledge. If you're trying to clarify something specific, go right to that chapter and section. If you're a real novice, start with Chapter 1 and go from there. If you know a little chemistry, I suggest reviewing Chapter 3 and then going on to Part II. Chapter 10 on the mole is essential, and so is Chapter 13 on gases.

If you're just interested in knowing about chemistry in your everyday life, read Chapter 1 and then skip to Chapters 16 and 17. If you're most interested in environmental chemistry, go on to Chapters 18 and 19. You really can't go wrong. I hope that you enjoy your chemistry trip.

Part I
Basic Concepts of Chemistry

In this part . . .

If you are new to chemistry, it may seem a little frightening. I see students every day who've psyched themselves out by saying so often that they can't do chemistry.

Anyone can figure out chemistry. Anyone can *do* chemistry. If you cook, clean, or simply exist, you're part of the chemical world.

I work with a lot of elementary school children, and they love science. I show them chemical reactions (vinegar plus baking soda, for example), and they go wild. And that's what I hope happens to you.

The chapters of Part I give you a background in chemistry basics. I tell you about matter and the states it can exist in. I talk a little about energy, including the different types and how it's measured. I discuss the microscopic world of the atom and its basic parts. I explain the periodic table, the most useful tool for a chemist. And I cover radioactivity, nuclear reactors, and bombs.

This part takes you on a fun ride, so get your motor running!

Chapter 1

What Is Chemistry, and Why Do I Need to Know Some?

In This Chapter

- Defining the science of chemistry
- Checking out the general areas of chemistry
- Discovering how chemistry is all around you

If you're taking a course in chemistry, you may want to skip this chapter and go right to the area you're having trouble with. But if you bought this book to help you decide whether to take a course in chemistry or to have fun discovering something new, I encourage you to read this chapter. I set the stage for the rest of the book here by showing you what chemistry is, what chemists do, and why you should be interested in chemistry.

I really enjoy chemistry. It's far more than a simple collection of facts and a body of knowledge. I think it's fascinating to watch chemical changes take place, to figure out unknowns, to use instruments, to extend my senses, and to make predictions and figure out why they were right or wrong. It all starts here — with the basics — so welcome to the interesting world of chemistry.

What Exactly Is Chemistry?

Simply put, this whole branch of science is all about *matter,* which is anything that has mass and occupies space. *Chemistry* is the study of the composition and properties of matter and the changes it undergoes.

A lot of chemistry comes into play with that last part — the changes matter undergoes. Matter is made up of either pure substances or mixtures of pure substances. The change from one substance into another is what chemists call a *chemical change,* or *chemical reaction,* and it's a big deal because when it occurs, a brand-new substance is created (see Chapter 2 for the nitty-gritty details).

What is science?

Science is far more than a collection of facts, figures, graphs, and tables. Science is a method for examining the physical universe. It's a way of asking and answering questions. Science is best described by the attitudes of scientists themselves: They're skeptical — they must be able to test phenomena. And they hold onto the results of their experiments tentatively, waiting for another scientist to disprove them. If it can't be tested, it's not science. Scientists wonder, they question, they strive to find out *why*, and they experiment — they have exactly the same attitudes that most small children have before they grow up. Maybe this is a good definition of scientists — they are adults who've never lost that wonder of nature and the desire to know.

Branches in the tree of chemistry

The general field of chemistry is so huge that it was originally subdivided into a number of different areas of specialization. But there's now a tremendous amount of overlap between the different areas of chemistry, just as there is among the various sciences. Here are the traditional fields of chemistry:

- **Analytical chemistry:** This branch is highly involved in the analysis of substances. Chemists from this field of chemistry may be trying to find out what substances are in a mixture *(qualitative analysis)* or how much of a particular substance is present *(quantitative analysis)* in something. A lot of instrumentation is used in analytical chemistry.
- **Biochemistry:** This branch specializes in living organisms and systems. Biochemists study the chemical reactions that occur at the *molecular level* of an organism — the level where items are so small that people can't directly see them. Biochemists study processes such as digestion, metabolism, reproduction, respiration, and so on. Sometimes it's difficult to distinguish between a biochemist and a molecular biologist because they both study living systems at a microscopic level. However, a biochemist really concentrates more on the reactions that are occurring.
- **Biotechnology:** This is a relatively new area of science that is commonly placed with chemistry. It's the application of biochemistry and biology when creating or modifying genetic material or organisms for specific purposes. It's used in such areas as cloning and the creation of disease-resistant crops, and it has the potential for eliminating genetic diseases in the future.
- **Inorganic chemistry:** This branch is involved in the study of inorganic compounds such as salts. It includes the study of the structure and properties of these compounds. It also commonly involves the study of the individual elements of the compounds. Inorganic chemists would probably say that it is the study of everything except carbon, which they leave to the organic chemists.

So what are compounds and elements? Just more of the anatomy of matter. Matter is made up of either pure substances or mixtures of pure substances, and substances themselves are made up of either elements or compounds. (Chapter 2 dissects the anatomy of matter. And, as with all matters of dissection, it's best to be prepared — with a nose plug and an empty stomach.)

- **Organic chemistry:** This is the study of carbon and its compounds. It's probably the most organized of the areas of chemistry — with good reason. There are millions of organic compounds, with thousands more discovered or created each year. Industries such as the polymer industry, the petrochemical industry, and the pharmaceutical industry depend on organic chemists.
- **Physical chemistry:** This branch figures out how and why a chemical system behaves as it does. Physical chemists study the physical properties and behavior of matter and try to develop models and theories that describe this behavior.

The scientific method

Scientific method is normally described as the way scientists go about examining the physical world around them. In fact, there is no one scientific method that everyone uses every time, but the one I cover here describes most of the critical steps scientists go through sooner or later.

Scientists make observations and note facts regarding something in the physical universe. The observations may raise a question or problem that the researcher wants to solve. He or she comes up with a *hypothesis,* a tentative explanation that's consistent with the observations. The researcher then designs an *experiment* to test the hypothesis. This experiment generates observations or facts that can then be used to generate another hypothesis or modify the current one. Then more experiments are designed, and the loop continues.

In good science, this loop never ends. As scientists become more sophisticated in their scientific skills and build better and better instruments, their hypotheses are tested over and over. But a couple of things can come out of this loop. First, a law may be created. A *law* is a generalization of *what* happens in the scientific system being studied. And like the laws that have been created for the judicial system, scientific laws sometimes have to be modified based on new facts. A theory or model may also be proposed. A *theory* or *model* attempts to explain *why* something happens. It's similar to a hypothesis except that it has much more evidence to support it. The power of the theory or model is prediction. If the scientist can use the model to gain a good understanding of the system, then he or she can make predictions based on the model and then check them out with more experimentation. The observations from this experimentation can be used to refine or modify the theory or model, thus establishing another loop in the process. When does it end? Never.

Macroscopic versus microscopic viewpoints

Most chemists that I know operate quite comfortably in two worlds. One is the *macroscopic* world that you and I see, feel, and touch. This is the world of stained lab coats — of weighing out things like sodium chloride to create things like hydrogen gas. This is the world of experiments, or what some nonscientists call the "real world."

But chemists also operate quite comfortably in the *microscopic* world that you and I can't directly see, feel, or touch. Here, chemists work with theories and models. They may measure the volume and pressure of a gas in the macroscopic world, but they have to mentally translate the measurements into how close the gas particles are in the microscopic world.

Scientists often become so accustomed to slipping back and forth between these two worlds that they do so without even realizing it. An occurrence or observation in the macroscopic world generates an idea related to the microscopic world, and vice versa. You may find this flow of ideas disconcerting at first. But as you study chemistry, you'll soon adjust so that it becomes second nature.

Pure versus applied chemistry

In *pure chemistry,* chemists are free to carry out whatever research interests them — or whatever research they can get funded. There is no real expectation of practical application at this point. The researcher simply wants to know for the sake of knowledge. This type of research (often called *basic research*) is most commonly conducted at colleges and universities. The chemist uses undergraduate and graduate students to help conduct the research. The work becomes part of the professional training of the student. The researcher publishes his or her results in professional journals for other chemists to examine and attempt to refute. Funding is almost always a problem, because the experimentation, chemicals, and equipment are quite expensive.

In *applied chemistry,* chemists normally work for private corporations. Their research is directed toward a very specific short-term goal set by the company — product improvement or the development of a disease-resistant strain of corn, for example. Normally, more money is available for equipment and instrumentation with applied chemistry, but there's also the pressure of meeting the company's goals.

These two types of chemistry, pure and applied, share the same basic differences as science and technology. In *science,* the goal is simply the basic acquisition of knowledge. There doesn't need to be any apparent practical application. Science is simply knowledge for knowledge's sake. *Technology* is the application of science toward a very specific goal.

There's a place in our society for science *and* technology — likewise for the two types of chemistry. The pure chemist generates data and information that is then used by the applied chemist. Both types of chemists have their own sets of strengths, problems, and pressures. In fact, because of the dwindling federal research dollars, many universities are becoming much more involved in gaining patents, and they're being paid for technology transfers into the private sector.

So What Does a Chemist Do All Day?

You can group the activities of chemists into these major categories:

- **Chemists analyze substances.** They determine what is in a substance, how much of something is in a substance, or both. They analyze solids, liquids, and gases. They may try to find the active compound in a substance found in nature, or they may analyze water to see how much lead is present.
- **Chemists create, or *synthesize,* new substances.** They may try to make the synthetic version of a substance found in nature, or they may create an entirely new and unique compound. They may try to find a way to synthesize insulin. They may create a new plastic, pill, or paint. Or they may try to find a new, more efficient process to use for the production of an established product.
- **Chemists create models and test the predictive power of theories.** This area of chemistry is referred to as *theoretical chemistry.* Chemists who work in this branch of chemistry use computers to model chemical systems. Theirs is the world of mathematics and computers. Some of these chemists don't even own a lab coat.
- **Chemists measure the physical properties of substances.** They may take new compounds and measure the melting points and boiling points. They may measure the strength of a new polymer strand or determine the octane rating of a new gasoline.

And Where Do Chemists Actually Work?

You may be thinking that all chemists can be found deep in a musty lab, working for some large chemical company, but chemists hold a variety of jobs in a variety of places:

- **Quality control chemist:** These chemists analyze raw materials, intermediate products, and final products for purity to make sure that they fall within specifications. They may also offer technical support for the customer or analyze returned products. Many of these chemists often solve problems when they occur within the manufacturing process.

- **Industrial research chemist:** Chemists in this profession perform a large number of physical and chemical tests on materials. They may develop new products, and they may work on improving existing products. They may work with particular customers to formulate products that meet specific needs. They may also supply technical support to customers.
- **Sales representative:** Chemists may work as sales representatives for companies that sell chemicals or pharmaceuticals. They may call on their customers and let them know of new products being developed. They may also help their customers solve problems.
- **Forensic chemist:** These chemists may analyze samples taken from crime scenes or analyze samples for the presence of drugs. They may also be called to testify in court as expert witnesses.
- **Environmental chemist:** These chemists may work for water purification plants, the Environmental Protection Agency, the Department of Energy, or similar agencies. This type of work appeals to people who like chemistry but also like to get out in nature. They often go out to sites to collect their own samples.
- **Preservationist of art and historical works:** Chemists may work to restore paintings or statues, or they may work to detect forgeries. With air and water pollution destroying works of art daily, these chemists work to preserve our heritage.
- **Chemical educator:** Chemists working as educators may teach physical science and chemistry in public schools. They may also teach at the college or university level. University chemistry teachers often conduct research and work with graduate students. Chemists may even become chemical education specialists for organizations such as the American Chemical Society.

These are just a few of the professions chemists may find themselves in. I didn't even get into law, medicine, technical writing, governmental relations, and consulting. Chemists are involved in almost every aspect of society. Some chemists even write books.

If you aren't interested in becoming a chemist, why should you be interested in chemistry? (The quick answer is probably "to pass a course.") Chemistry is an integral part of our everyday world, and knowing something about chemistry will help you interact more effectively with our chemical environment.

Chapter 2

Matter and Energy

In This Chapter

- Understanding the states of matter and their changes
- Differentiating between pure substances and mixtures
- Finding out about the metric system
- Examining the properties of chemical substances
- Discovering the different types of energy
- Measuring the energy in chemical bonds

Walk into a room and turn on the light. Look around — what do you see? There might be a table, some chairs, a lamp, a computer humming away. But really all you see is matter and energy. There are many kinds of matter and many kinds of energy, but when all is said and done, you're left with these two things. Scientists used to believe that these two were separate and distinct, but now they realize that matter and energy are linked. In an atomic bomb or nuclear reactor, matter is converted into energy. Perhaps someday the science fiction of *Star Trek* will become a reality and converting the human body into energy and back in a transporter will be commonplace. But in the meantime, I'll stick to the basics of matter and energy.

In this chapter, I cover the two basic components of the universe — matter and energy. I examine the different states of matter and what happens when matter goes from one state to another. I show you how the metric system is used to make matter and energy measurements, and I examine the different types of energy and see how energy is measured.

States of Matter: Macroscopic and Microscopic Views

Look around you. All the stuff you see — your chair, the water you're drinking, the paper this book is printed on — is matter. Matter is the material part of the universe. It's anything that has mass and occupies space. (Later in this chapter, I introduce you to energy, the other part of the universe.) Matter can exist in one of three states: solid, liquid, and gas.

Solids

At the *macroscopic level,* the level at which we directly observe with our senses, a solid has a definite shape and occupies a definite volume. Think of an ice cube in a glass — it's a solid. You can easily weigh the ice cube and measure its volume. At the *microscopic level* (where items are so small that people can't directly observe them), the particles that make up the ice are very close together and aren't moving around very much (see Figure 2-1a).

The reason the particles that make up the ice (also known as *water molecules)* are close together and have little movement is because, in many solids, the particles are pulled into a rigid, organized structure of repeating patterns called a *crystal lattice.* The particles that are contained in the crystal lattice are still moving, but barely — it's more of a slight vibration. Depending on the particles, this crystal lattice may be of different shapes.

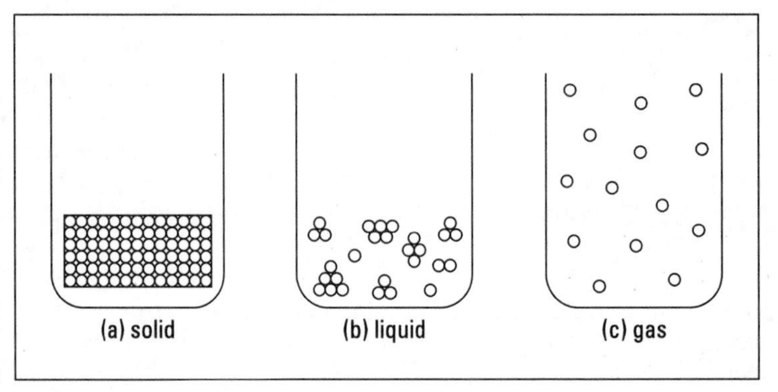

Figure 2-1: Solid, liquid, and gaseous states of matter.

Liquids

When an ice cube melts, it becomes a liquid. Unlike solids, liquids have no definite shape, but they do have a definite volume, just like solids do. For example, a cup of water in a tall skinny glass has a different shape than a cup of water in a pie pan, but in both cases, the volume of water is the same — one cup. Why? The particles in liquids are much farther apart than the particles in solids, and they're also moving around much more (see Figure 2-1b.). Even though the particles are farther apart in liquids than in solids, some particles in liquids may still be near each other, clumped together in small groups.

Because the particles are farther apart in liquids, the attractive forces among them aren't as strong as they are in solids — which is why liquids don't have a definite shape. However, these attractive forces are strong enough to keep the substance confined in one large mass — a liquid — instead of going all over the place.

Gases

If you heat water, you can convert it to *steam,* the gaseous form of water. A gas has no definite shape and no definite volume. In a gas, particles are much farther apart than they are in solids or liquids (see Figure 2-1c), and they're moving relatively independent of each other. Because of the distance between the particles and the independent motion of each of them, the gas expands to fill the area that contains it (and thus it has no definite shape).

Ice in Alaska, Water in Texas: Matter Changes States

When a substance goes from one state of matter to another, we call the process a *change of state.* Some rather interesting things occur during this process.

I'm melting away! Oh, what a world!

Imagine taking a big chunk of ice out of your freezer and putting it into a large pot on your stove. If you measure the temperature of that chunk of ice, you may find it to be –5° Celsius or so. If you take temperature readings while heating the ice, you find that the temperature of the ice begins to rise as the heat from the stove causes the ice particles to begin vibrating faster and faster in the crystal lattice. After a while, some of the particles move so fast that they break free of the lattice, and the crystal lattice (which keeps a solid solid) eventually breaks apart. The solid begins to go from a solid state to a liquid state — a process called *melting.* The temperature at which melting occurs is called the *melting point (mp)* of the substance. The melting point for ice is 32° Fahrenheit, or 0° Celsius.

If you watch the temperature of ice as it melts, you see that the temperature remains steady at 0°C until all the ice has melted. During changes of state (*phase changes*), the temperature remains constant even though the liquid contains more energy than the ice (because the particles in liquids move faster than the particles in solids, as mentioned in the previous section).

Boiling point

If you heat a pot of cool water (or if you continue to heat the pot of now-melted ice cubes mentioned in the preceding section), the temperature of the water rises and the particles move faster and faster as they absorb the heat. The temperature rises until the water reaches the next change of state — boiling. As the particles move faster and faster as they heat up, they begin to break the attractive forces between each other and move freely as steam — a gas. The process by which a substance moves from the liquid state to the gaseous state is called *boiling.* The temperature at which a liquid begins to boil is called the *boiling point (bp).* The bp is dependent on atmospheric pressure, but for water at sea level, it's 212°F, or 100°C. The temperature of the boiling water will remain constant until all the water has been converted to steam.

You can have both water and steam at 100°C. They will have the same temperature, but the steam will have a lot more energy (because the particles move independently and pretty quickly). Because steam has more energy, steam burns are normally a lot more serious than boiling water burns — much more energy is transferred to your skin. I was reminded of this one morning while trying to iron a wrinkle out of a shirt that I was still wearing. My skin and I can attest — steam contains a *lot* of energy!

I can summarize the process of water changing from a solid to a liquid in this way:

ice→water→steam

Because the basic particle in ice, water, and steam is the water molecule (written as H_2O), the same process can also be shown as

$$H_2O(s)\rightarrow H_2O(l)\rightarrow H_2O(g)$$

Here the *(s)* stands for solid, the *(l)* stands for liquid, and the *(g)* stands for gas. This second depiction is much better, because unlike H_2O, most chemical substances don't have different names for the solid, liquid, and gas forms.

Freezing point: The miracle of ice cubes

If you cool a gaseous substance, you can watch the phase changes that occur. The phase changes are

- **Condensation** — going from a gas to a liquid
- **Freezing** — going from a liquid to a solid

The gas particles have a high amount of energy, but as they're cooled, that energy is reduced. The attractive forces now have a chance to draw the particles closer together, forming a liquid. This process is called *condensation.* The particles are now in clumps (as is characteristic of particles in a liquid state), but as more energy is removed by cooling, the particles start to align themselves, and a solid is formed. This is known as *freezing.* The temperature at which this occurs is called the *freezing point (fp)* of the substance.

The freezing point is the same as the melting point — it's the point at which the liquid is able to become a gas or solid.

I can represent water changing states from a gas to a solid like this:

$$H_2O(g) \rightarrow H_2O(l) \rightarrow H_2O(s)$$

Sublimate this!

Most substances go through the logical progression from solid to liquid to gas as they're heated — or vice versa as they're cooled. But a few substances go directly from the solid to the gaseous state without ever becoming a liquid. Scientists call this process *sublimation.* Dry ice — solid carbon dioxide, written as $CO_2(s)$ — is the classic example of sublimation. You can see dry ice particles becoming smaller as the solid begins to turn into a gas, but no liquid is formed during this phase change. (If you've seen dry ice, then you remember that a white cloud usually accompanies it — magicians and theater productions often use dry ice for a cloudy or foggy effect. The white cloud you normally see isn't the carbon dioxide gas — the gas itself is colorless. The white cloud is the condensation of the water vapor in the air due to the cold of the dry ice.)

The process of sublimation is represented as

$$CO_2(s) \rightarrow CO_2(g)$$

In addition to dry ice, mothballs and certain solid air fresheners also go through the process of sublimation. The reverse of sublimation is *deposition* — going directly from a gaseous state to a solid state.

Pure Substances and Mixtures

One of the basic processes in science is classification. As discussed in the preceding section, chemists can classify matter as solid, liquid, or gas. But there are other ways to classify matter, as well. In this section, I discuss how all matter can be classified as either a pure substance or a mixture (see Figure 2-2).

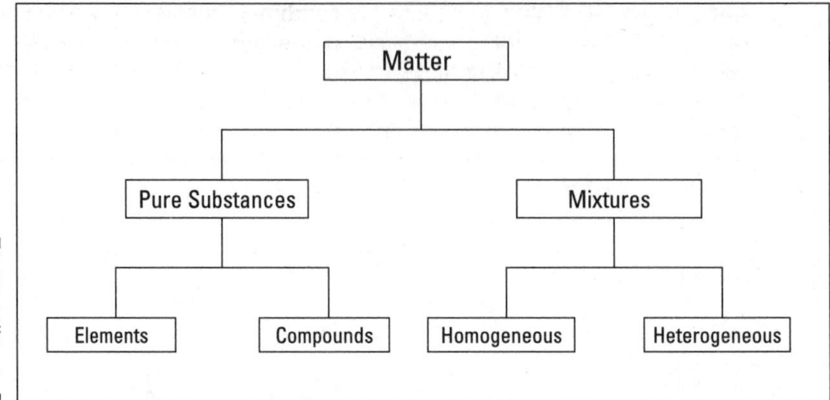

Figure 2-2: Classification of matter.

Pure substances

A *pure substance* has a definite and constant composition or make-up — like salt or sugar.

A pure substance can be either an element or a compound, but the composition of a pure substance doesn't vary.

Elementary, my dear reader

An *element* is composed of a single kind of atom. An *atom* is the smallest particle of an element that still has all the properties of the element. Here's an example: Gold is an element. If you slice and slice a chunk of gold until only one tiny particle is left that can't be chopped any more without losing the properties that make gold *gold,* then you've got an atom.

The atoms in an element all have the same number of protons. *Protons* are subatomic particles — particles of an atom. There are three major subatomic particles, which Chapter 3 covers in great, gory detail.

The important thing to remember right now is that elements are the building blocks of matter. And they're represented in a strange table you may have seen at one time or another — the periodic table. (If you haven't seen such a table before, it's just a list of elements. Chapter 3 contains one if you want to take a peek.)

Compounding the problem

A *compound* is composed of two or more elements in a specific ratio. For example, water (H_2O) is a compound made up of two elements, hydrogen (H) and oxygen (O). These elements are combined in a very specific way — in a ratio of two hydrogen atoms to one oxygen atom (hence H_20). A lot of compounds contain hydrogen and oxygen, but only one has that special 2 to 1 ratio we call water. Even though water is made up of hydrogen and oxygen, the compound water has physical and chemical properties different from both hydrogen and oxygen — water's properties are a unique combination of the two elements.

Chemists can't easily separate the components of a compound: They have to resort to some type of chemical reaction.

Throwing mixtures into the mix

Mixtures are physical combinations of pure substances that have no definite or constant composition — the composition of a mixture varies according to who prepares the mixture. Suppose I asked two people to prepare me a margarita (a delightful mixture). Unless these two people used exactly the same recipe, these mixtures would vary somewhat in their relative amounts of tequila, triple sec, and so on. They would have produced two slightly different mixtures. However, each component of a mixture (that is, each pure substance that makes up the mixture — in the drink example, each *ingredient*) retains its own set of physical and chemical characteristics. Because of this, it's relatively easy to separate the various substances in a mixture.

Although chemists have a difficult time separating compounds into their specific elements, the different parts of a mixture can be easily separated by physical means, such as filtration. For example, suppose you have a mixture of salt and sand, and you want to purify the sand by removing the salt. You can do this by adding water, dissolving the salt, and then filtering the mixture. You then end up with pure sand.

Mixtures can be either homogeneous or heterogeneous.

Homogeneous mixtures, sometimes called *solutions,* are relatively uniform in composition; every portion of the mixture is like every other portion. If you dissolve sugar in water and mix it really well, your mixture is basically the same no matter where you sample it.

But if you put some sugar in a jar, add some sand, and then give the jar a couple of shakes, your mixture doesn't have the same composition throughout the jar. Because the sand is heavier, there's probably more sand at the bottom of the jar and more sugar at the top. In this case, you have a *heterogeneous mixture,* a mixture whose composition varies from position to position within the sample.

Measuring Matter

Scientists are often called on to make measurements, which may include such things as mass (weight), volume, and temperature. If each nation had its own measurement system, communication among scientists would be tremendously hampered, so a worldwide measurement system has been adopted to ensure that scientists can speak the same language.

The SI system

The *SI system* (from the French *Systeme International*) is a worldwide measurement system based on the older metric system that most of us learned in school. There are minor differences between the SI and metric systems, but, for purposes of this book, they're interchangeable.

SI is a decimal system with basic units for things like mass, length, and volume, and prefixes that modify the basic units. For example, the prefix *kilo- (k)* means 1,000. So a kilogram (kg) is 1,000 grams and a kilometer (km) is 1,000 meters. Two other very useful SI prefixes are *centi- (c)* and *milli- (m),* which mean 0.01 and 0.001, respectively. So a milligram (mg) is 0.001 grams — or you can say that there are 1,000 milligrams in a gram. (Check out Appendix A for the most useful SI prefixes.)

SI/English conversions

Many years ago, there was a movement in the United States to convert to the metric system. But, alas, Americans are still buying their potatoes by the pound and their gasoline by the gallon. Don't worry about it. Most professional chemists I know use both the U.S. and SI systems without any trouble. It's necessary to make conversions when using two systems, but I show you how to do that right here.

The basic unit of length in the SI system is the *meter (m).* A meter is a little longer than a yard; there are 1.094 yards in a meter, to be exact. But that's not a really useful conversion. The most useful SI/English conversion for length is

2.54 centimeters = 1 inch

The basic unit of mass in the SI system for chemists is the *gram (g).* And the most useful conversion for mass is

454 grams = 1 pound

The basic unit for volume in the SI system is the *liter (L).* The most useful conversion is

0.946 liter = 1 quart

By using the preceding conversions and the unit conversion method I describe in Appendix C, you'll be able to handle most SI/English conversions you need to do.

For example, suppose that you have a 5-pound sack of potatoes and you want to know its weight in kilograms. Write down 5 pounds (lbs) as a fraction by placing it over 1.

$$\frac{5.0\,\text{lbs}}{1}$$

Because you need to cancel the unit *lbs* in the numerator, you must find a relationship between *lbs* and something else — and then express that something else with *lbs* in the denominator. You know the relationship between pounds and grams, so you can use that.

$$\frac{5.0\,\cancel{\text{lbs}}}{1} \times \frac{454\,\text{g}}{1\,\cancel{\text{lb}}}$$

Now simply convert from grams to kilograms in the same way.

$$\frac{5.0\,\cancel{\text{lbs}}}{1} \times \frac{454\,\cancel{\text{g}}}{1\,\cancel{\text{lb}}} \times \frac{1\,\text{kg}}{1000\,\cancel{\text{g}}} = 2.3\,\text{kg}$$

Nice Properties You've Got There

When chemists study chemical substances, they examine two types of properties:

- **Chemical properties:** These properties enable a substance to change into a brand-new substance, and they describe how a substance reacts with other substances. Does a substance change into something completely new when water is added — like sodium metal changes to sodium hydroxide? Does it burn in air?
- **Physical properties:** These properties describe the physical characteristics of a substance. The mass, volume, and color of a substance are physical properties, and so is its ability to conduct electricity.

Some physical properties are *extensive properties,* properties that depend on the amount of matter present. Mass and volume are extensive properties. *Intensive properties,* however, don't depend on the amount of matter present. Color is an intensive property. A large chunk of gold, for example, is the same color as a small chunk of gold. The mass and volume of these two chunks are different (extensive properties), but the color is the same. Intensive properties are especially useful to chemists because they can use intensive properties to identify a substance.

How dense are you?

Density is one of the most useful intensive properties of a substance, enabling chemists to more easily identify substances. For example, knowing the differences between the density of quartz and diamond allows a jeweler to check out that engagement ring quickly and easily. *Density (d)* is the ratio of the mass (m) to volume (v) of a substance. Mathematically, it looks like this:

$$d = m/v$$

Usually, mass is described in grams (g) and volume in milliliters (mL), so density is g/mL. Because the volumes of liquids vary somewhat with temperature, chemists also usually specify the temperature at which a density measurement is made. Most reference books report densities at 20°C, because it's close to room temperature and easy to measure without a lot of heating or cooling. The density of water at 20°C, for example, is 1g/mL.

Another term you may sometimes hear is *specific gravity (sg),* which is the ratio of the density of a substance to the density of water at the same temperature. Specific gravity is just another way for you to get around the problem of volumes of liquids varying with the temperature. Specific gravity is used with urinalysis in hospitals and to describe automobile battery fluid in auto repair shops. Note that specific gravity has no units of measure associated with it, because the units g/mL appear in both the numerator and denominator, canceling each other out (see the "SI/English conversions" section, earlier in this chapter, for info about canceling out units of measure). In most cases, the density and specific gravity are almost the same, so it's common to simply use the density.

You may sometimes see density reported as g/cm^3 or g/cc. These examples are the same as g/mL. A cube measuring 1 centimeter on each edge (written as 1 cm^3) has a volume of 1 milliliter (1 mL). Because 1 mL = 1 cm^3, g/mL and g/cm^3 are interchangeable. And because a cubic centimeter (cm^3) is commonly abbreviated *cc,* g/cc also means the same thing. (You hear *cc* a lot in the medical profession. When you receive a 10cc injection, you're getting 10 milliliters of liquid.)

Measuring density

Calculating density is pretty straightforward. You measure the mass of an object by using a balance or scale, determine the object's volume, and then divide the mass by the volume.

Determining the volume of liquids is easy, but solids can be tricky. If the object is a regular solid, like a cube, you can measure its three dimensions and calculate the volume by multiplying the length by the width by the height (volume = $l \times w \times h$). But if the object is an irregular solid, like a rock, determining the volume is more difficult. With irregular solids, you can measure the volume by using something called the Archimedes Principle.

The *Archimedes Principle* states that the volume of a solid is equal to the volume of water it displaces. The Greek mathematician Archimedes discovered this concept in the third century B.C., and finding an object's density is greatly simplified by using it. Say that you want to measure the volume of a small rock in order to determine its density. First, put some water into a graduated cylinder with markings for every mL and read the volume. (The example in Figure 2-3 shows 25 mL.) Next, put the rock in, making sure that it's totally submerged, and read the volume again (29 mL in Figure 2-3). The difference in volume (4 mL) is the volume of the rock.

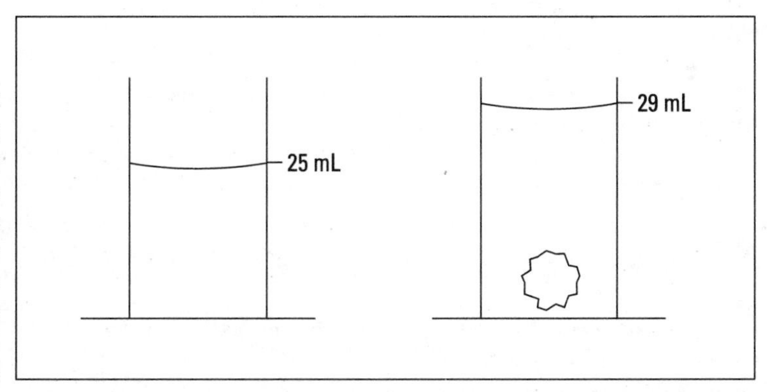

Figure 2-3: Determining the volume of an irregular solid: The Archimedes Principle.

Anything with a density lower than water will float when put into water, and anything with a density greater than 1 g/mL will sink.

For your pondering pleasure, Table 2-1 lists the density of some common materials.

Table 2-1 Densities of Typical Solids and Liquids in g/mL

Substance	*Density*
Gasoline	0.68
Ice	0.92
Water	1.00
Table Salt	2.16
Iron	7.86
Lead	11.38
Mercury	13.55
Gold	19.3

Energy (Wish I Had More)

Matter is one of two components of the universe. Energy is the other. *Energy* is the ability to do work. And if you're like I am, at about 5 p.m. your ability to do work — and your energy level — are pretty low.

Energy can take several forms — such as heat energy, light energy, electrical energy, and mechanical energy. But two general categories of energy are especially important to chemists — kinetic energy and potential energy.

Kinetic energy — moving right along

Kinetic energy is energy of motion. A baseball flying through the air toward a batter has a large amount of kinetic energy. Ask anyone who's ever been hit with a baseball, and I'm sure that they'll agree! Chemists sometimes study moving particles, especially gases, because the kinetic energy of these particles helps determine whether a particular reaction may take place. The reason is that collisions between particles and the transfer of energy cause chemical reactions to occur.

The kinetic energy of moving particles can be transferred from one particle to another. Have you ever shot pool? You transfer kinetic energy from your moving pool stick to the cue ball to (hopefully) the ball you're aiming at.

Kinetic energy can be converted into other types of energy. In a hydroelectric dam, the kinetic energy of the falling water is converted into electrical energy. In fact, a scientific law — *The Law of Conservation of Energy* — states that in ordinary chemical reactions (or physical processes), energy is neither created nor destroyed but can be converted from one form to another. (This law doesn't hold in nuclear reactions, though. Chapter 5 tells you why.)

Potential energy — sitting pretty

Suppose you take a ball and throw it up into a tree where it gets stuck. You gave that ball kinetic energy — energy in motion — when you threw it. But where's that energy now? It's been converted into the other major category of energy — potential energy.

Potential energy is stored energy. Objects may have potential energy stored in terms of their position. That ball up in the tree has potential energy due to its height. If the ball were to fall, that potential energy would be converted to kinetic energy. (Watch out!)

Potential energy due to position isn't the only type of potential energy. In fact, chemists really aren't all that interested in potential energy due to position. Chemists are far more interested in the energy stored (potential energy) in *chemical bonds,* which are the forces that hold atoms together in compounds.

It takes a lot of energy to run a human body. What if there were no way to store the energy you extract from food? You'd have to eat all the time just to keep your body going. (My wife claims I eat all the time, anyway!) But humans can store energy in terms of chemical bonds. And then later, when we need that energy, our bodies can break those bonds and release it.

The same is true of the fuels we commonly use to heat our homes and run our automobiles. Energy is stored in these fuels — gasoline, for example — and is released when chemical reactions take place.

Measuring Energy

Measuring potential energy can be a difficult task. The potential energy of a ball stuck up in a tree is related to the mass of the ball and its height above the ground. The potential energy contained in chemical bonds is related to the type of bond and the number of bonds that can potentially break.

It's far easier to measure kinetic energy. You can do that with a relatively simple instrument — a thermometer.

Temperature and temperature scales

When you measure, say, the air temperature in your backyard, you're really measuring the average kinetic energy (the energy of motion) of the gas particles in your backyard. The faster those particles are moving, the higher the temperature is.

Now all the particles aren't moving at the same speed. Some are going very fast, and some are going relatively slow, but most are moving at a speed between the two extremes. The temperature reading from your thermometer is related to the *average* kinetic energy of the particles.

You probably use the Fahrenheit scale to measure temperatures, but most scientists and chemists use either the Celsius (°C) or Kelvin (K) temperature scale. (There's no degree symbol associated with K.) Figure 2-4 compares the three temperature scales using the freezing point and boiling point of water as reference points.

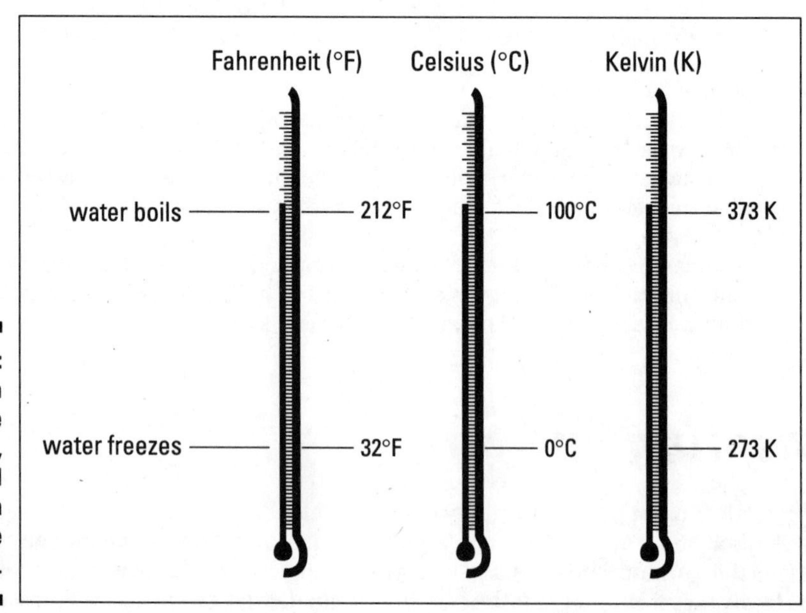

Figure 2-4: Comparison of the Fahrenheit, Celsius, and Kelvin temperature scales.

As you can see from Figure 2-4, water boils at 100°C (373K) and freezes at 0°C (273K). To get the Kelvin temperature, you take the Celsius temperature and add 273. Mathematically, it looks like this:

$$K = {}^{\circ}C + 273$$

You may want to know how to convert from Fahrenheit to Celsius (because most of us still think in °F). Here are the equations you need:

$$^{\circ}C = 5/9\ (^{\circ}F - 32)$$

Be sure to subtract 32 from your Fahrenheit temperature before multiplying by 5/9.

$$^{\circ}F = 9/5\ (^{\circ}C) + 32$$

Be sure to multiply your Celsius temperature by 9/5 and *then* add 32.

Go ahead — try these equations out by confirming that the normal body temperature of 98.6°F equals 37°C.

Most of the time in this book, I use the Celsius scale. But when I describe the behavior of gases, I use the Kelvin scale.

Feel the heat

Heat is not the same as temperature. When you measure the temperature of something, you're measuring the average kinetic energy of the individual particles. *Heat,* on the other hand, is a measure of the total amount of energy a substance possesses. For example, a glass of water and a swimming pool may be the same temperature, but they contain vastly different amounts of heat. It takes much more energy to raise the temperature of a swimming pool 5°C than it does a glass of water, because there's so much more water in the swimming pool.

Counting calories

When you hear the word *calories,* you may think about food and counting calories. Food contains energy (heat). The measure of that energy is the nutritional Calorie (which is commonly capitalized), which is really a kilocalorie (kcal). That candy bar you just ate contained 300 nutritional Calories, which is 300 kcal or 300,000 calories. Thinking of it that way may make it a little easier to resist temptation.

The unit of heat in the SI system is the *joule (J)*. Most of us still use the metric unit of heat, the *calorie (cal)*. Here's the relationship between the two:

1 calorie = 4.184 joule

The calorie is a fairly small amount of heat — the amount it takes to raise the temperature of 1 gram of water 1°C. I often use the *kilocalorie (kcal),* which is 1,000 calories, as a convenient unit of heat. If you burn a large kitchen match completely, it produces about 1 kilocalorie (1,000 cal) of heat.

Chapter 3

Something Smaller Than an Atom? Atomic Structure

In This Chapter

- Taking a look at the particles that make up an atom
- Finding out about isotopes and ions
- Coming to understand electron configurations
- Discovering the importance of valence electrons

I remember learning about atoms as a child in school. My teachers called them building blocks and, in fact, we used blocks and Legos to represent atoms. I also remember being told that atoms were so small that nobody would ever see one. Imagine my surprise years later when the first pictures of atoms appeared. They weren't very detailed, but they did make me stop and think how far science had come. I am still amazed when I see pictures of atoms.

In this chapter, I tell you about atoms, the fundamental building blocks of the universe. I cover the three basic particles of an atom — protons, neutrons, and electrons — and show you where they're located. And I use a slew of pages on electrons themselves, because chemical reactions (where a lot of chemistry comes into play) depend on the loss, gain, or sharing of them.

Subatomic Particles: So That's What's in an Atom

The *atom* is the smallest part of matter that represents a particular element. For quite a while, the atom was thought to be the smallest part of matter that could exist. But in the latter part of the nineteenth century and early part of the twentieth, scientists discovered that atoms are composed of certain subatomic particles and that, no matter what the element, the same subatomic particles make up the atom. The number of the various subatomic particles is the only thing that varies.

Scientists now recognize that there are many subatomic particles (this really makes physicists salivate). But in order to be successful in chemistry, you really only need to be concerned with the three major subatomic particles:

- Protons
- Neutrons
- Electrons

Table 3-1 summarizes the characteristics of these three subatomic particles.

Table 3-1 The Three Major Subatomic Particles

Name	*Symbol*	*Charge*	*Mass (g)*	*Mass (amu)*	*Location*
Proton	p^+	+1	1.673×10^{-24}	1	Nucleus
Neutron	n^0	0	1.675×10^{-24}	1	Nucleus
Electron	e^-	–1	9.109×10^{-28}	0.0005	Outside Nucleus

In Table 3-1, the masses of the subatomic particles are listed in two ways: grams and *amu,* which stands for *atomic mass units.* Expressing mass in amu is much easier than using the gram equivalent.

Atomic mass units are based on something called the Carbon 12 scale, a worldwide standard that's been adopted for atomic weights. By international agreement, a carbon atom that contains 6 protons and 6 neutrons has an atomic weight of exactly 12 amu, so 1 amu is $\frac{1}{12}$ of this carbon atom. I know, what do carbon atoms and the number 12 have to do with anything? Just trust me. Because the mass in grams of protons and neutrons are almost exactly the same, both protons and neutrons are said to have a mass of 1 amu. Notice that the mass of an electron is much smaller than that of either a proton or neutron. It takes almost 2,000 electrons to equal the mass of a single proton.

Table 3-1 also shows the electrical charge associated with each subatomic particle. Matter can be electrically charged in one of two ways: positive or negative. The proton carries one unit of positive charge, the electron carries one unit of negative charge, and the neutron has no charge — it's neutral.

Scientists have discovered through observation that objects with like charges, whether positive or negative, repel each other, and objects with unlike charges attract each other.

The atom itself has no charge. It's neutral. (Well, actually, Chapter 6 explains that certain atoms can gain or lose electrons and acquire a charge. Atoms that gain a charge, either positive or negative, are called *ions.*) So how can an atom be neutral if it contains positively charged protons and negatively charged electrons? Ah, good question. The answer is that there are *equal* numbers of protons and electrons — equal numbers of positive and negative charges — so they cancel each other out.

The last column in Table 3-1 lists the location of the three subatomic particles. Protons and neutrons are located in the *nucleus,* a dense central core in the middle of the atom, while the electrons are located outside the nucleus (see "Where Are Those Electrons?" later in this chapter).

The Nucleus: Center Stage

In 1911, Ernest Rutherford discovered that atoms have a nucleus — a center — containing protons. Scientists later discovered that the nucleus also houses the neutron.

The nucleus is very, very small and very, very dense when compared to the rest of the atom. Typically, atoms have diameters that measure around 10^{-10} meters. (That's small!) Nuclei are around 10^{-15} meters in diameter. (That's *really* small!) For example, if the Superdome in New Orleans represented a hydrogen atom, the nucleus would be about the size of a pea.

The protons of an atom are all crammed together inside the nucleus. Now some of you may be thinking, "Okay, each proton carries a positive charge, and like charges repel each other. So if all the protons are repelling each other, why doesn't the nucleus simply fly apart?" It's *The Force,* Luke. Forces in the nucleus counteract this repulsion and hold the nucleus together. (Physicists call these forces *nuclear glue.* But sometimes this "glue" isn't strong enough, and the nucleus does break apart. This process is called *radioactivity.*)

Not only is the nucleus very small, but it also contains most of the mass of the atom. In fact, for all practical purposes, the mass of the atom is the sum of the masses of the protons and neutrons. (I ignore the minute mass of the electrons unless I'm doing very, very precise calculations.)

The sum of the number of protons plus the number of neutrons in an atom is called the *mass number.* And the number of protons in a particular atom is given a special name, the *atomic number.* Chemists commonly use the symbolization shown in Figure 3-1 to represent these things for a particular element.

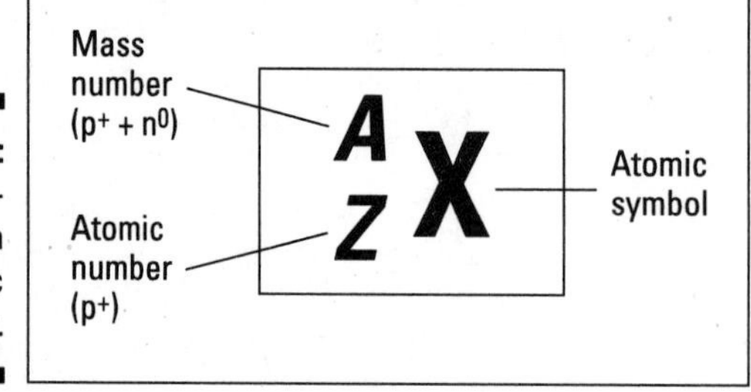

Figure 3-1: Representing a specific element.

As shown in Figure 3-1, chemists use the placeholder *X* to represent the chemical symbol. You can find an element's chemical symbol on the periodic table or a list of elements (see Table 3-2 for a list of elements). The placeholder *Z* represents the atomic number — the number of protons in the nucleus. And *A* represents the mass number, the sum of the number of protons plus neutrons. The mass number (also called the *atomic weight*) is listed in amu.

Suppose you want to represent uranium. You can refer to a periodic table or a list of elements, such as the one shown in Table 3-2, and find that the symbol for uranium is U, its atomic number is 92, and its mass number is 238.

Table 3-2 The Elements

Name	*Symbol*	*Atomic Number*	*Mass Number*	*Name*	*Symbol*	*Atomic Number*	*Mass Number*
Actinium	Ac	89	227.028	Cerium	Ce	58	140.115
Aluminum	Al	13	26.982	Cesium	Cs	55	132.905
Americium	Am	95	243	Chlorine	Cl	17	35.453
Antimony	Sb	51	121.76	Chromium	Cr	24	51.996
Argon	Ar	18	39.948	Cobalt	Co	27	58.933
Arsenic	As	33	74.922	Copper	Cu	29	63.546
Astatine	At	85	210	Curium	Cm	96	247
Barium	Ba	56	137.327	Dubnium	Db	105	262
Berkelium	Bk	97	247	Dysprosium	Dy	66	162.5
Beryllium	Be	4	9.012	Einsteinium	Es	99	252
Bismuth	Bi	83	208.980	Erbium	Er	68	167.26
Bohrium	Bh	107	262	Europium	Eu	63	151.964
Boron	B	5	10.811	Fermium	Fm	100	257
Bromine	Br	35	79.904	Fluorine	F	9	18.998
Cadmium	Cd	48	112.411	Francium	Fr	87	223
Calcium	Ca	20	40.078	Gadolinium	Gd	64	157.25
Californium	Cf	98	251	Gallium	Ga	31	69.723
Carbon	C	6	12.011	Germanium	Ge	32	72.61

(continued)

Table 3-2 (continued)

Name	Symbol	Atomic Number	Mass Number	Name	Symbol	Atomic Number	Mass Number
Gold	Au	79	196.967	Mendelevium	Md	101	258
Hafnium	Hf	72	178.49	Mercury	Hg	80	200.59
Hassium	Hs	108	265	Molybdenum	Mo	42	95.94
Helium	He	2	4.003	Neodymium	Nd	60	144.24
Holmfum	Ho	67	164.93	Neon	Ne	10	20.180
Hydrogen	H	1	1.0079	Neptunium	Np	93	237.048
Indium	In	49	114.82	Nickel	Ni	28	58.69
Iodine	I	53	126.905	Niobium	Nb	41	92.906
Iridium	Ir	77	192.22	Nitrogen	N	7	14.007
Iron	Fe	26	55.845	Nobelium	No	102	259
Krypton	Kr	36	83.8	Osmiun	Os	76	190.23
Lanthanum	La	57	138.906	Oxygen	O	8	15.999
Lawrencium	Lr	103	262	Palladium	Pd	46	106.42
Lead	Pb	82	207.2	Phosphorus	P	15	30.974
Lithium	Li	3	6.941	Platinum	Pt	78	195.08
Lutetium	Lu	71	174.967	Plutonium	Pu	94	244
Magnesium	Mg	12	24.305	Polonium	Po	84	209
Manganese	Mn	25	54.938	Potassium	K	19	39.098
Meitnerium	Mt	109	266	Praseodymium	Pr	59	140.908

Name	**Symbol**	**Atomic Number**	**Mass Number**	**Name**	**Symbol**	**Atomic Number**	**Mass Number**
Promethium	Pm	61	145	Tantalum	Ta	73	180.948
Protactinium	Pa	91	231.036	Technetium	Tc	43	98
Radium	Ra	88	226.025	Tellurium	Te	52	127.60
Radon	Rn	86	222	Terbium	Tb	65	158.925
Rhenium	Re	75	186.207	Thallium	Tl	81	204.383
Rhodium	Rh	45	102.906	Thorium	Th	90	232.038
Rubidium	Rb	37	85.468	Thulium	Tm	69	168.934
Ruthenium	Ru	44	101.07	Tin	Sn	50	118.71
Rutherfordium	Rf	104	261	Titanium	Ti	22	47.88
Samarium	Sm	62	150.36	Tungsten	W	74	183.84
Scandium	Sc	21	44.956	Uranium	U	92	238.029
Seaborgium	Sg	106	263	Vanadium	V	23	50.942
Selenium	Se	34	78.96	Xenon	Xe	54	131.29
Silicon	Si	14	28.086	Ytterbium	Yb	70	173.04
Silver	Ag	47	107.868	Yttrium	Y	39	88.906
Sodium	Na	11	22.990	Zinc	Zn	30	65.39
Strontium	Sr	38	87.62	Zirconium	Zr	40	91.224
Sulfur	S	16	32.066				

So you can represent uranium as shown in Figure 3-2.

$^{238}_{92}U$

Figure 3-2: Representing uranium.

You know that uranium has an atomic number of 92 (number of protons) and mass number of 238 (protons plus neutrons). So if you want to know the number of neutrons in uranium, all you have to do is subtract the atomic number (92 protons) from the mass number (238 protons plus neutrons). The resulting number shows that uranium has 146 neutrons.

But how many electrons does uranium have? Because the atom is neutral (it has no electrical charge), there must be equal numbers of positive and negative charges inside it, or equal numbers of protons and electrons. So there are 92 electrons in each uranium atom.

Where Are Those Electrons?

Early models of the atom had electrons spinning around the nucleus in a random fashion. But as scientists learned more about the atom, they found that this representation probably wasn't accurate. Today, two models of atomic structure are used: the Bohr model and the quantum mechanical model. The Bohr model is simple and relatively easy to understand; the quantum mechanical model is based on mathematics and is more difficult to understand. Both, though, are helpful in understanding the atom, so I explain each in the following sections (without resorting to a lot of math).

A model is useful because it helps you understand what's observed in nature. It's not unusual to have more than one model represent and help people understand a particular topic.

The Bohr model — it's really not boring

Have you ever bought color crystals for your fireplace — to make flames of different colors? Or have you ever watched fireworks and wondered where the colors came from?

Color comes from different elements. If you sprinkle table salt — or any salt containing sodium — on a fire, you get a yellow color. Salts that contain copper give a greenish-blue flame. And if you look at the flames through a *spectroscope,* an instrument that uses a prism to break up light into its various components, you see a number of lines of various colors. Those distinct lines of color make up a *line spectrum.*

Niels Bohr, a Danish scientist, explained this line spectrum while developing a model for the atom.

The Bohr model shows that the electrons in atoms are in orbits of differing energy around the nucleus (think of planets orbiting around the sun). Bohr used the term *energy levels* (or *shells*) to describe these orbits of differing energy. And he said that the energy of an electron is *quantized,* meaning electrons can have one energy level or another but nothing in between.

The energy level an electron normally occupies is called its *ground state.* But it can move to a higher-energy, less-stable level, or shell, by absorbing energy. This higher-energy, less-stable state is called the electron's *excited state.*

After it's done being excited, the electron can return to its original ground state by releasing the energy it has absorbed (see Figure 3-3). And here's where the line spectrum explanation comes in. Sometimes the energy released by electrons occupies the portion of the *electromagnetic spectrum* (the range of wavelengths of energy) that humans detect as visible light. Slight variations in the amount of the energy is seen as light of different colors.

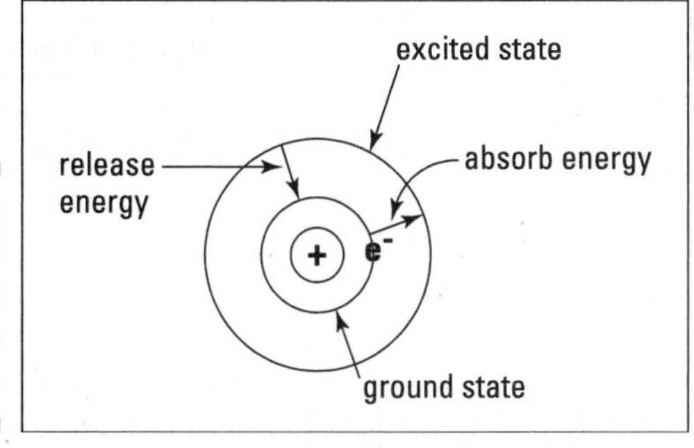

Figure 3-3: Ground and excited states in the Bohr model.

Bohr found that the closer an electron is to the nucleus, the less energy it needs, but the farther away it is, the more energy it needs. So Bohr numbered the electron's energy levels. The higher the energy-level number, the farther away the electron is from the nucleus — and the higher the energy.

Bohr also found that the various energy levels can hold differing numbers of electrons: energy level 1 may hold up to 2 electrons, energy level 2 may hold up to 8 electrons, and so on.

The Bohr model worked well for very simple atoms such as hydrogen (which has 1 electron) but not for more complex atoms. Although the Bohr model is still used today, especially in elementary textbooks, a more sophisticated (and complex) model — the quantum mechanical model — is used much more frequently.

Quantum mechanical model

The simple Bohr model was unable to explain observations made on complex atoms, so a more complex, highly mathematical model of atomic structure was developed — the quantum mechanical model.

This model is based on *quantum theory,* which says matter also has properties associated with waves. According to quantum theory, it's impossible to know the exact position and *momentum* (speed and direction) of an electron at the same time. This is known as the *Uncertainty Principle.* So scientists had to replace Bohr's orbits with *orbitals* (sometimes called *electron clouds*), volumes of space in which there is *likely* to be an electron. In other words, certainty was replaced with probability.

The quantum mechanical model of the atom uses complex shapes of orbitals rather than Bohr's simple circular orbits. Without resorting to a lot of math (you're welcome), this section shows you some aspects of this newest model of the atom.

Four numbers, called *quantum numbers,* were introduced to describe the characteristics of electrons and their orbitals. You'll notice that they were named by totally top-rate techno-geeks:

- Principal quantum number n
- Angular momentum quantum number l
- Magnetic quantum number m_l
- Spin quantum number m_s

Table 3-3 summarizes the four quantum numbers. When they're all put together, theoretical chemists have a pretty good description of the characteristics of a particular electron.

Table 3-3 Summary of the Quantum Numbers

Name	*Symbol*	*Description*	*Allowed Values*
Principal	n	Orbital energy	Positive integers (1, 2, 3, and so on)
Angular momentum	l	Orbital shape	Integers from 0 to n–1

Name	Symbol	Description	Allowed Values
Magnetic	m_l	Orientation	Integers from –*l* to 0 to +*l*
Spin	m_s	Electron spin	+½ or –½

The principal quantum number n

The principal quantum number n describes the average distance of the orbital from the nucleus — and the energy of the electron in an atom. It's really about the same as Bohr's energy-level numbers. It can have positive integer (whole number) values: 1, 2, 3, 4, and so on. The larger the value of n, the higher the energy and the larger the orbital. Chemists sometimes call the orbitals *electron shells.*

The angular momentum quantum number l

The angular momentum quantum number *l* describes the shape of the orbital, and the shape is limited by the principal quantum number n: The angular momentum quantum number *l* can have positive integer values from 0 to n–1. For example, if the n value is 3, three values are allowed for *l*: 0, 1, and 2.

The value of *l* defines the shape of the orbital, and the value of n defines the size.

Orbitals that have the same value of n but different values of *l* are called *subshells.* These subshells are given different letters to help chemists distinguish them from each other. Table 3-4 shows the letters corresponding to the different values of *l*.

Table 3-4 Letter Designation of the Subshells

Value of l (subshell)	Letter
0	s
1	p
2	d
3	f
4	g

When chemists describe one particular subshell in an atom, they can use both the n value and the subshell letter — 2p, 3d, and so on. Normally, a subshell value of 4 is the largest needed to describe a particular subshell. If chemists ever need a larger value, they can create subshell numbers and letters.

Figure 3-4 shows the shapes of the s, p, and d orbitals.

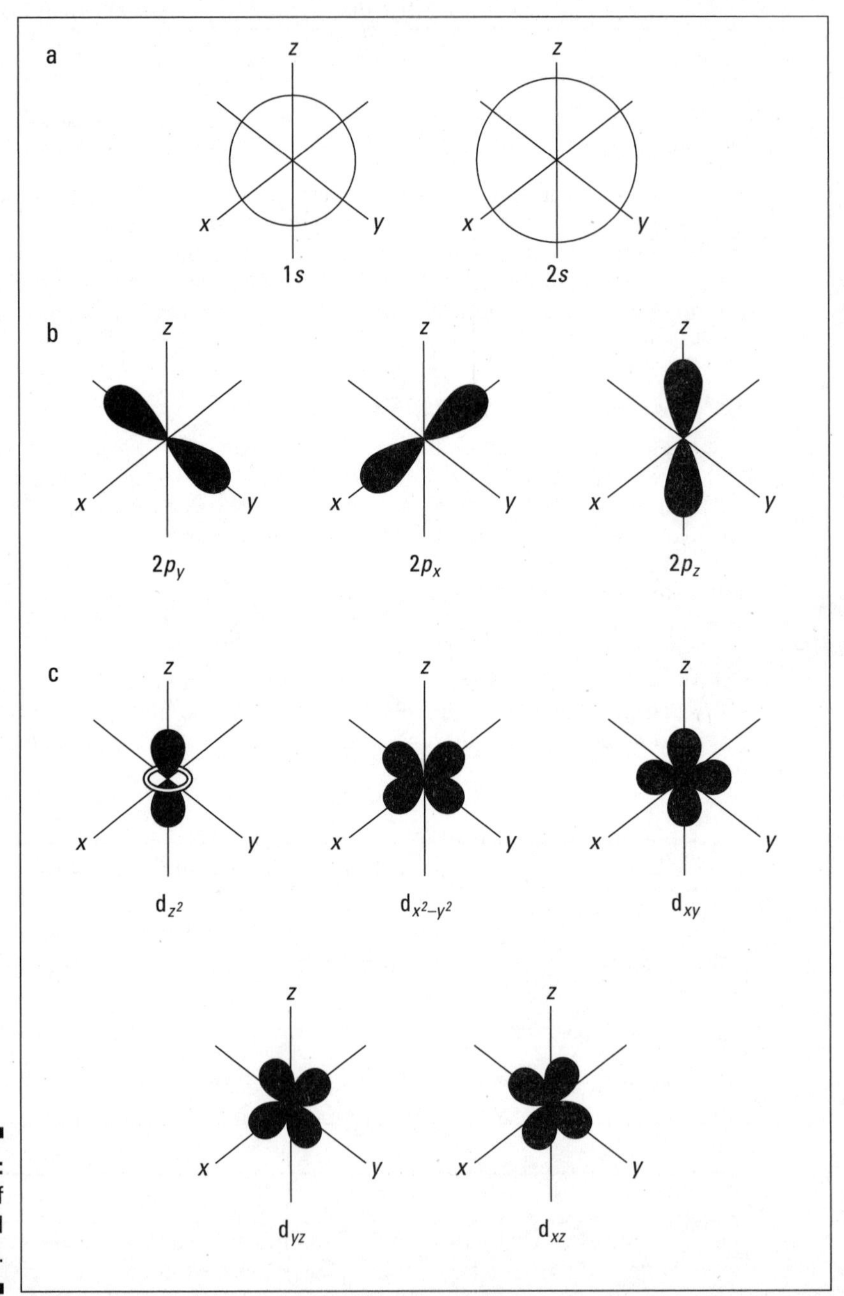

Figure 3-4: Shapes of the s, p, and d orbitals.

In Figure 3-4 (a), there are two s orbitals — one for energy level 1 (1s) and the other for energy level 2 (2s). S orbitals are spherical with the nucleus at the center. Notice that the 2s orbital is larger in diameter than the 1s orbital. In large atoms, the 1s orbital is nestled inside the 2s, just like the 2p is nestled inside the 3p.

Figure 3-4 (b) shows the shapes of the p orbitals, and Figure 3-4 (c) shows the shapes of the d orbitals. Notice that the shapes get progressively more complex.

The magnetic quantum number m_l

The magnetic quantum number m_l describes how the various orbitals are oriented in space. The value of m_l depends on the value of l. The values allowed are integers from $-l$ to 0 to $+l$. For example, if the value of $l = 1$ (p orbital — see Table 3-4), you can write three values for m_l: –1, 0, and +1. This means that there are three different p subshells for a particular orbital. The subshells have the same energy but different orientations in space.

Figure 3-4 (b) shows how the p orbitals are oriented in space. Notice that the three p orbitals correspond to m_l values of –1, 0, and +1, oriented along the x, y, and z axes.

The spin quantum number m_s

The fourth and final (I know you're glad — techie stuff, eh?) quantum number is the spin quantum number m_s. This one describes the direction the electron is spinning in a magnetic field — either clockwise or counterclockwise. Only two values are allowed for m_s: +½ or –½. For each subshell, there can be only two electrons, one with a spin of +½ and another with a spin of –½.

Put all the numbers together and whaddya get? (A pretty table)

I know. Quantum number stuff makes science nerds drool and normal people yawn. But, hey, sometime if the TV's on the blink and you've got some time to kill, take a peek at Table 3-5. You can check out the quantum numbers for each electron in the first two energy levels (oh boy, oh boy, oh boy).

Table 3-5 Quantum Numbers for the First Two Energy Levels

n	*l*	*Subshell Notation*	m_l	m_s
1	0	1s	0	+1/2, –1/2
2	0	2s	0	+1/2, –1/2
	1	2p	–1	+1/2, –1/2
			0	+1/2, –1/2
			+1	+1/2, –1/2

Table 3-5 shows that in energy level 1 (n=1) there's only an s orbital. There's no p orbital because an *l* value of 1 (p orbital) is not allowed. And notice that there can be only two electrons in that 1s orbital (m_s of +½ and –½). In fact, there can be only two electrons in any s orbital, whether it's 1s or 5s.

When you move from energy level 1 to energy level 2 (n=2), there can be both s and p orbitals. If you write out the quantum numbers for energy level 3, you see s, p, and d orbitals. Each time you move higher in a major energy level, you add another orbital type.

Notice also that there are three subshells (m*l*) for the 2p orbital (see Figure 3-4 (b)) and that each holds a maximum of two electrons. The three 2p subshells can hold a maximum of six electrons.

There's an energy difference in the major energy levels (energy level 2 is higher in energy than energy level 1), but there's also a difference in the energies of the different orbitals within an energy level. At energy level 2, both s and p orbitals are present. But the 2s is lower in energy than the 2p. The three subshells of the 2p orbital have the same energy. Likewise, the five subshells of the d orbitals (see Figure 3-4 (c)) have the same energy.

Okay. Enough already.

Electron configurations (Bed Check for Electrons)

Chemists find quantum numbers useful when they're looking at chemical reactions and bonding (and those are things many chemists like to study). But they find two other representations for electrons *more* useful and easier to work with:

- Energy level diagrams
- Electron configurations

Chemists use both of these things to represent which energy level, subshell, and orbital are occupied by electrons in any particular atom. Chemists use this information to predict what type of bonding will occur with a particular element and show exactly which electrons are being used. These representations are also useful in showing why certain elements behave in similar ways.

In this section, I show you how to use an energy level diagram and write electron configurations.

The dreaded energy level diagram

Figure 3-5 is a blank energy level diagram you can use to depict electrons for any particular atom. Not all the known orbitals and subshells are shown. But with this diagram, you should be able to do most anything you need to. (If you don't have a clue what orbitals, subshells, or all those numbers and letters in the figure have to do with the price of beans, check out the "Quantum mechanical model" section, earlier in this chapter. Fun read, lemme tell ya.)

I represent orbitals with dashes in which you can place a maximum of two electrons. The 1s orbital is closest to the nucleus, and it has the lowest energy. It's also the only orbital in energy level 1 (refer to Table 3-5). At energy level 2, there are both s and p orbitals, with the 2s having lower energy than the 2p. The three 2p subshells are represented by three dashes of the same energy. Energy levels 3, 4, and 5 are also shown. Notice that the 4s has lower energy than the 3d: This is an exception to what you may have thought, but it's what's observed in nature. Go figure. Speaking of which, Figure 3-6 shows the *Aufbau Principle,* a method for remembering the order in which orbitals fill the vacant energy levels.

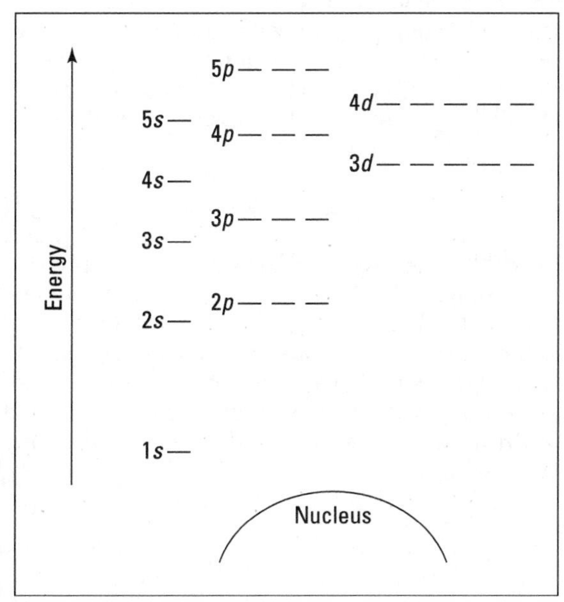

Figure 3-5: Energy level diagram.

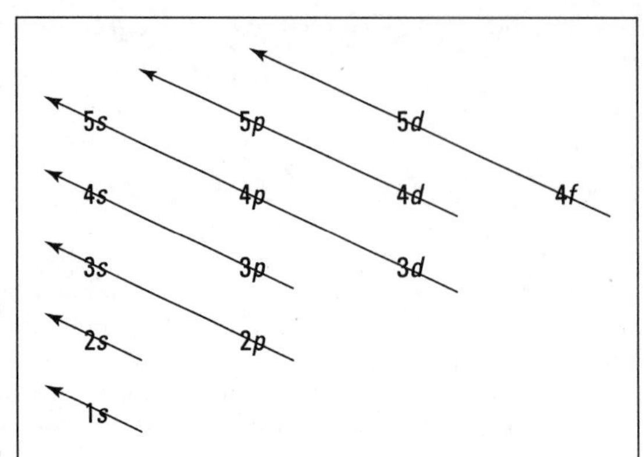

Figure 3-6: Aufbau filling chart.

In using the energy level diagram, remember two things:

- Electrons fill the lowest vacant energy levels first.
- When there's more than one subshell at a particular energy level, such as at the 3p or 4d levels (see Figure 3-5), only one electron fills each subshell until each subshell has one electron. Then electrons start pairing up in each subshell. This rule is named *Hund's Rule.*

Suppose you want to draw the energy level diagram of oxygen. You look on the periodic table or an element list and find that oxygen is atomic number 8. This number means that oxygen has 8 protons in its nucleus and 8 electrons. So you put 8 electrons into your energy level diagram. You can represent electrons as arrows (see Figure 3-7). Note that if two electrons end up in the same orbital, one arrow faces up and the other faces down. (This is called *spin pairing.* It corresponds to the +½ and –½ of m_s.— see "The spin quantum number m_s"section earlier in this chapter.)

The first electron goes into the 1s orbital, filling the lowest energy level first, and the second one spin pairs with the first one. Electrons 3 and 4 spin pair in the next lowest vacant orbital — the 2s. Electron 5 goes into one of the 2p subshells (no, it doesn't matter which one — they all have the same energy), and electrons 6 and 7 go into the other two totally vacant 2p orbitals (see the two things you're supposed to remember about the energy level diagram, a little ways back from here). The last electron spin pairs with one of the electrons in the 2p subshells (again, it doesn't matter which one you pair it with). Figure 3-7 shows the completed energy level diagram for oxygen.

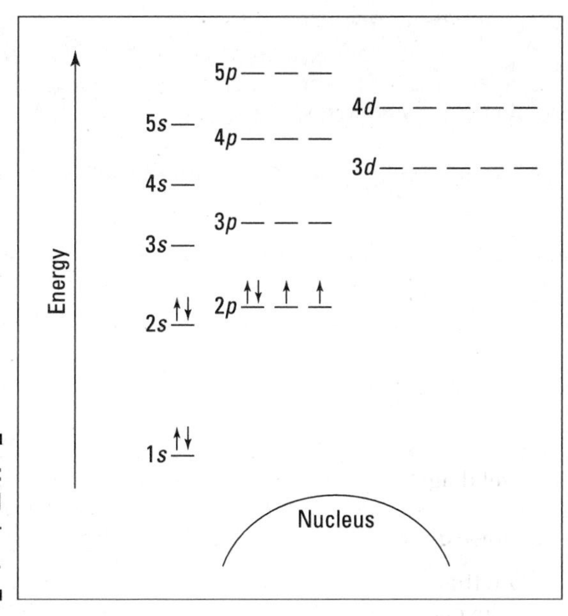

Figure 3-7: Energy level diagram for oxygen.

Electron configurations: Easy and space efficient

Energy level diagrams are useful when you need to figure out chemical reactions and bonding, but they're very bulky to work with. Wouldn't it be nice if there were another representation that gives just about the same information but in a much more concise, shorthand-notation form? Well, there is. It's called the *electron configuration.*

The electron configuration for oxygen is $1s^2 2s^2 2p^4$. Compare that notation with the energy level diagram for oxygen in Figure 3-7. Doesn't the electron configuration take up a lot less space? You can derive the electron configuration from the energy level diagram. The first two electrons in oxygen fill the 1s orbital, so you show it as $1s^2$ in the electron configuration. The 1 is the energy level, the s represents the type of orbital, and the superscript 2 represents the number of electrons in that orbital. The next two electrons are in the 2s orbital, so you write $2s^2$. And, finally, you show the 4 electrons in the 2p orbital as $2p^4$. Put it all together, and you get $1s^2 2s^2 2p^4$.

Some people use a more expanded form, showing how the individual p_x, p_y, and p_z orbitals are oriented along the x,y, and z axes and the number of electrons in each orbital. (The section "The magnetic quantum number m_l," earlier in this chapter, explains how orbitals are oriented in space.) The expanded form is nice if you're really looking at the finer details, but most of the time you won't need that amount of detail in order to show bonding situations and such, so I'm not going to explain the expanded form here.

The sum of the superscript numbers equals the atomic number, or the number of electrons in the atom.

Here are a couple of electron configurations you can use to check your conversions from energy level diagrams:

Chlorine (Cl): $1s^2 2s^2 2p^6 3s^2 3p^5$

Iron (Fe): $1s^2 2s^2 2p^6 3s^2 3p^6 4s^2 3d^6$

Although I've showed you how to use the energy level diagram to write the electron configuration, with a little practice, you can omit doing the energy level diagram altogether and simply write the electron configuration by knowing the number of electrons and the orbital filling pattern. Anything to save a little precious time, right?

Valence electrons: Living on the edge

When chemists study chemical reactions, they study the transfer or sharing of electrons. The electrons more loosely held by the nucleus — the electrons in the energy level farthest away from the nucleus — are the ones that are gained, lost, or shared.

Electrons are negatively charged, while the nucleus has a positive charge due to the protons. The protons attract and hold the electrons, but the farther away the electrons are, the less the attractive force.

The electrons in the outermost energy level are commonly called *valence electrons.* Chemists really only consider the electrons in the s and p orbitals in the energy level that is currently being filled as valence electrons. In the electron configuration for oxygen, $1s^2 2s^2 2p^4$, energy level 1 is filled, and there are 2 electrons in the 2s orbital and 4 electrons in the 2p orbital for a total of 6 valence electrons. Those valence electrons are the ones lost, gained, or shared.

Being able to determine the number of valence electrons in a particular atom gives you a big clue as to how that atom will react. In Chapter 4, which gives an overview of the periodic table, I show you a quick way to determine the number of valence electrons without writing the electron configuration of the atom.

Isotopes and Ions: These Are a Few of My Favorite Things

But then again, I'm a nerd. The atoms in a particular element have an identical number of protons and electrons but can have varying numbers of neutrons. If they have different numbers of neutrons, then the atoms are called *isotopes*.

Isolating the isotope

Hydrogen is a common element here on earth. Hydrogen's atomic number is 1 — its nucleus contains 1 proton. The hydrogen atom also has 1 electron. Because it has the same number of protons as electrons, the hydrogen atom is neutral (the positive and negative charges have canceled each other out).

Most of the hydrogen atoms on earth contain no neutrons. You can use the symbolization shown in Figure 3-2 to represent hydrogen atoms that don't contain neutrons, as shown in Figure 3-8 (a).

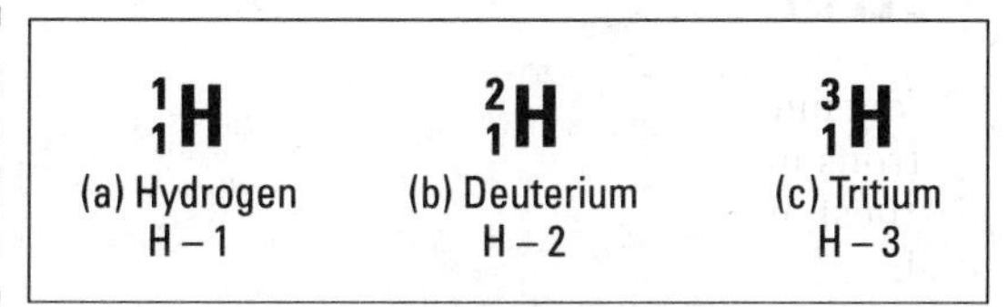

Figure 3-8: The isotopes of hydrogen.

However, approximately one hydrogen atom out of 6,000 contains a neutron in its nucleus. These atoms are still hydrogen, because they have one proton and one electron; they simply have a neutron that most hydrogen atoms lack. So these atoms are called isotopes. Figure 3-8 (b) shows an isotope of hydrogen, commonly called *deuterium*. It's still hydrogen, because it contains only one proton, but it's different from the hydrogen in Figure 3-8 (a), because it also has one neutron. Because it contains one proton and one neutron, its mass number is two.

There's even an isotope of hydrogen containing two neutrons. This one's called *tritium*, and it's represented in Figure 3-8 (c). Tritium doesn't occur naturally on earth, but it can easily be created.

Now take another look at Figure 3-8. It also shows an alternative way of representing isotopes: Write the element symbol, a dash, and then the mass number.

Now you may be wondering, "If I'm doing a calculation involving the atomic mass of hydrogen, which isotope do I use?" Well, you use an average of all the naturally occurring isotopes of hydrogen. But not a simple average. (You have to take into consideration that there's a *lot* more H-1 than H-2, and you don't even consider H-3, because it's not naturally occurring.) You use a *weighted average,* which takes into consideration the abundances of the naturally occurring isotopes. That's why the atomic mass of hydrogen in Table 3-2 isn't a whole number: It's 1.0079 amu. The number shows that there's a lot more H-1 than H-2.

Many elements have several isotopic forms. You can find out more about them in Chapter 5.

Keeping an eye on ions

Because an atom itself is neutral, throughout this book I say that the number of protons and electrons in atoms are equal. But there are cases in which an atom can acquire an electrical charge. For example, in the compound sodium chloride — table salt — the sodium atom has a positive charge and the chlorine atom has a negative charge. Atoms (or groups of atoms) in which there are unequal numbers of protons and electrons are called *ions.*

The neutral sodium atom has 11 protons and 11 electrons, which means it has 11 positive charges and 11 negative charges. Overall, the sodium atom is neutral, and it's represented like this: Na. But the sodium *ion* contains one more positive charge than negative charge, so it's represented like this: Na^+ (the $^+$ represents its net positive electrical charge).

This unequal number of negative and positive charges can occur in one of two ways: An atom can gain a proton (a positive charge) or lose an electron (a negative charge). So which process is more likely to occur? Well, a rough guideline says that it's easy to gain or lose electrons but very difficult to gain or lose protons.

So atoms become ions by gaining or losing electrons. And ions that have a positive charge are called *cations.* The progression goes like this: The Na^+ ion is formed from the loss of one electron. Because it lost an electron, it has more protons than electrons, or more *positive* charges than negative charges, which means it's now called the Na^+ cation. Likewise, The Mg^{2+} cation is formed when the neutral magnesium atom loses two electrons.

Now consider the chlorine atom in sodium chloride. The neutral chlorine atom has acquired a negative charge by gaining an electron. Because it has unequal numbers of protons and electrons, it's now an ion, represented like this: Cl^-. And because ions that have a negative charge are called *anions,* it's now called the Cl^- anion. (You can get the full scoop on ions, cations, and anions in Chapter 6, if you're interested. This here's just a teaser.)

Just for kicks, here are some extra tidbits about ions for your reading pleasure:

- You can write electron configurations and energy level diagrams for ions. The neutral sodium atom (11 protons) has an electron configuration of $1s^2 2s^2 2p^6 3s^1$. The sodium cation has lost an electron — the valence electron, which is *farthest* away from the nucleus (the 3s electron, in this case). The electron configuration of Na^+ is $1s^2 2s^2 2p^6$.
- The electron configuration of the chloride ion (Cl^-) is $1s^2 2s^2 2p^6 3s^2 3p^6$. This is the same electron configuration as the neutral Argon atom. If two chemical species have the same electron configuration, they're said to be *isoelectronic.* Figuring out chemistry requires learning a whole new language, eh?
- This section has been discussing *monoatomic* (one atom) ions. But *polyatomic* (many atom) ions do exist. The ammonium ion, NH_4^+, is a polyatomic ion, or, specifically, a polyatomic cation. The nitrate ion, NO_3^-, is also a polyatomic ion, or, specifically, a polyatomic anion.
- Ions are commonly found in a class of compounds called *salts,* or *ionic solids.* Salts, when melted or dissolved in water, yield solutions that conduct electricity. A substance that conducts electricity when melted or dissolved in water is called an *electrolyte.* Table salt — sodium chloride — is a good example. On the other hand, when table sugar (sucrose) is dissolved in water, it becomes a solution that doesn't conduct electricity. So sucrose is a *nonelectrolyte.* Whether a substance is an electrolyte or a nonelectrolyte gives clues to the type of bonding in the compound. If the substance is an electrolyte, the compound is probably *ionically bonded* (see Chapter 6). If it's a nonelectrolyte, it's probably *covalently bonded* (see Chapter 7).

Chapter 4

The Periodic Table (But No Chairs)

In This Chapter

- Understanding periodicity
- Figuring out how elements are organized in the periodic table

In this chapter, I introduce you to the second most important tool a chemist possesses — the periodic table. (The most important? The beaker and Bunsen burner he or she brews coffee with.)

Chemists are a little lazy, as are most scientists. They like to put things together into groups based on similar properties. This process, called *classification,* makes it much easier to study a particular system. Scientists have grouped the elements together in the periodic table so they don't have to learn the properties of individual elements. With the periodic table, they can just learn the properties of the various groups. So in this chapter, I show you how the elements are arranged in the table, and I show you some important groups. I also explain how chemists and other scientists go about using the periodic table.

Repeating Patterns of Periodicity

In nature, as well as in things that mankind invents, you may notice some repeating patterns. The seasons repeat their pattern of fall, winter, spring, and summer. The tides repeat their pattern of rising and falling. Tuesday follows Monday, December follows November, and so on. This pattern of repeating order is called *periodicity.*

In the mid-1800s, Dmitri Mendeleev, a Russian chemist, noticed a repeating pattern of chemical properties in the elements that were known at the time. Mendeleev arranged the elements in order of increasing atomic mass (see Chapter 3 for a description of atomic mass), to form something that fairly closely resembles our modern periodic table. He was even able to predict the properties of some of the then-unknown elements. Later, the elements were rearranged in order of increasing *atomic number,* the number of protons in the nucleus of the atom (again, see Chapter 3). Figure 4-1 shows the modern periodic table.

PERIODIC TABLE OF THE ELEMENTS

	1 IA	2 IIA	3 IIIB	4 IVB	5 VB	6 VIB	7 VIIB	8 VIIIB	9 VIIIB
1	1 H Hydrogen 1.00797								
2	3 Li Lithium 6.939	4 Be Beryllium 9.0122							
3	11 Na Sodium 22.9898	12 Mg Magnesium 24.312							
4	19 K Potassium 39.102	20 Ca Calcium 40.08	21 Sc Scandium 44.956	22 Ti Titanium 47.90	23 V Vanadium 50.942	24 Cr Chromium 51.996	25 Mn Manganese 54.9380	26 Fe Iron 55.847	27 Co Cobalt 58.9332
5	37 Rb Rubidium 85.47	38 Sr Strontium 87.62	39 Y Yttrium 88.905	40 Zr Zirconium 91.22	41 Nb Niobium 92.906	42 Mo Molybdenum 95.94	43 Tc Technetium (99)	44 Ru Ruthenium 101.07	45 Rh Rhodium 102.905
6	55 Cs Cesium 132.905	56 Ba Barium 137.34	57 La Lanthanum 138.91	72 Hf Hafnium 179.49	73 Ta Tantalum 180.948	74 W Tungsten 183.85	75 Re Rhenium 186.2	76 Os Osmium 190.2	77 Ir Iridium 192.2
7	87 Fr Francium (223)	88 Ra Radium (226)	89 Ac Actinium (227)	104 Rf Rutherfordium (261)	105 Db Dubnium (262)	106 Sg Seaborgium (266)	107 Bh Bohrium (264)	108 Hs Hassium (269)	109 Mt Meitnerium (268)

Lanthanide Series	58 Ce Cerium 140.12	59 Pr Praseodymium 140.907	60 Nd Neodymium 144.24	61 Pm Promethium (145)	62 Sm Samarium 150.35	63 Eu Europium 151.96
Actinide Series	90 Th Thorium 232.038	91 Pa Protactinium (231)	92 U Uranium 238.03	93 Np Neptunium (237)	94 Pu Plutonium (242)	95 Am Americium (243)

Figure 4-1: The periodic table.

10 VIIIB	11 IB	12 IIB	13 IIIA	14 IVA	15 VA	16 VIA	17 VIIA	18 0
								He Helium 4.0026
			5 B Boron 10.811	6 C Carbon 12.01115	7 N Nitrogen 14.0067	8 O Oxygen 15.9994	9 F Flourine 18.9984	10 Ne Neon 20.183
			13 Al Aluminum 26.9815	14 Si Silicon 28.086	15 P Phosphorus 30.9738	16 S Sulfur 32.064	17 Cl Chlorine 35.453	18 Ar Argon 39.948
28 Ni Nickel 58.71	29 Cu Copper 63.546	30 Zn Zinc 65.37	31 Ga Gallium 69.72	32 Ge Germanium 72.59	33 As Arsenic 74.9216	34 Se Selenium 78.96	35 Br Bromine 79.904	36 Kr Krypton 83.80
46 Pd Palladium 106.4	47 Ag Silver 107.868	48 Cd Cadmium 112.40	49 In Indium 114.82	50 Sn Tin 118.69	51 Sb Antimony 121.75	52 Te Tellurium 127.60	53 I Iodine 126.9044	54 Xe Xenon 131.30
78 Pt Platinum 195.09	79 Au Gold 196.967	80 Hg Mercury 200.59	81 Tl Thallium 204.37	82 Pb Lead 207.19	83 Bi Bismuth 208.980	84 Po Polonium (210)	85 At Astatine (210)	86 Rn Radon (222)
110 Uun Ununnilium (269)	111 Uuu Unununium (272)	112 Uub Ununbium (277)	113 Uut §	114 Uuq Ununquadium (285)	115 Uup §	116 Uuh Ununhexium (289)	117 Uus §	118 Uuo Ununoctium (293)

64 Gd Gadolinium 157.25	65 Tb Terbium 158.924	66 Dy Dysprosium 162.50	67 Ho Holmium 164.930	68 Er Erbium 167.26	69 Tm Thulium 168.934	70 Yb Ytterbium 173.04	71 Lu Lutetium 174.97
96 Cm Curium (247)	97 Bk Berkelium (247)	98 Cf Californium (251)	99 Es Einsteinium (254)	100 Fm Fermium (257)	101 Md Mendelevium (258)	102 No Nobelium (259)	103 Lr Lawrencium (260)

§ Note: Elements 113, 115, and 117 are not known at this time, but are included in the table to show their expected positions.

Chemists can't imagine doing much of anything without having access to the periodic table. Instead of learning the properties of 109+ elements (more are created almost every year), chemists — and chemistry students — can simply learn the properties of families of elements, thus saving a lot of time and effort. They can find the relationships among elements and figure out the formulas of many different compounds by referring to the periodic table. The table readily provides atomic numbers, mass numbers, and information about the number of valence electrons.

I remember reading a science fiction story many years ago about an alien life based on the element silicon. Silicon was the logical choice for this story because it's in the same family as carbon, the element that's the basis for life on earth. So the periodic table is an absolute necessity for chemists, chemistry students, and science fiction novelists. Don't leave home without it!

Understanding How Elements Are Arranged in the Periodic Table

Look at the periodic table in Figure 4-1. The elements are arranged in order of increasing atomic number. The atomic number (number of protons) is located right above the element symbol. Under the element symbol is the *atomic mass,* or *atomic weight* (sum of the protons and neutrons). Atomic mass is a *weighted average* of all naturally occurring isotopes. (And if that's Greek to you, just flip to Chapter 3 for tons of fun with atomic mass and isotopes.) Notice also that two rows of elements — Ce-Lu (commonly called the Lanthanides) and Th-Lr (the Actinides) — have been pulled out of the main body of the periodic table. If they were included in the main body of the periodic table, the table would be much larger.

The periodic table is composed of horizontal rows called *periods.* The periods are numbered 1 through 7 on the left-hand side of the table. The vertical columns are called *groups,* or *families.* Members of these families have similar properties (see the section "Families and periods," later in this chapter). The families may be labeled at the top of the columns in one of two ways. The older method uses Roman numerals and letters. Many chemists (especially old ones like me) prefer and still use this method. The newer method simply uses the numbers 1 through 18. I use the older method in describing the features of the table.

Using the periodic table, you can classify the elements in many ways. Two quite useful ways are

- Metals, nonmetals, and metalloids
- Families and periods

Metals, nonmetals, and metalloids

If you look carefully at Figure 4-1, you can see a stair-stepped line starting at Boron (B), atomic number 5, and going all the way down to Polonium (Po), atomic number 84. Except for Germanium (Ge) and Antimony (Sb), all the elements to the left of that line can be classified as *metals.* Figure 4-2 shows the metals.

These metals have properties that you normally associate with the metals you encounter in everyday life. They are solid (with the exception of mercury, Hg, a liquid), shiny, good conductors of electricity and heat, *ductile* (they can be drawn into thin wires), and *malleable* (they can be easily hammered into very thin sheets). And all these metals tend to lose electrons easily (see Chapter 6). As you can see, the vast majority of the elements on the periodic table are classified as metals.

Except for the elements that border the stair-stepped line (more on those in a second), the elements to the right of the line are classified as *nonmetals* (along with hydrogen). These elements are shown in Figure 4-3.

Nonmetals have properties opposite those of the metals. The nonmetals are brittle, not malleable or ductile, poor conductors of both heat and electricity, and tend to gain electrons in chemical reactions. Some nonmetals are liquids.

The elements that border the stair-stepped line are classified as *metalloids,* and they're shown in Figure 4-4.

The metalloids, or *semimetals,* have properties that are somewhat of a cross between metals and nonmetals. They tend to be economically important because of their unique conductivity properties (they only partially conduct electricity), which make them valuable in the semiconductor and computer chip industry. (Did you think the term *silicon valley* referred to a valley covered in sand? Nope. Silicon, one of the metalloids, is used in making computer chips.)

Figure 4-2: The metals.

	IIA (2)	IIIB (3)	IVB (4)	VB (5)	VIB (6)	VIIB (7)	VIIIB (8)	(9)	(10)	IB (11)	IIB (12)				
3 Li Lithium 6.939	4 Be Beryllium 9.0122														
11 Na Sodium 22.9898	12 Mg Magnesium 24.312											13 Al Aluminum 26.9815			
19 K Potassium 39.102	20 Ca Calcium 40.08	21 Sc Scandium 44.956	22 Ti Titanium 47.90	23 V Vanadium 50.942	24 Cr Chromium 51.996	25 Mn Manganese 54.9380	26 Fe Iron 55.847	27 Co Cobalt 58.9332	28 Ni Nickel 58.71	29 Cu Copper 63.546	30 Zn Zinc 65.37	31 Ga Gallium 69.72			
37 Rb Rubidium 85.47	38 Sr Strontium 87.62	39 Y Yttrium 88.905	40 Zr Zirconium 91.22	41 Nb Niobium 92.906	42 Mo Molybdenum 95.94	43 Tc Technetium (99)	44 Ru Ruthenium 101.07	45 Rh Rhodium 102.905	46 Pd Palladium 106.4	47 Ag Silver 107.868	48 Cd Cadmium 112.40	49 In Indium 114.82	50 Sn Tin 118.69		
55 Cs Cesium 132.905	56 Ba Barium 137.34	57 La Lanthanum 138.91	72 Hf Hafnium 179.49	73 Ta Tantalum 180.948	74 W Tungsten 183.85	75 Re Rhenium 186.2	76 Os Osmium 190.2	77 Ir Iridium 192.2	78 Pt Platinum 195.09	79 Au Gold 196.967	80 Hg Mercury 200.59	81 Tl Thallium 204.37	82 Pb Lead 207.19	83 Bi Bismuth 208.980	84 Po Polonium (210)
87 Fr Francium (223)	88 Ra Radium (226)	89 Ac Actinium (227)	104 Rf Rutherfordium (261)	105 Db Dubnium (262)	106 Sg Seaborgium (266)	107 Bh Bohrium (264)	108 Hs Hassium (269)	109 Mt Meitnerium (268)	110 Uun Ununnilium (269)	111 Uuu Unununium (272)	112 Uub Ununbium (277)				

58 Ce Cerium 140.12	59 Pr Praseodymium 140.907	60 Nd Neodymium 144.24	61 Pm Promethium (145)	62 Sm Samarium 150.35	63 Eu Europium 151.96	64 Gd Gadolinium 157.25	65 Tb Terbium 158.924	66 Dy Dysprosium 162.50	67 Ho Holmium 164.930	68 Er Erbium 167.26	69 Tm Thulium 168.934	70 Yb Ytterbium 173.04	71 Lu Lutetium 174.97
90 Th Thorium 232.038	91 Pa Protactinium (231)	92 U Uranium 238.03	93 Np Neptunium (237)	94 Pu Plutonium (242)	95 Am Americium (243)	96 Cm Curium (247)	97 Bk Berkelium (247)	98 Cf Californium (251)	99 Es Einsteinium (254)	100 Fm Fermium (257)	101 Md Mendelevium (258)	102 No Nobelium (259)	103 Lr Lawrencium (260)

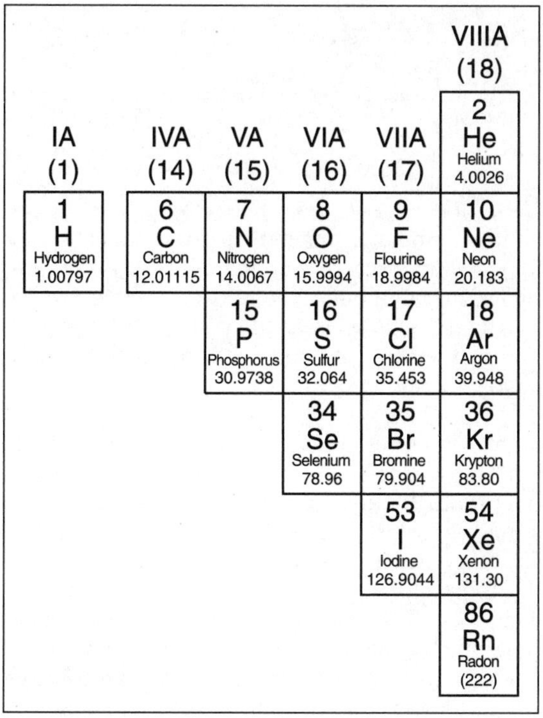

VIIIA
(18)
IA
(1)
IVA
(14)
VA
(15)
VIA
(16)
VIIA
(17)
2 He Helium 4.0026
1 H Hydrogen 1.00797
6 C Carbon 12.01115
7 N Nitrogen 14.0067
8 O Oxygen 15.9994
9 F Flourine 18.9984
10 Ne Neon 20.183
15 P Phosphorus 30.9738
16 S Sulfur 32.064
17 Cl Chlorine 35.453
18 Ar Argon 39.948
34 Se Selenium 78.96
35 Br Bromine 79.904
36 Kr Krypton 83.80
53 I Iodine 126.9044
54 Xe Xenon 131.30
86 Rn Radon (222)

Figure 4-3: The nonmetals.

IIIA
(13)
5 B Boron 10.811
14 Si Silicon 28.086
32 Ge Germanium 72.59
33 As Arsenic 74.9216
51 Sb Antimony 121.75
52 Te Tellurium 127.60
85 At Astatine (210)

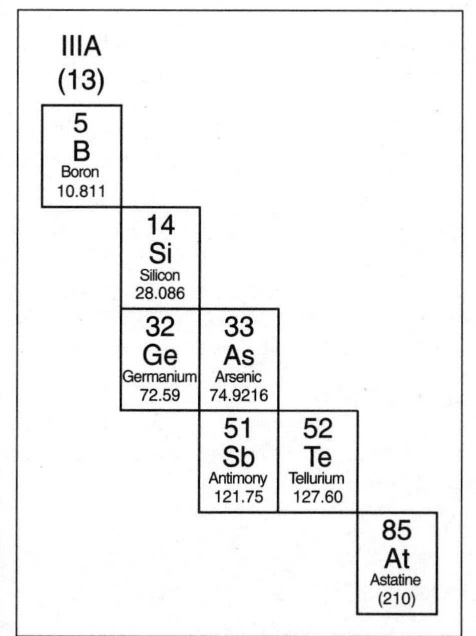

Figure 4-4: The metalloids.

Families and periods

If you refer to the periodic table shown in Figure 4-1, you see seven horizontal rows of elements called periods. In each period, the atomic numbers increase from left to right.

Even though they're in the same period, these elements have chemical properties that are not all that similar. Consider the first two members of period 3: sodium (Na) and magnesium (Mg). In reactions, they both tend to lose electrons (after all, they are metals), but sodium loses one electron, while magnesium loses two. Chlorine (Cl), down near the end of the period, tends to gain an electron (it's a nonmetal). So what you need to remember is that members of a period don't have very similar properties.

The members of a family do have similar properties. Consider the IA family, starting with Lithium (Li) — don't worry about hydrogen, because it's unique, and it doesn't really fit anywhere — and going through Francium (Fr). All these elements tend to lose only one electron in reactions. And all the members of the VIIA family tend to gain one electron.

So why do the elements in the same family have similar properties? And why do some families have the particular properties of electron loss or gain? To find out, you can examine four specific families on the periodic table and look at the electron configurations for a few elements in each family.

My family name is special

Take a look at Figure 4-5, which lists some important families that are given special names:

- The IA family is made up of the *alkali metals.* In reactions, these elements all tend to lose a single electron. This family contains some important elements, such as sodium (Na) and potassium (K). Both of these elements play an important role in the chemistry of the body and are commonly found in salts.
- The IIA family is made up of the *alkaline earth metals.* All these elements tend to lose two electrons. Calcium (Ca) is an important member of the IIA family (you need calcium for healthy teeth and bones).
- The VIIA family is made up of the *halogens.* They all tend to gain a single electron in reactions. Important members in the family include chlorine (Cl), used in making table salt and bleach, and iodine (I). Ever use tincture of iodine as a disinfectant?

- The VIIIA family is made up of the *noble gases.* These elements are *very* unreactive. For a long time, the noble gases were called the inert gases, because people thought that these elements wouldn't react at all. Later, a scientist named Neil Bartlett showed that at least some of the inert gases could be reacted, but they required very special conditions. After Bartlett's discovery, the gases were then referred to as noble gases.

What valence electrons have to do with families

Chapter 3 explains that an *electron configuration* shows the number of electrons in each orbital in a particular atom. The electron configuration forms the basis of the concept of bonding and molecular geometry and other important stuff that I cover in the various chapters of this book.

IA (1)	IIA (2)	VIIA (17)	VIIIA (18)
3 Li Lithium 6.939	4 Be Beryllium 9.0122		2 He Helium 4.0026
11 Na Sodium 22.9898	12 Mg Magnesium 24.312	9 F Flourine 18.9984	10 Ne Neon 20.183
19 K Potassium 39.102	20 Ca Calcium 40.08	17 Cl Chlorine 35.453	18 Ar Argon 39.948
37 Rb Rubidium 85.47	38 Sr Strontium 87.62	35 Br Bromine 79.904	36 Kr Krypton 83.80
55 Cs Cesium 132.905	56 Ba Barium 137.34	53 I Iodine 126.9044	54 Xe Xenon 131.30
87 Fr Francium (223)	88 Ra Radium (226)	85 At Astatine (210)	86 Rn Radon (222)
Alkali Metals	Alkaline Earth Metals	Halogens	Noble Gases

Figure 4-5: Some important chemical families.

Tables 4-1 through 4-4 show the electron configurations for the first three members of the families IA, IIA, VIIA, and VIIIA.

Table 4-1 Electron Configurations for Members of IA (alkali metals)

Element	*Electron Configuration*
Li	$1s^2\mathbf{2s^1}$
Na	$1s^22s^22p^6\mathbf{3s^1}$
K	$1s^22s^22p^63s^23p^6\mathbf{4s^1}$

Table 4-2 Electron Configurations for Members of IIA (alkaline earth metals)

Element	*Electron Configuration*
Be	$1s^2\mathbf{2s^2}$
Mg	$1s^22s^22p^6\mathbf{3s^2}$
Ca	$1s^22s^22p^63s^23p^6\mathbf{4s^2}$

Table 4-3 Electron Configurations for Members of VIIA (halogens)

Element	*Electron Configuration*
F	$1s^2\mathbf{2s^22p^5}$
Cl	$1s^22s^22p^6\mathbf{3s^23p^5}$
Br	$1s^22s^22p^63s^23p^6\mathbf{4s^2}3d^{10}\mathbf{4p^5}$

Table 4-4 Electron Configurations for Members of VIIIA (noble gases)

Element	*Electron Configuration*
Ne	$1s^2\mathbf{2s^22p^6}$
Ar	$1s^22s^22p^6\mathbf{3s^23p^6}$
Kr	$1s^22s^22p^63s^23p^6\mathbf{4s^2}3d^{10}\mathbf{4p^6}$

These electron configurations show that some similarities among each group of elements are in terms of their valence electrons. *Valence electrons* are the s and p electrons in the outermost energy level of an atom (see Chapter 3).

Look at the electron configurations for the alkali metals (Table 4-1). In lithium, energy level 1 is filled, and a single electron is in the 2s orbital. In sodium, energy levels 1 and 2 are filled, and a single electron is in energy level 3. All these elements have one valence electron in an s orbital. The alkaline earth elements (Table 4-2) each have two valence electrons. The halogens (Table 4-3) each have seven valence electrons (in s and p orbitals — d orbitals don't count), and the noble gases (Table 4-4) each have eight valence electrons, which fill their valence orbitals.

So how do you remember all this stuff?

Here's something to keep in mind about the number of valence electrons and the Roman numeral column number: The *I*A family has *1* valence electron; the *II*A family has *2* valence electrons; the *VII*A family has *7* valence electrons; and the *VIII*A family has *8* valence electrons. So for the families labeled with a Roman numeral and an A, the Roman numeral gives the number of valence electrons. Pretty cool, eh?

The Roman numeral makes it very easy to determine that oxygen (O) has six valence electrons (it's in the VIA family), that silicon (Si) has four, and so on. You don't even have to write the electronic configuration or the energy diagram to determine the number of valence electrons.

Noble and gassy

The fact that the noble gases have eight valence electrons, filling their valence, or outermost energy level, explains why the noble gases are extremely hard to react. They are stable, or "satisfied," with a filled (complete) valence energy level. They don't easily lose, gain, or share electrons.

A lot of stability in nature seems to be associated with this condition. Chemists observe that the other elements in the *A* families on the periodic table tend to lose, gain, or share valence electrons in order to achieve the goal of having a filled valence shell of eight electrons: This is sometimes called the *octet rule.* For example, look at the electron configuration for sodium (Na): $1s^2 2s^2 2p^6 \mathbf{3s^1}$. It has one valence electron — the $3s^1$. If it lost that electron, its valence shell would be energy level 2, which is filled. Without the $3s^1$, it would become *isoelectronic* (have the same electronic configuration) as Neon (Ne) and achieve stability. As I show you in Chapters 6 and 7, this is the driving force in chemical bonding: achieving stability by having a filled valence shell.

But what about elements that are labeled with a Roman numeral and a B? These elements, found in the middle of the periodic table, are commonly called the *transition metals;* their electrons are progressively filling the d orbitals. Scandium (Sc) is the first member of the transition metals, and it has an electronic configuration of $1s^22s^22p^63s^23p^64s^23d^1$. Titanium (Ti), the next transition metal, has a configuration $1s^22s^22p^63s^23p^64s^23d^2$. Notice that the number of electrons in the s and p orbitals is not changing. The progressively added electrons fill the d orbitals. Lanthanides and Actinides, the two groups of elements that are pulled out of the main body of the periodic table and shown below it, are classified as *inner transition metals.* In these elements, the electrons are progressively filling the f orbitals in much the same way that the electrons of the transition metals fill the d orbitals.

Chapter 5

Nuclear Chemistry: It'll Blow Your Mind

In This Chapter

- Understanding radioactivity and radioactive decay
- Figuring out half-lives
- The basics of nuclear fission
- Taking a look at nuclear fusion
- Tracing the effects of radiation

Most of this book deals, in one way or the other, with chemical reactions. And when I talk about these reactions, I'm really talking about how the valence electrons (the electrons in the outermost energy levels of atoms) are lost, gained, or shared. I mention very little about the nucleus of the atom because, to a very large degree, it's not involved in chemical reactions.

But, in this chapter, I do discuss the nucleus and the changes it can undergo. I talk about radioactivity and the different ways an atom can decay. I discuss half-lives and show you why they are important in the storage of nuclear waste products. I also discuss nuclear fission in terms of bombs, power plants, and the hope that nuclear fusion holds for mankind.

Like most of you reading this book, I'm a child of the Atomic Age. I actually remember open air testing of nuclear weapons. I remember being warned not to eat snow because it might contain fallout. I remember friends building fallout shelters. I remember A-bomb drills at school. I remember x-ray machines in shoe stores. (I never did order those x-ray glasses, though!) And I remember radioactive Fiesta stoneware and radium watch hands. When I was growing up, atomic energy was new, exciting, and scary. And it still is.

It All Starts with the Atom

To understand nuclear chemistry, you need to know the basics of atomic structure. Chapter 3 drones on and on *(and on)* about atomic structure, if you're interested. This section just provides a quickie brain dump.

The *nucleus,* that dense central core of the atom, contains both protons and neutrons. Electrons are outside the nucleus in energy levels. Protons have a positive charge, neutrons have no charge, and electrons have a negative charge. A neutral atom contains equal numbers of protons and electrons. But the number of neutrons within an atom of a particular element can vary. Atoms of the same element that have differing numbers of neutrons are called *isotopes.* Figure 5-1 shows the symbolization chemists use to represent a specific isotope of an element.

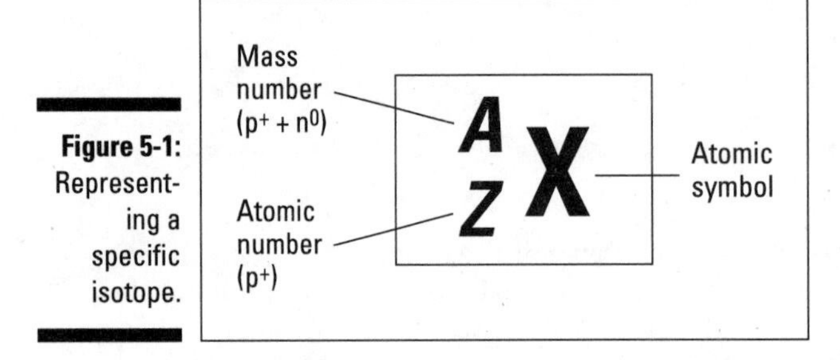

Figure 5-1: Representing a specific isotope.

In the figure, X represents the symbol of the element found on the periodic table, Z represents the *atomic number* (the number of protons in the nucleus), and A represents the *mass number* (the sum of the protons and neutrons in that particular isotope). If you subtract the atomic number from the mass number (A – Z), you get the number of neutrons in that particular isotope. A short way to show the same information is to simply use the element symbol (X) and the mass number (A) — for example, U-235.

Radioactivity and Man-Made Radioactive Decay

For purposes of this book, I define *radioactivity* as the spontaneous decay of an unstable nucleus. An unstable nucleus may break apart into two or more other particles with the release of some energy (see "Gone (Nuclear) Fission," later in this chapter, for more info on this process). This breaking apart can occur in a number of ways, depending on the particular atom that's decaying.

You can often predict one of the particles of a radioactive decay by knowing the other particle. Doing so involves something called *balancing the nuclear reaction.* (A *nuclear reaction* is any reaction involving a change in nuclear structure.)

Balancing a nuclear reaction is really a fairly simple process. But before I explain it, I want to show you how to represent a reaction:

Reactants → Products

Reactants are the substances you start with, and products are the new substances being formed. The arrow, called a *reaction arrow,* indicates that a reaction has taken place.

For a nuclear reaction to be balanced, the sum of all the atomic numbers on the left-hand side of the reaction arrow must equal the sum of all the atomic numbers on the right-hand side of the arrow. The same is true for the sums of the mass numbers. Here's an example: Suppose you're a scientist performing a nuclear reaction by bombarding a particular isotope of chlorine (Cl-35) with a neutron. (Work with me here. I'm just trying to get to a point.) You observe that an isotope of hydrogen, H-1, is created along with another isotope, and you want to figure out what the other isotope is. The equation for this example is

$$^{35}_{17}Cl + \underset{neutron}{^{1}_{0}n} \rightarrow \underline{\ \mathrm{Pr?}\ } + ^{1}_{1}H$$

Now to figure out the unknown isotope (represented by Pr), you need to balance the equation. The sum of the atomic numbers on the left is 17 (17 + 0), so you want the sum of the atomic numbers on the right to equal 17 too. Right now, you've got an atomic number of 1 on the right; 17 – 1 is 16, so that's the atomic number of the unknown isotope. This atomic number identifies the element as Sulfur (S).

Now look at the mass numbers in the equation. The sum of the mass numbers on the left is 36 (35 + 1), and you want the sum of the mass numbers on the right to equal 36, too. Right now, you've got a mass number of 1 on the right; 36 – 1 is 35, so that's the mass number of the unknown isotope. Now you know that the unknown isotope is a Sulfur isotope (S-35). And here's what the balanced nuclear equation looks like:

$$^{35}_{17}Cl + \underset{neutron}{^{1}_{0}n} \rightarrow ^{35}_{16}S + ^{1}_{1}H$$

This equation represents a nuclear *transmutation,* the conversion of one element into another. Nuclear transmutation is a process human beings control. S-35 is an isotope of sulfur that doesn't exist in nature. It's a *man-made isotope.* Alchemists, those ancient predecessors of chemists, dreamed of converting one element into another (usually lead into gold), but they were never able to master the process. Chemists are now able, sometimes, to convert one element into another.

Natural Radioactive Decay: How Nature Does It

Certain isotopes are unstable: Their nucleus breaks apart, undergoing nuclear decay. Sometimes the product of that nuclear decay is unstable itself and undergoes nuclear decay, too. For example, when U-238 (one of the radioactive isotopes of uranium) initially decays, it produces Th-234, which decays to Pa-234. The decay continues until, finally, after a total of 14 steps, Pb-206 is produced. Pb-206 is stable, and the decay sequence, or series, stops.

Before I show you *how* radioactive isotopes decay, I want to briefly explain *why* a particular isotope decays. The nucleus has all those positively charged protons shoved together in an extremely small volume of space. All those protons are repelling each other. The forces that normally hold the nucleus together, the "nuclear glue," sometimes can't do the job, and so the nucleus breaks apart, undergoing nuclear decay.

All elements with 84 or more protons are unstable; they eventually undergo decay. Other isotopes with fewer protons in their nucleus are also radioactive. The radioactivity corresponds to the neutron/proton ratio in the atom. If the neutron/proton ratio is too high (there are too many neutrons or too few protons), the isotope is said to be *neutron rich* and is, therefore, unstable. Likewise, if the neutron/proton ratio is too low (there are too few neutrons or too many protons), the isotope is unstable. The neutron/proton ratio for a certain element must fall within a certain range for the element to be stable. That's why some isotopes of an element are stable and others are radioactive.

There are three primary ways that naturally occurring radioactive isotopes decay:

- Alpha particle emission
- Beta particle emission
- Gamma radiation emission

In addition, there are a couple of less common types of radioactive decay:

- Positron emission
- Electron capture

Alpha emission

An *alpha particle* is defined as a positively charged particle of a helium nuclei. I hear ya: *Huh?* Try this: An alpha particle is composed of two protons and two neutrons, so it can be represented as a Helium-4 atom. As an alpha particle breaks away from the nucleus of a radioactive atom, it has no electrons, so it has a +2 charge. Therefore and to-wit, it's a positively charged particle of a helium nuclei. (Well, it's really a *cation,* a positively charged ion — see Chapter 3.)

But electrons are basically free — easy to lose and easy to gain. So normally, an alpha particle is shown with no charge because it very rapidly picks up two electrons and becomes a neutral helium atom instead of an ion.

Large, heavy elements, such as uranium and thorium, tend to undergo alpha emission. This decay mode relieves the nucleus of two units of positive charge (two protons) and four units of mass (two protons + two neutrons). What a process. Each time an alpha particle is emitted, four units of mass are lost. I wish I could find a diet that would allow me to lose four pounds at a time!

Radon-222 (Rn-222) is another alpha particle emitter, as shown in the following equation:

$$^{222}_{86}Rn \rightarrow ^{218}_{84}Po + \underset{\text{alpha particle}}{^{4}_{2}He}$$

Here, Radon-222 undergoes nuclear decay with the release of an alpha particle. The other remaining isotope must have a mass number of 218 (222 – 4) and an atomic number of 84 (86 – 2), which identifies the element as Polonium (Po). (If this subtraction stuff confuses you, check out how to balance equations in the section "Radioactivity and Man-Made Radioactive Decay," earlier in this chapter.)

Beta emission

A *beta particle* is essentially an electron that's emitted from the nucleus. (Now I know what you're thinking — electrons aren't in the nucleus. Keep on reading to find out how they can be formed in this nuclear reaction.) Iodine-131 (I-131), which is used in the detection and treatment of thyroid cancer, is a beta particle emitter:

$$^{131}_{53}I \rightarrow ^{131}_{54}Xe + \underset{\text{beta particle}}{^{0}_{-1}e}$$

Here, the Iodine-131 gives off a beta particle (an electron), leaving an isotope with a mass number of 131 (131 – 0) and an atomic number of 54 (53 – (-1)). An atomic number of 54 identifies the element as Xenon (Xe).

Notice that the mass number doesn't change in going from I-131 to Xe-131, but the atomic number increases by one. In the iodine nucleus, a neutron was converted (decayed) into a proton and an electron, and the electron was emitted from the nucleus as a beta particle. Isotopes with a high neutron/proton ratio often undergo beta emission, because this decay mode allows the number of neutrons to be decreased by one and the number of protons to be increased by one, thus lowering the neutron/proton ratio.

Gamma emission

Alpha and beta particles have the characteristics of matter: They have definite masses, occupy space, and so on. However, because there is no mass change associated with gamma emission, I refer to gamma emission as *gamma radiation emission.* Gamma radiation is similar to x-rays — high energy, short wavelength radiation. Gamma radiation commonly accompanies both alpha and beta emission, but it's usually not shown in a balanced nuclear reaction. Some isotopes, such as Cobalt-60 (Co-60), give off large amounts of gamma radiation. Co-60 is used in the radiation treatment of cancer. The medical personnel focus gamma rays on the tumor, thus destroying it.

Positron emission

Although positron emission doesn't occur with naturally occurring radioactive isotopes, it does occur naturally in a few man-made ones. A *positron* is essentially an electron that has a positive charge instead of a negative charge. A positron is formed when a proton in the nucleus decays into a neutron and a positively charged electron. The positron is then emitted from the nucleus. This process occurs in a few isotopes, such as Potassium-40 (K-40), as shown in the following equation:

$$^{40}_{19}K \rightarrow ^{40}_{18}Ar + \underset{positron}{^{0}_{+1}e}$$

The K-40 emits the positron, leaving an element with a mass number of 40 (40 – 0) and an atomic number of 18 (19 – 1). An isotope of argon (Ar), Ar-40, has been formed.

If you watch *Star Trek,* you may have heard about antimatter. The positron is a tiny bit of antimatter. When it comes in contact with an electron, both particles are destroyed with the release of energy. Luckily, not many positrons are produced: If a lot of them were produced, you'd probably have to spend a lot of time ducking explosions.

Electron capture

Electron capture is a rare type of nuclear decay in which an electron from the innermost energy level (the 1s — see Chapter 3) is captured by the nucleus. This electron combines with a proton to form a neutron. The atomic number decreases by one, but the mass number stays the same. The following equation shows the electron capture of Polonium-204 (Po-204):

$$^{204}_{84}Po + {}^{0}_{-1}e \rightarrow {}^{204}_{83}Bi + x\text{-}rays$$

The electron combines with a proton in the polonium nucleus, creating an isotope of bismuth (Bi-204).

The capture of the 1s electron leaves a vacancy in the 1s orbitals. Electrons drop down to fill the vacancy, releasing energy not in the visible part of the electromagnetic spectrum but in the X-ray portion.

Half-Lives and Radioactive Dating

If you could watch a single atom of a radioactive isotope, U-238, for example, you wouldn't be able to predict when that particular atom might decay. It might take a millisecond, or it might take a century. There's simply no way to tell.

But if you have a large enough sample — what mathematicians call a *statistically significant sample size* — a pattern begins to emerge. It takes a certain amount of time for half the atoms in a sample to decay. It then takes the same amount of time for half the remaining radioactive atoms to decay, and the same amount of time for half of those remaining radioactive atoms to decay, and so on. The amount of time it takes for one-half of a sample to decay is called the *half-life* of the isotope, and it's given the symbol $t_{1/2}$. This process is shown in Table 5-1.

Table 5-1 Half-Life Decay of a Radioactive Isotope

Half-Life	*Percent of Radioactive Isotope Remaining*
0	100.00
1	50.00
2	25.00
3	12.50
4	6.25
5	3.12
6	1.56
7	0.78
8	0.39
9	0.19
10	0.09

It's important to realize that the half-life decay of radioactive isotopes is not linear. For example, you can't find the remaining amount of an isotope as 7.5 half-lives by finding the midpoint between 7 and 8 half-lives. This decay is an example of an exponential decay, shown in Figure 5-2.

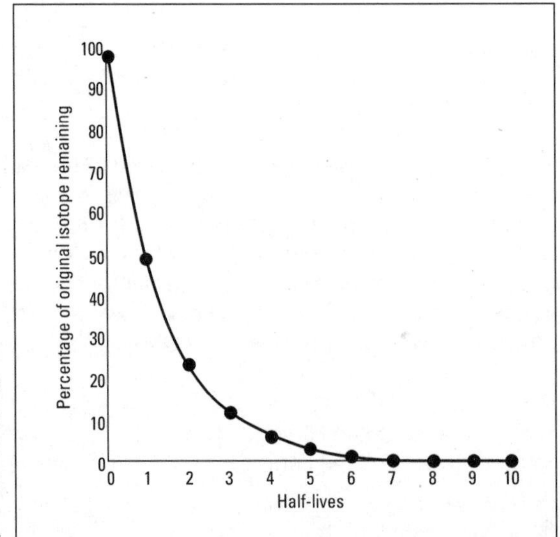

Figure 5-2: Decay of a radioactive isotope.

If you want to find times or amounts that are not associated with a simple multiple of a half-life, you can use this equation:

$$\ln\left(\frac{N_0}{N}\right) = \left(\frac{0.6963}{t_{1/2}}\right)t$$

In the equation, *ln* stands for the *natural logarithm* (not the base 10 log; it's that ln button on your calculator, not the log button), N_0 is the amount of radioactive isotope that you start with, N is the amount of radioisotope left at some time *(t)*, and $t_{1/2}$ is the half-life of the radioisotope. If you know the half-life and the amount of the radioactive isotope that you start with, you can use this equation to calculate the amount remaining radioactive at any time. But we are going to keep it simple.

Half-lives may be very short or very long. Table 5-2 shows the half-lives of some typical radioactive isotopes.

Table 5-2 Half-Lives of Some Radioactive Isotopes

Radioisotope	*Radiation Emitted*	*Half-Life*
Kr-94	Beta	1.4 seconds
Rn-222	Alpha	3.8 days
I-131	Beta	8 days
Co-60	Gamma	5.2 years
H-3	Beta	12.3 years
C-14	Beta	5,730 years
U-235	Alpha	4.5 billion years
Re-187	Beta	70 billion years

Safe handling

Knowing about half-lives is important because it enables you to determine when a sample of radioactive material is safe to handle. The rule is that a sample is safe when its radioactivity has dropped below detection limits. And that occurs at 10 half-lives. So, if radioactive iodine-131 ($t_{1/2}$ = 8 days) is injected into the body to treat thyroid cancer, it'll be "gone" in 10 half-lives, or 80 days.

This stuff is important to know when using radioactive isotopes as *medical tracers,* which are taken into the body to allow doctors to trace a pathway or find a blockage, or in cancer treatments. They need to be active long enough to treat the condition, but they should also have a short enough half-life so that they don't injure healthy cells and organs.

Radioactive dating

A useful application of half-lives is *radioactive dating.* No, radioactive dating has nothing to do with taking an X-ray tech to the movies. It has to do with figuring out the age of ancient things.

Carbon-14 (C-14), a radioactive isotope of carbon, is produced in the upper atmosphere by cosmic radiation. The primary carbon-containing compound in the atmosphere is carbon dioxide, and a very small amount of carbon dioxide contains C-14. Plants absorb C-14 during photosynthesis, so C-14 is incorporated into the cellular structure of plants. Plants are then eaten by animals, making C-14 a part of the cellular structure of all living things.

As long as an organism is alive, the amount of C-14 in its cellular structure remains constant. But when the organism dies, the amount of C-14 begins to decrease. Scientists know the half-life of C-14 (5,730 years, listed in Table 5-2), so they can figure out how long ago the organism died.

Radioactive dating using C-14 has been used to determine the age of skeletons found at archeological sites. Recently, it was used to date the *Shroud of Turin,* a piece of linen in the shape of a burial cloth that contains an image of a man. Many thought that it was the burial cloth of Jesus, but in 1988, radiocarbon dating determined that the cloth dated from around A.D. 1200–1300. Even though we don't know how the image of the man was placed on the Shroud, C-14 dating has proven that it's not the death cloth of Jesus.

Carbon-14 dating can only be used to determine the age of something that was once alive. It can't be used to determine the age of a moon rock or a meteorite. For nonliving substances, scientists use other isotopes, such as potassium-40.

Gone (Nuclear) Fission

In the 1930s, scientists discovered that some nuclear reactions can be initiated and controlled (see "Radioactivity and Man-Made Radioactive Decay," earlier in this chapter). Scientists usually accomplished this task by bombarding a large isotope with a second, smaller one — commonly a neutron.

The collision caused the larger isotope to break apart into two or more elements, which is called *nuclear fission.* The nuclear fission of uranium-235 is shown in the following equation:

$$^{235}_{92}U + ^{1}_{0}n \rightarrow ^{142}_{56}Ba + ^{91}_{36}Kr + 3\,^{1}_{0}n$$

Reactions of this type also release a lot of energy. Where does the energy come from? Well, if you make *very* accurate measurement of the masses of all the atoms and subatomic particles you start with and all the atoms and subatomic particles you end up with, and then compare the two, you find that there's some "missing" mass. Matter disappears during the nuclear reaction. This loss of matter is called the *mass defect.* The missing matter is converted into energy.

You can actually calculate the amount of energy produced during a nuclear reaction with a fairly simple equation developed by Einstein: $E = mc^2$. In this equation, E is the amount of energy produced, m is the "missing" mass, or the mass defect, and c is the speed of light, which is a rather large number. The speed of light is squared, making that part of the equation a *very* large number that, even when multiplied by a small amount of mass, yields a *large* amount of energy.

Chain reactions and critical mass

Take a look at the equation for the fission of U-235 in the preceding section. Notice that one neutron was used, but three were produced. These three neutrons, if they encounter other U-235 atoms, can initiate other fissions, producing even more neutrons. It's the old domino effect. In terms of nuclear chemistry, it's a continuing cascade of nuclear fissions called a *chain reaction.* The chain reaction of U-235 is shown in Figure 5-3.

This chain reaction depends on the release of more neutrons than were used during the nuclear reaction. If you were to write the equation for the nuclear fission of U-238, the more abundant isotope of uranium, you'd use one neutron and only get one back out. You can't have a chain reaction with U-238. But isotopes that produce an *excess* of neutrons in their fission support a chain reaction. This type of isotope is said to be *fissionable,* and there are only two main fissionable isotopes used during nuclear reactions — uranium-235 and plutonium-239.

A certain minimum amount of fissionable matter is needed to support a self-sustaining chain reaction, and it's related to those neutrons. If the sample is small, then the neutrons are likely to shoot out of the sample before hitting a U-235 nucleus. If they don't hit a U-235 nucleus, no extra electrons and no energy are released. The reaction just fizzles. The minimum amount of fissionable material needed to ensure that a chain reaction occurs is called the *critical mass.* Anything less than this amount is called *subcritical.*

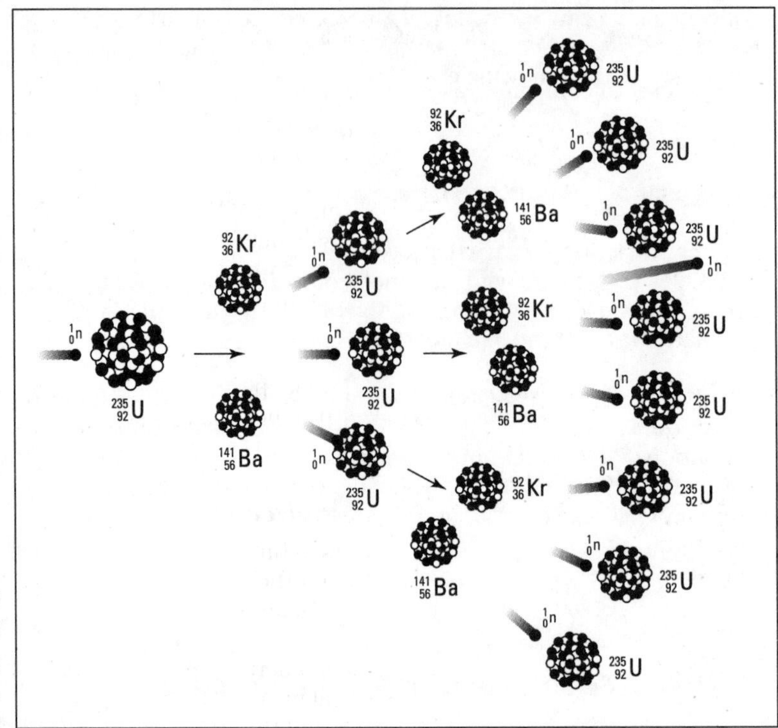

Figure 5-3: Chain reaction.

Atomic bombs (big bangs that aren't theories)

Because of the tremendous amount of energy released in a fission chain reaction, the military implications of nuclear reactions were immediately realized. The first atomic bomb was dropped on Hiroshima, Japan, on August 6, 1945.

In an atomic bomb, two pieces of a fissionable isotope are kept apart. Each piece, by itself, is subcritical. When it's time for the bomb to explode, conventional explosives force the two pieces together to cause a critical mass. The chain reaction is uncontrolled, releasing a tremendous amount of energy almost instantaneously.

The real trick, however, is to control the chain reaction, releasing its energy slowly so that ends other than destruction might be achieved.

Nuclear power plants

The secret to controlling a chain reaction is to control the neutrons. If the neutrons can be controlled, then the energy can be released in a controlled way. That's what scientists have done with nuclear power plants.

In many respects, a nuclear power plant is similar to a conventional fossil fuel power plant. In this type of plant, a fossil fuel (coal, oil, natural gas) is burned, and the heat is used to boil water, which, in turn, is used to make steam. The steam is then used to turn a turbine that is attached to a generator that produces electricity.

The big difference between a conventional power plant and a nuclear power plant is that the nuclear power plant produces heat through nuclear fission chain reactions.

How do nuclear power plants make electricity?

Most people believe that the concepts behind nuclear power plants are tremendously complex. That's really not the case. Nuclear power plants are very similar to conventional fossil fuel plants.

The fissionable isotope is contained in fuel rods in the reactor core. All the fuel rods together comprise the critical mass. Control rods, commonly made of boron or cadmium, are in the core, and they act like neutron sponges to control the rate of radioactive decay. Operators can stop a chain reaction completely by pushing the control rods all the way into the reactor core, where they absorb all the neutrons. The operators can then pull out the control rods a little at a time to produce the desired amount of heat.

A liquid (water or, sometimes, liquid sodium) is circulated through the reactor core, and the heat generated by the fission reaction is absorbed. The liquid then flows into a steam generator, where steam is produced as the heat is absorbed by water. This steam is then piped through a steam turbine that's connected to an electric generator. The steam is condensed and recycled through the steam generator. This forms a closed system; that is, no water or steam escapes — it's all recycled.

The liquid that circulates through the reactor core is also part of a closed system. This closed system helps ensure that no contamination of the air or water takes place. But sometimes problems do arise.

Oh, so many problems

In the United States, there are approximately 100 nuclear reactors, producing a little more than 20 percent of the country's electricity. In France, almost

80 percent of the country's electricity is generated through nuclear fission. Nuclear power plants have certain advantages. No fossil fuels are burned (saving fossil-fuel resources for producing plastics and medicines), and there are no combustion products, such as carbon dioxide, sulfur dioxide, and so on, to pollute the air and water. But problems are associated with nuclear power plants.

One is cost. Nuclear power plants are expensive to build and operate. The electricity that's generated by nuclear power costs about twice as much as electricity generated through fossil fuel or hydroelectric plants. Another problem is that the supply of fissionable uranium-235 is limited. Of all the naturally occurring uranium, only about 0.75 percent is U-235. A vast majority is nonfissionable U-238. At current usage levels, we'll be out of naturally occurring U-235 in fewer than 100 years. A little bit more time can be gained through the use of breeder reactors (see "Breeder reactors: Making more nuclear stuff," later in this chapter). But there's a limit to the amount of nuclear fuel available in the earth, just as there's a limit to the amount of fossil fuels.

However, the two major problems associated with nuclear fission power are accidents (safety) and disposal of nuclear wastes.

Accidents: Three Mile Island and Chernobyl

Although nuclear power reactors really do have a good safety record, the distrust and fear associated with radiation make most people sensitive to safety issues and accidents. The most serious accident to occur in the United States happened in 1979 at the Three Mile Island Plant in Pennsylvania. A combination of operator error and equipment failure caused a loss of reactor core coolant. The loss of coolant led to a partial meltdown and the release of a small amount of radioactive gas. There was no loss of life or injury to plant personnel or the general population.

This was not the case at Chernobyl, Ukraine, in 1986. Human error, along with poor reactor design and engineering, contributed to a tremendous overheating of the reactor core, causing it to rupture. Two explosions and a fire resulted, blowing apart the core and scattering nuclear material into the atmosphere. A small amount of this material made its way to Europe and Asia. The area around the plant is *still* uninhabitable. The reactor has been encased in concrete, and it must remain that way for hundreds of years. Hundreds of people died. Many others felt the effect of radiation poisoning. Instances of thyroid cancer, possibly caused by the release of I-13, have risen dramatically in the towns surrounding Chernobyl. It will be many more years until the effects of this disaster will be fully known.

How do you get rid of this stuff: Nuclear wastes

The fission process produces large amounts of radioactive isotopes. If you look at Table 5-2, you'll notice that some of the half-lives of radioactive isotopes are rather long. Those isotopes are safe after ten half-lives. The length of ten half-lives presents a problem when dealing with the waste products of a fission reactor.

Eventually, all reactors must have their nuclear fuel replenished. And as we disarm nuclear weapons, we must deal with their radioactive material. Many of these waste products have long half-lives. How do we safely store the isotopes until their residual radioactivity has dropped to safe limits (ten half-lives)? How do we protect the environment and ourselves, and our children for generations to come, from this waste? These questions are undoubtedly the most serious problem associated with the peaceful use of nuclear power.

Nuclear waste is divided into low-level and high-level material, based on the amount of radioactivity being emitted. In the United States, low-level wastes are stored at the site of generation or at special storage facilities. The wastes are basically buried and guarded at the sites. High-level wastes pose a much larger problem. They're temporarily being stored at the site of generation, with plans to eventually seal the material in glass and then in drums. The material will then be stored underground in Nevada. At any rate, the waste must be kept safe and undisturbed for at least 10,000 years. Other countries face the same problems. There has been some dumping of nuclear material into deep trenches in the sea, but this practice has been discouraged by many nations.

Breeder reactors: Making more nuclear stuff

Only the U-235 isotope of uranium is fissionable because it's the only isotope of uranium that produces the excess of neutrons needed to maintain a chain-reaction. The far more plentiful U-238 isotope doesn't produce those extra neutrons.

The other commonly used fissionable isotope, plutonium-239 (Pu-239), is very rare in nature. But there's a way to make Pu-239 from U-238 in a special fission reactor called a *breeder reactor.* Uranium-238 is first bombarded with a neutron to produce U-239, which decays to Pu-239. The process is shown in Figure 5-4.

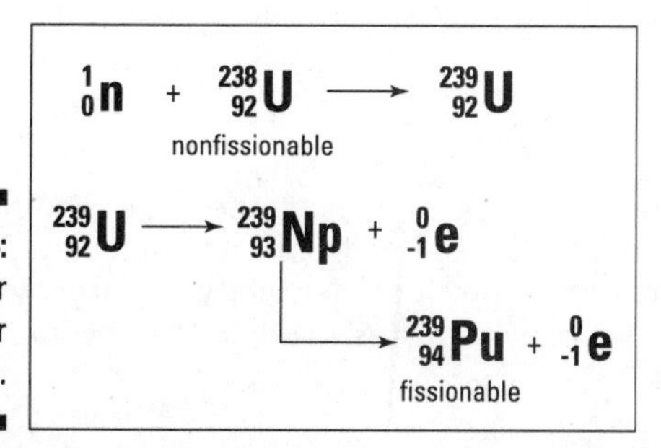

Figure 5-4: The breeder reactor process.

Breeder reactors can extend the supply of fissionable fuels for many, many years, and they're currently being used in France. But the United States is moving *slowly* with the construction of breeder reactors because of several problems associated with them. First, they're extremely expensive to build. Second, they produce large amounts of nuclear wastes. And, finally, the plutonium that's produced is much more hazardous to handle than uranium and can easily be used in an atomic bomb.

Nuclear Fusion: The Hope for Our Energy Future

Soon after the fission process was discovered, another process, called *fusion,* was discovered. Fusion is essentially the opposite of fission. In fission, a heavy nucleus is split into smaller nuclei. With fusion, lighter nuclei are fused into a heavier nucleus.

The fusion process is the reaction that powers the sun. On the sun, in a series of nuclear reactions, four isotopes of hydrogen-1 are fused into a helium-4 with the release of a tremendous amount of energy. Here on earth, two other isotopes of hydrogen are used: H-2, called deuterium, and H-3, called tritium. Deuterium is a minor isotope of hydrogen, but it's still relatively abundant. Tritium doesn't occur naturally, but it can easily be produced by bombarding deuterium with a neutron. The fusion reaction is shown in the following equation:

$$^{3}_{1}H + ^{2}_{1}H \rightarrow ^{4}_{2}He + ^{1}_{0}n$$

The first demonstration of nuclear fusion — the hydrogen bomb — was conducted by the military. A hydrogen bomb is approximately 1,000 times as powerful as an ordinary atomic bomb.

The isotopes of hydrogen needed for the hydrogen bomb fusion reaction were placed around an ordinary fission bomb. The explosion of the fission bomb released the energy needed to provide the *activation energy* (the energy necessary to initiate, or start, the reaction) for the fusion process.

Control issues

The goal of scientists for the last 45 years has been the controlled release of energy from a fusion reaction. If the energy from a fusion reaction can be released slowly, it can be used to produce electricity. It will provide an unlimited supply of energy that has no wastes to deal with or contaminants to harm the atmosphere — simply non-polluting helium. But achieving this goal requires overcoming three problems:

- Temperature
- Time
- Containment

Temperature

The fusion process requires an extremely high activation energy. Heat is used to provide the energy, but it takes a *lot* of heat to start the reaction. Scientists estimate that the sample of hydrogen isotopes must be heated to approximately 40,000,000 K. (K represents the Kelvin temperature scale. To get the Kelvin temperature, you add 273 to the Celsius temperature. Chapter 2 explains all about Kelvin and his pals Celsius and Fahrenheit.)

Now 40,000,000 K is hotter than the sun! At this temperature, the electrons have long since left the building; all that's left is a positively-charged *plasma,* bare nuclei heated to a tremendously high temperature. Presently, scientists are trying to heat samples to this high temperature through two ways — magnetic fields and lasers. Neither one has yet achieved the necessary temperature.

Time

Time is the second problem scientists must overcome to achieve the controlled release of energy from fusion reactions. The charged nuclei must be held together close enough and long enough for the fusion reaction to start. Scientists estimate that the plasma needs to be held together at 40,000,000 K for about one second.

Containment

Containment is the major problem facing fusion research. At 40,000,000 K, everything is a gas. The best ceramics developed for the space program would vaporize when exposed to this temperature. Because the plasma has a charge, magnetic fields can be used to contain it — like a magnetic bottle. But if the bottle leaks, the reaction won't take place. And scientists have yet to create a magnetic field that won't allow the plasma to leak. Using lasers to zap the hydrogen isotope mixture and provide the necessary energy bypasses the containment problem. But scientists have not figured out how to protect the lasers themselves from the fusion reaction.

What the future holds

The latest estimates indicate that science is 5 to 10 years away from showing that fusion can work: This is the so-called *break-even point,* where we get out more energy than we put in. It will then be another 20 to 30 years before a functioning fusion reactor is developed. But scientists are optimistic that controlled fusion power will be achieved. The rewards are great — an unlimited source of nonpolluting energy.

An interesting by-product of fusion research is the *fusion torch* concept. With this idea, the fusion plasma, which must be cooled in order to produce steam, is used to incinerate garbage and solid wastes. Then the individual atoms and small molecules that are produced are collected and used as raw materials for industry. It seems like an ideal way to close the loop between waste and raw materials. Time will tell if this concept will eventually make it into practice.

Am I Glowing? The Effects of Radiation

Radiation can have two basic effects on the body:

- It can destroy cells with heat.
- It can ionize and fragment cells.

Radiation generates heat. This heat can destroy tissue, much like a sunburn does. In fact, the term *radiation burn* is commonly used to describe the destruction of skin and tissue due to heat.

The other major way that radiation can affect the body is through the ionization and fragmentation of cells. Radioactive particles and radiation have a lot of *kinetic energy* (energy of motion — see Chapter 2) associated with them. When these particles strike cells within the body, they can *fragment* (destroy) the cells or *ionize* the cells — turn the cells into ions (charged atoms) by knocking off an electron. (Flip to Chapter 3 for the full scoop on ions.) Ionization weakens bonds and can lead to the damage, destruction, or mutation of the cells.

Radon: Hiding in our houses

Radon is a radioactive isotope that's been receiving a lot of publicity recently. Radon-222 is formed naturally as part of the decay of uranium. It's an unreactive noble gas, so it escapes from the ground into the air. Because it's heavier than air, it can accumulate in basements.

Radon itself has a short half-life of 3.8 days, but it decays to Polonium-218, a solid. So if radon is inhaled, solid Po-218 can accumulate in the lungs. Po-218 is an alpha emitter, and, even though this type of radiation is not very penetrating, it has been linked to increased instances of lung cancer. In many parts of the United States, radon testing is performed before selling a house. Commercial test kits can be opened, left in the basement area for a specified amount of time, and then sent to a lab for analysis. The question of whether radon represents a serious problem is still being investigated and debated.

Part II
Blessed Be the Bonds That Tie

"Okay, now that the paramedic is here with the defibrillator and smelling salts, prepare to learn about covalent bonds..."

In this part . . .

Mention chemistry, and most people immediately think of chemical reactions. Scientists use chemical reactions to make new drugs, plastics, cleaners, fabrics — the list is endless. They also use chemical reactions to analyze samples and find out what and how much is in them. Chemical reactions power our bodies, our sun, and our universe. Chemistry is all reactions and the bonding that occurs in them. And that's what this part is all about.

These chapters introduce you to the two main types of bonding found in nature: ionic bonding and covalent bonding. I show you how to predict the formulas of ionic compounds (salts) and how to name them. I explain covalent bonding, how to draw Lewis structural formulas, and how to predict the shapes of simple molecules. I tell you about chemical reactions and show you the various general types. In addition, I cover chemical equilibrium, kinetics, and electrochemistry — batteries, cells, and electroplating.

I think you'll get a charge out of the material in this part. In fact, I don't see how you can fail to react to it.

Chapter 6

Opposites Do Attract: Ionic Bonds

In This Chapter

- Finding out why and how ions are formed
- Discovering how cations and ions are formed
- Understanding polyatomic ions
- Deciphering the formulas of ionic compounds
- Naming ionic compounds
- Clarifying the difference between electrolytes and nonelectrolytes

If I had to point to the one thing that made me want to major in chemistry, it would be the reactions of salts. I remember the day clearly: It was the second half of general chemistry, and I was doing *qualitative analysis* (finding out what's in a sample) of salts. I really enjoyed the colors of the compounds formed in the reactions I was doing, and the labs were fun and challenging. I was hooked.

In this chapter, I introduce you to ionic bonding, the type of bonding that holds salts together. I discuss simple ions and polyatomic ions: how they form and how they combine. I also show you how to predict the formulas of ionic compounds and how chemists detect ionic bonds.

The Magic of an Ionic Bond: Sodium + Chlorine = Table Salt

Sodium is a fairly typical metal. It's silvery, soft, and a good conductor. It's also highly reactive: Sodium is normally stored under oil to keep it from reacting with the water in the atmosphere. If you melt a freshly cut piece of sodium and put it into a beaker filled with greenish-yellow chlorine gas, something very impressive happens. The molten sodium begins to glow with a white light that gets brighter and brighter. The chlorine gas swirls, and soon the color of the gas begins to disappear. In a couple of minutes, the reaction is over, and the beaker can be safely uncovered. You find table salt, or NaCl, deposited on the inside of the beaker.

Understanding the components

If you really stop and think about it, the process of creating table salt is pretty remarkable. You take two substances that are both very hazardous (chlorine was used by the Germans against Allied troops during World War I), and from them you make a substance that's necessary for life. In this section, I show you what happens during the chemical reaction to create salt and, more importantly, why it occurs.

Sodium is an alkali metal, a member of the IA family on the periodic table. The Roman numerals at the top of the A families show the number of valence electrons (s and p electrons in the outermost energy level) in the particular element (see Chapter 4 for details). So sodium has 1 valence electron and 11 total electrons because its atomic number is 11.

You can use an energy level diagram to represent the distribution of electrons in an atom. Sodium's energy level diagram is shown in Figure 6-1. (If energy level diagrams are new to you, check out Chapter 3. There are a number of minor variations that are commonly used in writing energy level diagrams, so don't worry if the diagrams in Chapter 3 are slightly different than the ones I show you here.)

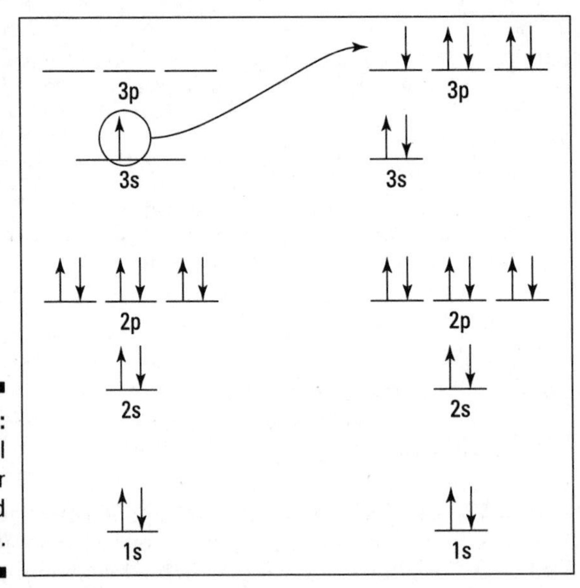

Figure 6-1: Energy level diagram for sodium and chlorine.

Chlorine is a member of the halogen family — the VIIA family on the periodic table. It has 7 valence electrons and a total of 17 electrons. The energy level diagram for chlorine is also shown in Figure 6-1.

If you want, instead of using the bulky energy level diagram to represent the distribution of electrons in an atom, you can use the electron configuration. (For a complete discussion of electron configurations, see Chapter 3.) Write, *in order,* the energy levels being used, the orbital types (s, p, d, and so on), and — in superscript — the number of electrons in each orbital. Here are the electronic configurations for sodium and chlorine:

Sodium (Na) $1s^2 2s^2 2p^6 3s^1$

Chlorine (Cl) $1s^2 2s^2 2p^6 3s^2 3p^5$

Understanding the reaction

The noble gases are the VIIIA elements on the periodic table. They're extremely unreactive because their valence energy level (outermost energy level) is filled. Achieving a filled (complete) valence energy level is a driving force in nature in terms of chemical reactions, because that's when elements become stable, or "satisfied." They don't lose, gain, or share electrons.

The other elements in the *A* families on the periodic table do gain, lose, or share valence electrons in order to fill their valence energy level and become satisfied. Because this process, in most cases, involves filling the outermost s and p orbitals, it's sometimes called the *octet rule* — elements gain, lose, or share electrons to reach a full octet (8 valence electrons: 2 in the s orbital and 6 in the p orbital).

Sodium's role

Sodium has one valence electron; by the octet rule, it becomes stable when it has eight valence electrons. Two possibilities exist for sodium to become stable: It can gain seven more electrons to fill energy level 3, or it can lose the one 3s electron so that energy level 2 (which is filled at eight electrons) becomes the valence energy level. In general, the loss or gain of one, two, or sometimes even three electrons can occur, but an element doesn't lose or gain more than three electrons. So to gain stability, sodium loses its 3s electron. At this point, it has 11 protons (11 positive charges) and 10 electrons (10 negative charges). The once neutral sodium atom now has a single positive charge [11(+) plus 10(-) equals 1+]. It's now an *ion,* an atom that has a charge due to the loss or gain of electrons. And ions that have a positive charge (such as sodium) due to the loss of electrons are called *cations.* You can write an electron configuration for the sodium cation:

Na^+ $1s^2 2s^2 2p^6$

The sodium ion (cation) has the same electron configuration as neon, so it's *isoelectronic* with neon. So has sodium become neon by losing an electron? No. Sodium still has 11 protons, and the number of protons determines the identity of the element.

There's a difference between the neutral sodium atom and the sodium cation — one electron. In addition, their chemical reactivities are different *and* their sizes are different. The cation is smaller. The filled energy level determines the size of an atom or ion (or, in this case, cation). Because sodium loses an entire energy level to change from an atom to a cation, the cation is smaller.

Chlorine's role

Chlorine has seven valence electrons. To obtain its full octet, it must lose the seven electrons in energy level 3 or gain one at that level. Because elements don't gain or lose more than three electrons, chlorine must gain a single electron to fill energy level 3. At this point, chlorine has 17 protons (17 positive charges) and 18 electrons (18 negative charges). So chlorine becomes an ion with a single negative charge (Cl^-). The neutral chlorine atom becomes the chloride ion. Ions with a negative charge due to the gain of electrons are called *anions.* The electronic configuration for the chloride anion is

Cl^- $1s^2 2s^2 2p^6 3s^2 3p^6$

The chloride anion is isoelectronic with argon. The chloride anion is also slightly larger than the neutral chlorine atom. To complete the octet, the one electron gained went into energy level 3, but now there are 17 protons attracting 18 electrons. The attractive force has been reduced slightly, and the electrons are free to move outward a little, making the anion a little larger. In general, a cation is smaller than its corresponding atom, and an anion is slightly larger.

Ending up with a bond

Sodium can achieve its full octet and stability by losing an electron. Chlorine can fill its octet by gaining an electron. If the two are in the same container, then the electron sodium loses can be the same electron chlorine gains. I show this process in Figure 6-1, indicating that the 3s electron in sodium is transferred to the 3p orbital of chlorine.

The transfer of an electron creates ions — cations (positive charge) and anions (negative charge) — and opposite charges attract each other. The Na^+ cation attracts the Cl^- anion and forms the compound NaCl, or table salt. This is an example of an *ionic bond,* which is a *chemical bond* (a strong attractive force that keeps two chemical elements together) that comes from the *electrostatic attraction* (attraction of opposite charges) between cations and anions.

The compounds that have ionic bonds are commonly called *salts.* In sodium chloride, a crystal is formed in which each sodium cation is surrounded by six different chloride anions, and each chloride anion is surrounded by six different sodium cations. The crystal structure is shown in Figure 6-2.

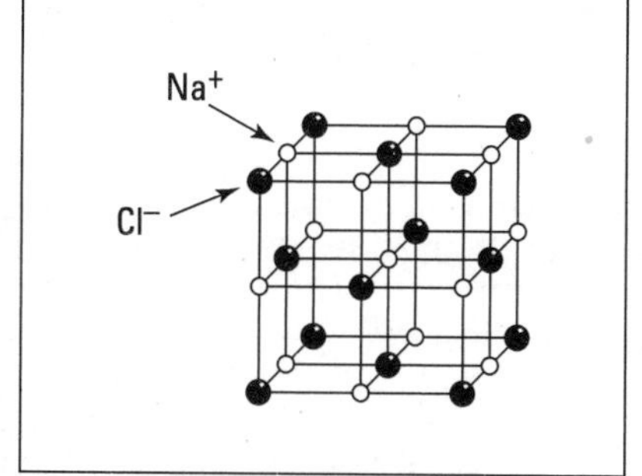

Figure 6-2: Crystal structure of sodium chloride.

Notice the regular, repeating structure. Different types of salts have different crystal structures. Cations and anions can have more than one unit of positive or negative charge if they lose or gain more than one electron. In this fashion, many different kinds of salts are possible.

Ionic bonding, the bonding that holds the cations and anions together in a salt, is one of the two major types of bonding in chemistry. The other type, *covalent bonding,* is described in Chapter 7. Grasping the concepts involved in ionic bonding makes understanding covalent bonding much easier.

Positive and Negative Ions: Cations and Anions

The basic process that occurs when sodium chloride is formed also occurs when other salts are formed. A metal loses electrons, and a nonmetal gains those electrons. Cations and anions are formed, and the electrostatic attraction between the positives and negatives brings the particles together and creates the ionic compound.

A metal reacts with a nonmetal to form an ionic bond.

You can often determine the charge an ion normally has by the element's position on the periodic table. For example, all the alkali metals (the IA elements) lose a single electron to form a cation with a 1+ charge. In the same way, the alkaline earth metals (IIA elements) lose two electrons to form a 2+ cation. Aluminum, a member of the IIIA family, loses three electrons to form a 3+ cation.

By the same reasoning, the halogens (VIIA elements) all have seven valence electrons. All the halogens gain a single electron to fill their valence energy level. And all of them form an anion with a single negative charge. The VIA elements gain two electrons to form anions with a 2- charge, and the VA elements gain three electrons to form anions with a 3- charge.

Table 6-1 shows the family, element, ion name, and ion symbol for some common monoatomic (one atom) cations, and Table 6-2 gives the same information for some common monoatomic anions.

Table 6-1 Some Common Monoatomic Cations

Family	*Element*	*Ion Name*	*Ion Symbol*
IA	Lithium	Lithium cation	Li^{+}
	Sodium	Sodium cation	Na^{+}
	Potassium	Potassium cation	K^{+}
IIA	Beryllium	Beryllium cation	Be^{2+}
	Magnesium	Magnesium cation	Mg^{2+}
	Calcium	Calcium cation	Ca^{2+}
	Strontium	Strontium cation	Sr^{2+}
	Barium	Barium cation	Ba^{2+}
IB	Silver	Silver cation	Ag^{+}
IIB	Zinc	Zinc cation	Zn^{2+}
IIIA	Aluminum	Aluminum cation	Al^{3+}

Table 6-2 Some Common Monoatomic Anions

Family	*Element*	*Ion Name*	*Ion Symbol*
VA	Nitrogen	Nitride anion	N^{3-}
	Phosphorus	Phosphide anion	P^{3-}

Family	Element	Ion Name	Ion Symbol
VIA	Oxygen	Oxide anion	O^{2-}
	Sulfur	Sulfide anion	S^{2-}
VIIA	Fluorine	Fluoride anion	F^{-}
	Chlorine	Chloride anion	Cl^{-}
	Bromine	Bromide anion	Br^{-}
	Iodine	Iodide anion	I^{-}

It's more difficult to determine the number of electrons that members of the transition metals (the B families) lose. In fact, many of these elements lose a varying number of electrons so that they form two or more cations with different charges.

The electrical charge that an atom achieves is sometimes called its *oxidation state.* Many of the transition metal ions have varying oxidation states. Table 6-3 shows some common transition metals that have more than one oxidation state.

Table 6-3 Some Common Metals with More than One Oxidation State

Family	Element	Ion Name	Ion Symbol
VIB	Chromium	Chromium(II) or chromous	Cr^{2+}
		Chromium(III) or chromic	Cr^{3+}
VIIB	Manganese	Manganese(II) or manganous	Mn^{2+}
		Manganese(III) or manganic	Mn^{3+}
VIIIB	Iron	Iron(II) or ferrous	Fe^{2+}
		Iron(III) or ferric	Fe^{3+}
	Cobalt	Cobalt(II) or cobaltous	Co^{2+}
		Cobalt(III) or cobaltic	Co^{3+}
IB	Copper	Copper(I) or cuprous	Cu^{+}
		Copper(II) or cupric	Cu^{2+}
IIB	Mercury	Mercury(I) or mercurous	Hg_2^{2+}
		Mercury(II) or mercuric	Hg^{2+}

(continued)

Table 6-3 *(continued)*

Family	*Element*	*Ion Name*	*Ion Symbol*
IVA	Tin	Tin(II) or stannous	Sn^{2+}
		Tin(IV) or stannic	Sn^{4+}
	Lead	Lead(II) or plumbous	Pb^{2+}
		Lead(IV) or plumbic	Pb^{4+}

Notice that these cations can have more than one name. The current way of naming ions is to use the metal name, such as Chromium, followed in parentheses by the ionic charge written as a Roman numeral, such as (II). An older way of naming ions uses *-ous* and *-ic* endings. When an element has more than one ion — Chromium, for example — the ion with the lower oxidation state (lower numerical charge, ignoring the + or -) is given an *-ous* ending, and the ion with the higher oxidation state (higher numerical charge) is given an *-ic* ending. So for Chromium, the Cr^{2+} ion is named *chromous* and the Cr^{3+} ion is named *chromic*. (See the section "Naming Ionic Compounds," later in this chapter, for more on naming ions.)

Polyatomic Ions

Ions aren't always monoatomic, composed of just one atom. Ions can also be polyatomic, composed of a group of atoms. For example, take a look at Table 6-3. Notice anything about the Mercury(I) ion? Its ion symbol, Hg_2^{2+}, shows that two mercury atoms are bonded together. This group has a 2+ charge, with each mercury cation having a 1+ charge. The mercurous ion is classified as a polyatomic ion.

Polyatomic ions are treated the same as monoatomic ions (see "Naming Ionic Compounds," later in this chapter). Table 6-4 lists some important polyatomic ions.

Table 6-4 Some Important Polyatomic Ions

Ion Name	*Ion Symbol*
Sulfate	SO_4^{2-}
Sulfite	SO_3^{2-}
Nitrate	NO_3^-
Nitrite	NO_2^-

Ion Name	*Ion Symbol*
Hypochlorite	ClO^-
Chlorite	ClO_2^-
Chlorate	ClO_3^-
Perchlorate	ClO_4^-
Acetate	$C_2H_3O_2^-$
Chromate	CrO_4^{2-}
Dichromate	$Cr_2O_7^{2-}$
Arsenate	AsO_4^{3-}
Hydrogen phosphate	HPO_4^{2-}
Dihydrogen phosphate	$H_2PO_4^-$
Bicarbonate or hydrogen carbonate	HCO_3^-
Bisulfate or hydrogen sulfate	HSO_4^-
Mercury (I)	Hg_2^{2+}
Ammonium	NH_4^+
Phosphate	PO_4^{3-}
Carbonate	CO_3^{2-}
Permanganate	MnO_4^-
Cyanide	CN^-
Cyanate	OCN^-
Thiocyanate	SCN^-
Oxalate	$C_2O_4^{2-}$
Thiosulfate	$S_2O_3^{2-}$
Hydroxide	OH^-
Arsenite	AsO_3^{3-}
Peroxide	O_2^{2-}

The symbol for the sulfate ion, SO_4^{2-}, indicates that one sulfur atom and four oxygen atoms are bonded together and that the whole polyatomic ion has two extra electrons.

Putting Ions Together: Ionic Compounds

When an ionic compound is formed, the cation and anion attract each other, resulting in a salt (see "The Magic of an Ionic Bond: Sodium + Chlorine = Table Salt," earlier in this chapter). An important thing to remember is that the compound must be neutral — have equal numbers of positive and negative charges.

Putting magnesium and bromine together

Suppose you want to know the *formula,* or composition, of the compound that results from reacting magnesium with bromine. You start by putting the two atoms side by side, with the metal on the left, and then adding their charges. Figure 6-3 shows this process. (Forget about the crisscrossing lines for now. Well, if you're really curious, they're discussed in the "Using the crisscross rule" section, later in this chapter.)

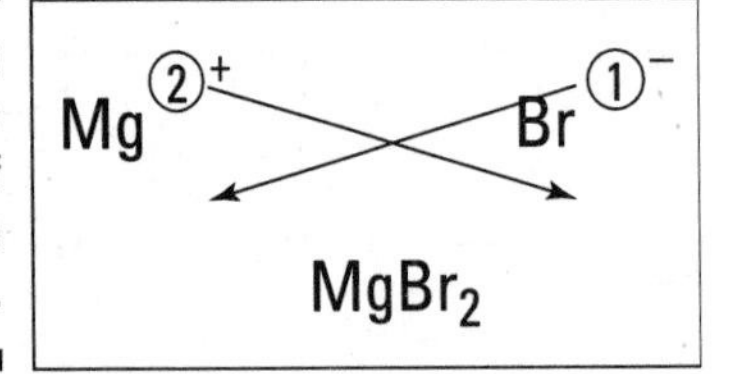

Figure 6-3: Figuring the formula of magnesium bromide.

The electron configurations for magnesium and bromine are

Magnesium (Mg)	$1s^2 2s^2 2p^6 3s^2$
Bromine (Br)	$1s^2 2s^2 2p^6 3s^2 3p^6 4s^2 3d^{10} 4p^5$

Magnesium, an alkaline earth metal, has two valence electrons that it loses to form a cation with a 2+ charge. The electron configuration for the magnesium cation is

Mg^{2+} $1s^2 2s^2 2p^6$

Bromine, a halogen, has seven valence electrons, so it gains one to complete its octet (eight valence electrons) and form the bromide anion with a 1- charge. The electron configuration for the bromide anion is

Br^{1-} $1s^2 2s^2 2p^6 3s^2 3p^6 4s^2 3d^{10} 4p^6$

Note that if the anion simply has 1 unit of charge, positive or negative, you normally don't write the 1; you just use the plus or minus symbol, with the 1 being understood. But for the example of the bromide ion, I use the 1.

The compound must be neutral; it must have the same number of positive and negative charges so that, overall, it has a zero charge. The magnesium ion has a 2+, so it requires 2 bromide anions, each with a single negative charge, to balance the 2 positive charges of magnesium. So the formula of the compound that results from reacting magnesium with bromine is $MgBr_2$.

Using the crisscross rule

There's a quick way to determine the formula of an ionic compound: Use the *crisscross rule.*

Look at Figure 6-3 for an example of using this rule. Take the numerical value of the metal ion's superscript (forget about the charge symbol) and move it to the bottom right-hand side of the nonmetal's symbol — as a subscript. Then take the numerical value of the nonmetal's superscript and make it the subscript of the metal. (Note that if the numerical value is 1, it's just understood and not shown.) So in this example, you make magnesium's 2 a subscript of bromine and make bromine's 1 a subscript of magnesium (but because it's 1, you don't show it), and you get the formula $MgBr_2$.

So what happens if you react aluminum and oxygen? Figure 6-4 shows the crisscross rule used for this reaction.

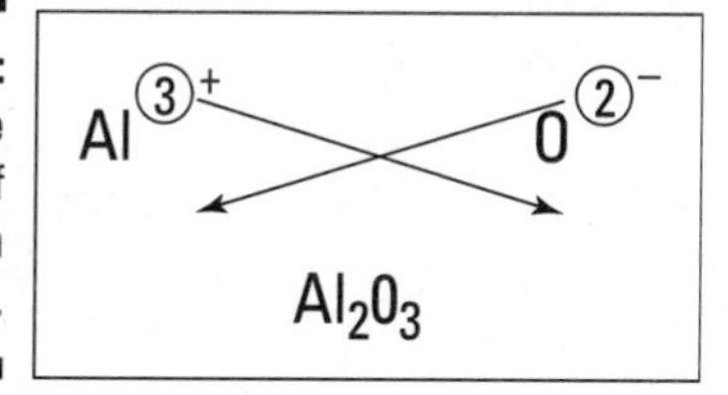

Figure 6-4: Figuring the formula of aluminum oxide.

Compounds involving polyatomic ions work exactly the same way. For example, here's the compound made from the ammonium cation and the sulfide anion:

$(NH_4)_2S$

Notice that because two ammonium ions (two positive charges) are needed to neutralize the two negative charges of the sulfide ion, the ammonium ion is enclosed in parentheses and a subscript 2 is added.

The crisscross rule works very well, but there's a situation where you have to be careful. Suppose that you want to write the compound formed when calcium reacts with oxygen. Calcium, an alkaline earth metal, forms a 2+ cation, and oxygen forms a 2- anion. So you might predict that the formula is

Mg_2O_2

But this formula is incorrect. After you use the crisscross rule, you need to reduce all the subscripts by a common factor, if possible. In this case, you divide each subscript by 2 and get the correct formula:

MgO

Naming Ionic Compounds

When you name inorganic compounds, you write the name of the metal first and then the nonmetal. Suppose, for example, that you want to name Li_2S, the compound that results from the reaction of lithium and sulfur. You first write the name of the metal, lithium, and then write the name of the nonmetal, adding an *-ide* ending so that *sulfur* becomes *sulfide*.

Li_2S	lithium sulfide

Ionic compounds involving polyatomic ions follow the same basic rule: Write the name of the metal first, and then simply add the name of the nonmetal (with the polyatomic anions, it is not necessary to add the -ide ending).

$(NH_4)_2CO_3$	Ammonium carbonate
K_3PO_4	Potassium phosphate

When the metal involved is a transition metal with more than one oxidation state (see "Positive and Negative Ions: Cations and Anions," earlier in the chapter, for more info about *that*), there can be more than one way to correctly name the compound. For example, suppose that you want to name the compound formed between the Fe^{3+} cation and the cyanide ion, CN^-. The preferred method is to use the metal name followed in parentheses by the ionic charge written as a Roman numeral: Iron(III). But an older naming method, which is still sometimes used (so it's a good idea to know it), is to use *-ous* and *-ic* endings. The ion with the lower oxidation state (lower numerical charge, ignoring the + or -) is given an *-ous* ending, and the ion with the higher oxidation state (higher numerical charge) is given an *-ic* ending. So, because

Fe^{3+} has a higher oxidation state than Fe^{2+}, it's called a *ferric ion*. So the compound can be named

$Fe(CN)_3$ Iron(III) cyanide or ferric cyanide

Sometimes figuring out the charge on an ion can be a little challenging (and fun), so now I want to show you how to name $FeNH_4(SO_4)_2$.

I show you in Table 6-4 that the sulfate ion has a 2- charge, and from the formula you can see that there are two of them. Therefore, you have a total of four negative charges. Table 6-4 also indicates that the ammonium ion has a 1+ charge, so you can figure out the charge on the iron cation.

Ion	***Charge***
Fe	?
NH_4	1+
$(SO_4)_2$	(2-)×2

Because you have a 4- for the sulfates and a 1+ for the ammonium, the iron must be a 3+ to make the compound neutral. So the iron is in the Iron(III), or ferric, oxidation state. You can name the compound

$FeNH_4(SO_4)_2$ Iron(III) ammonium sulfate or ferric ammonium sulfate

And, finally, if you have the name, you can derive the formula and the charge on the ions. For example, suppose that you're given the name *cuprous oxide*. You know that the cuprous ion is Cu^+ and the oxide ion is O^{2-}. Applying the crisscross rule, you get the following formula:

Cuprous oxide Cu_2O

Electrolytes and Nonelectrolytes

When an ionic compound such as sodium chloride is put into water, the water molecules attract both the cations and anions in the crystal (the crystal is shown in Figure 6-2) and pull them into the solution. (In Chapter 7, I talk a lot about water molecules and show you why they attract the NaCl ions.) The cations and anions get distributed throughout the solution. You can detect the presence of these ions by using an instrument called a *conductivity tester*.

A conductivity tester tests whether water solutions of various substances conduct electricity. It's composed of a light bulb with two electrodes attached. The light bulb is plugged into a wall outlet, but it doesn't light until some type of conductor (substance capable of transmitting electricity) between the electrodes completes the circuit. (A finger will complete the circuit, so this experiment should be done carefully. If you're not careful, it can be a shocking experience!)

When you place the electrodes in pure water, nothing happens, because there's no conductor between the electrodes. Pure water is a nonconductor. But if you put the electrodes in the NaCl solution, the light bulb lights, because the ions conduct the electricity (carry the electrons) from one electrode to the other.

In fact, you don't even really need the water. If you were to melt pure NaCl (it requires a lot of heat!) and then place the electrodes in it, you'd find that the molten table salt also conducts electricity. In the molten state, the NaCl ions are free to move and carry electrons, just as they are in the saltwater solution. Substances that conduct electricity in the molten state or when dissolved in water are called *electrolytes.* Substances that don't conduct electricity when in these states are called *nonelectrolytes.*

Scientists can get some good clues as to the type of bonding in a compound by discovering whether a substance is an electrolyte or a nonelectrolyte. Ionically bonded substances act as electrolytes. But covalently bonded compounds (see Chapter 7), in which no ions are present, are commonly nonelectrolytes. Table sugar, or sucrose, is a good example of a nonelectrolyte. You can dissolve sugar in water or melt it, but it won't have conductivity. No ions are present to transfer the electrons.

Chapter 7

Covalent Bonds: Let's Share Nicely

In This Chapter

- Seeing how one hydrogen atom bonds to another hydrogen atom
- Defining covalent bonding
- Finding out about the different types of chemical formulas
- Taking a look at polar covalent bonding and electronegativity
- Accepting the unusual properties of water

Sometimes when I'm cooking, I have one of my chemistry nerd moments and start reading the ingredients on food labels. I usually find lots of salts, such as sodium chloride, and lots of other compounds, such as potassium nitrate, that are all ionically bonded (see Chapter 6). But I also find many compounds, such as sugar, that aren't ionically bonded.

If no ions are holding a compound together, what *does* hold it together? What holds together sugar, vinegar, and even DNA? In this chapter, I discuss the other major type of bonding: covalent bonding. I explain the basics with an extremely simple covalent compound, hydrogen, and I tell you some cool stuff about one of the most unusual covalent compounds I know — water.

Covalent Bond Basics

An ionic bond is a chemical bond that comes from the transfer of electrons from a metal to a nonmetal, resulting in the formation of oppositely charged ions — cations (positive charge) and anions (negative charge) — and the attraction between those oppositely charged ions. The driving force in this whole process is achieving a filled valence energy level, completing the atom's octet. (For a more complete explanation of this concept, see Chapter 6.)

But many other compounds exist in which electron transfer hasn't occurred. The driving force is still the same: achieving a filled valence energy level. But instead of achieving it by gaining or losing electrons, the atoms in these compounds *share* electrons. That's the basis of a covalent bond.

A hydrogen example

Hydrogen is #1 on the periodic table — upper left corner. The hydrogen found in nature is often not comprised of an individual atom. It's primarily found as H_2, a *diatomic* (two atom) compound. (Taken one step further, because a *molecule* is a combination of two or more atoms, H_2 is called a *diatomic molecule.*)

Hydrogen has one valence electron. It'd love to gain another electron to fill its 1s energy level, which would make it *isoelectronic* with helium (because the two would have the same electronic configuration), the nearest noble gas. Energy level 1 can only hold two electrons in the 1s orbital, so gaining another electron fills it. That's the driving force of hydrogen — filling the valence energy level and achieving the same electron arrangement as the nearest noble gas.

Imagine one hydrogen atom transferring its single electron to another hydrogen atom. The hydrogen atom receiving the electron fills its valence shell and reaches stability while becoming an anion (H^-). However, the other hydrogen atom now has no electrons (H^+) and moves further away from stability. This process of electron loss and gain simply won't happen, because the driving force of both atoms is to fill their valence energy level. So the H_2 compound can't result from the loss or gain of electrons. What *can* happen is that the two atoms can share their electrons. At the atomic level, this sharing is represented by the electron orbitals (sometimes called electron clouds) overlapping. The two electrons (one from each hydrogen atom) "belong" to both atoms. Each hydrogen atom feels the effect of the two electrons; each has, in a way, filled its valence energy level. A *covalent bond* is formed — a chemical bond that comes from the sharing of one or more electron pairs between two atoms. The overlapping of the electron orbitals and the sharing of an electron pair is represented in Figure 7-1(a).

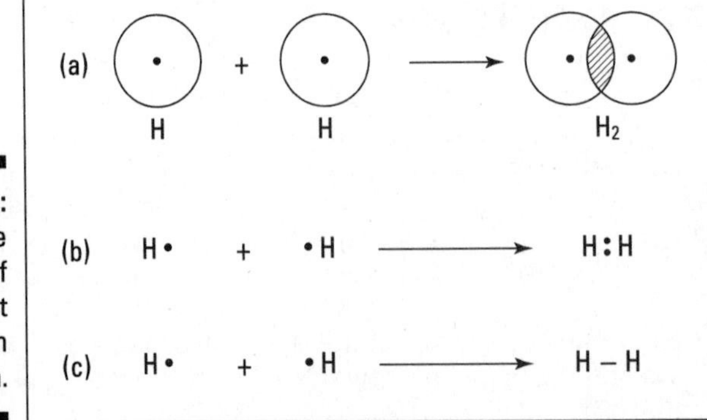

Figure 7-1: The formation of a covalent bond in hydrogen.

Another way to represent this process is through the use of an *electron-dot formula.* In this type of formula, valence electrons are represented as dots surrounding the atomic symbol, and the shared electrons are shown between the two atoms involved in the covalent bond. The electron-dot formula representations of H_2 are shown in Figure 7-1(b).

Most of the time, I use a slight modification of the electron-dot formula called the *Lewis structural formula;* it's basically the same as the electron-dot formula, but the shared pair of electrons (the covalent bond) is represented by a dash. The Lewis structural formula is shown in Figure 7-1(c). (Check out the section, "Structural formula: Add the bonding pattern," for more about writing structural formulas of covalent compounds.)

In addition to hydrogen, six other elements are found in nature in the diatomic form: oxygen (O_2), nitrogen (N_2), fluorine (F_2), chlorine (Cl_2), bromine (Br_2), and iodine (I_2). So when I talk about oxygen gas or liquid bromine, I'm talking about the diatomic compound (diatomic molecule).

Here's one more example of using the electron-dot formula to represent the shared electron pair of a diatomic compound: This time, look at bromine (Br_2), which is a member of the halogen family (see Figure 7-2). The two halogen atoms, each with seven valence electrons, share an electron pair and fill their octet.

$$:\ddot{\underset{\cdot\cdot}{Br}}\cdot \; + \; :\ddot{\underset{\cdot\cdot}{Br}}\cdot \longrightarrow :\ddot{\underset{\cdot\cdot}{Br}}:\ddot{\underset{\cdot\cdot}{Br}}:$$

$$\left(:\ddot{\underset{\cdot\cdot}{Br}}-\ddot{\underset{\cdot\cdot}{Br}}:\right)$$

Figure 7-2: The covalent bond formation of Br_2.

Comparing covalent bonds with other bonds

Ionic bonding occurs between a metal and a nonmetal. Covalent bonding, on the other hand, occurs between two nonmetals. The properties of these two types of compounds are different. Ionic compounds are usually solids at room temperature, while covalently bonded compounds can be solids, liquids, or gases. There's more. Ionic compounds (salts) usually have a much higher melting point than covalent compounds. In addition, ionic compounds tend to be electrolytes, and covalent compounds tend to be nonelectrolytes. (Chapter 6 explains all about ionic bonds, electrolytes, and nonelectrolytes.)

I know just what you're thinking: "If metals react with nonmetals to form ionic bonds, and nonmetals react with other nonmetals to form covalent bonds, do metals react with other metals?" The answer is yes and no.

Metals don't really react with other metals to form compounds. Instead, the metals combine to form *alloys,* solutions of one metal in another. But there is such a situation as metallic bonding, and it's present in both alloys and pure metals. In *metallic bonding,* the valence electrons of each metal atom are donated to an electron pool, commonly called a *sea of electrons,* and are shared by all the atoms in the metal. These valence electrons are free to move throughout the sample instead of being tightly bound to an individual metal nucleus. The ability of the valence electrons to flow throughout the entire metal sample is why metals tend to be conductors of electricity and heat.

Understanding multiple bonds

I define covalent bonding as the sharing of one *or more* electron pairs. In hydrogen and the other diatomic molecules, only one electron pair is shared. But in many covalent bonding situations, more than one electron pair is shared. This section shows you an example of a molecule in which more than one electron pair is shared.

Nitrogen (N_2) is a diatomic molecule in the *VA family* on the periodic table, meaning that it has five valence electrons (see Chapter 4 for a discussion of families on the periodic table). So nitrogen needs three more valence electrons to complete its octet.

A nitrogen atom can fill its octet by sharing three electrons with another nitrogen atom, forming three covalent bonds, a so-called *triple bond.* The triple bond formation of nitrogen is shown in Figure 7-3.

A triple bond isn't quite three times as strong as a single bond, but it's a very strong bond. In fact, the triple bond in nitrogen is one of the strongest bonds known. This strong bond is what makes nitrogen very stable and resistant to reaction with other chemicals. It's also why many explosive compounds (such as TNT and ammonium nitrate) contain nitrogen. When these compounds break apart in a chemical reaction, nitrogen gas (N_2) is formed, and a large amount of energy is released.

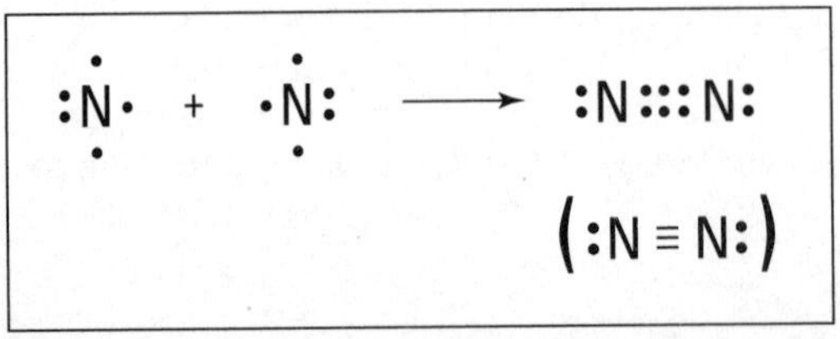

$$:\dot{\underset{\cdot}{N}}\cdot + \cdot\dot{\underset{\cdot}{N}}: \longrightarrow :N:::N:$$

$$(:N \equiv N:)$$

Figure 7-3: Triple bond formation in N_2.

There are no salt molecules!

A *molecule* is a compound that is covalently bonded. It's technically incorrect to refer to sodium chloride, which has ionic bonds, as a molecule, but lots of chemists do it anyway. It's kind of like using the wrong fork at a formal dinner. Some people may notice, but most don't notice or don't care. But just so you know, the correct term for ionic compounds is *formula unit.*

Carbon dioxide (CO_2) is another example of a compound containing a multiple bond. Carbon can react with oxygen to form carbon dioxide. Carbon has four valence electrons, and oxygen has six. Carbon can share two of its valence electrons with each of the two oxygen atoms, forming two double bonds. These double bonds are shown in Figure 7-4.

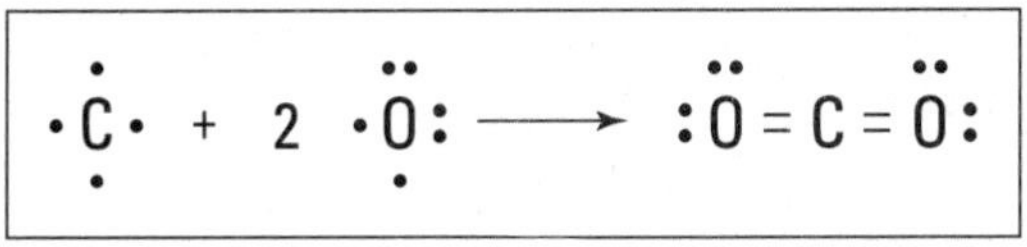

$$\cdot\dot{\underset{\cdot}{C}}\cdot + 2\ \cdot\ddot{\underset{\cdot}{O}}: \longrightarrow\ :\ddot{O}=C=\ddot{O}:$$

Figure 7-4: Formation of carbon dioxide.

Naming Binary Covalent Compounds

Binary compounds are compounds made up of only two elements, such as carbon dioxide (CO_2). Prefixes are used in the names of binary compounds to indicate the number of atoms of each nonmetal present. Table 7-1 lists the most common prefixes for binary covalent compounds.

Table 7-1 Common Prefixes for Binary Covalent Compounds

Number of Atoms	*Prefix*
1	mono-
2	di-
3	tri-
4	tetra-

(continued)

Table 7-1 *(continued)*

Number of Atoms	*Prefix*
5	penta-
6	hexa-
7	hepta-
8	octa-
9	nona-
10	deca-

In general, the prefix *mono-* is rarely used. Carbon monoxide is one of the few compounds that uses it.

Take a look at the following examples to see how to use the prefixes when naming binary covalent compounds (I've bolded the prefixes for you):

CO_2 carbon **di**oxide

P_4O_{10} **tetra**phosphorus **dec**oxide (Chemists try to avoid putting an *a* and an *o* together with the oxide name, as in dec**ao**xide, so they normally drop the *a* off the prefix.)

SO_3 sulfur **tri**oxide

N_2O_4 **di**nitrogen **tetr**oxide

This naming system is used only with binary, nonmetal compounds, with one exception — MnO_2 is commonly called manganese dioxide.

So Many Formulas, So Little Time

In Chapter 6, I show you how to predict the formula of an ionic compound, based on the loss and gain of electrons, to reach a noble gas configuration. (For example, if you react Ca with Cl, you can predict the formula of the resulting salt — $CaCl_2$.) You really can't make that type of prediction with covalent compounds, because they can combine in many ways, and many different possible covalent compounds may result.

Most of the time, you have to know the formula of the molecule you're studying. But you may have several different types of formulas, and each gives a slightly different amount of information. Oh joy.

Empirical formula: Just the elements

The *empirical formula* indicates the different types of elements in a molecule and the lowest whole-number ratio of each kind of atom in the molecule. For example, suppose that you have a compound with the empirical formula C_2H_6O. Three different kinds of atoms are in the compound, C, H, and O, and they're in the lowest whole-number ratio of 2 C to 6 H to 1 O. So the actual formula (called the *molecular formula* or *true formula*) may be C_2H_6O, $C_4H_{12}O_2$, $C_6H_{18}O_3$, $C_8H_{24}O_4$, or another multiple of 2:6:1

Molecular or true formula: Inside the numbers

The *molecular formula,* or *true formula,* tells you the kinds of atoms in the compound and the actual number of each atom. You may determine, for example, that the empirical formula C_2H_6O is actually the molecular formula, too, meaning that there are actually two carbon atoms, six hydrogen atoms, and one oxygen atom in the compound.

For ionic compounds, this formula is enough to fully identify the compound, but it's not enough to identify covalent compounds. Look at the Lewis formulas presented in Figure 7-5. Both compounds have the molecular formula of C_2H_6O.

```
    H     H                 H   H
    |  ..  |                 |   |   ..
H - C - O - C - H       H - C - C - O - H
    |  ..  |                 |   |   ..
    H     H                 H   H

  Dimethyl ether           Ethyl alcohol
```

Figure 7-5: Two possible compounds of C_2H_6O.

Both compounds in Figure 7-5 have two carbon atoms, six hydrogen atoms, and one oxygen atom. The difference is in the way the atoms are bonded, or what's bonded to what. These are two entirely different compounds with two entirely different sets of properties. The one on the left is called dimethyl ether. This compound is used in some refrigeration units and is highly flammable. The one on the right is ethyl alcohol, the drinking variety of alcohol. Simply knowing the molecular formula isn't enough to distinguish between the two compounds. Can you imagine going into a restaurant and ordering a shot of C_2H_6O and getting dimethyl ether instead of tequila?

It's always important to KISS

A lot of molecules obey the octet rule: Each atom in the compound ends up with a full octet of eight electrons filling its valence energy level. However, like most rules, the octet rule does have exceptions. Some stable molecules have atoms with just 6 electrons, and some have 10 or 12. I point out some examples of compounds that don't obey the octet rule in the section, "What Does Water Really Look Like? The VSEPR Theory," later in this chapter, but for the most part in this book, I concentrate on situations in which the octet rule is obeyed.

I pretty much stick to the KISS principle — Keep It Simple, Silly. Electron-dot formulas are used quite a bit by organic chemists in explaining why certain compounds react the way they do and are the first step in determining the molecular geometry of a compound.

Compounds that have the same molecular formula but different structures are called *isomers* of each other.

To identify the *exact* covalent compound, you need its structural formula.

Structural formula: Add the bonding pattern

To write a formula that stands for the exact compound you have in mind, you often must write the structural formula instead of the molecular formula. The *structural formula* shows the elements in the compound, the exact number of each atom in the compound, and the bonding pattern for the compound. The electron-dot formula and Lewis formula are examples of structural formulas.

Writing the electron-dot formula for water

The following steps explain how to write the electron-dot formula for a simple molecule — water — and provide some general guidelines and rules to follow:

1. **Write a skeletal structure showing a reasonable bonding pattern using just the element symbols.**

   Often, most atoms are bonded to a single atom. This atom is called the *central atom.* Hydrogen and the halogens are very rarely, if ever, central atoms. Carbon, silicon, nitrogen, phosphorus, oxygen, and sulfur are always good candidates, because they form more than one covalent bond in filling their valence energy level. In the case of water, H_2O, oxygen is the central element and the hydrogen atoms are both bonded to it. The bonding pattern looks like this:

O H
H

It doesn't matter where you put the hydrogen atoms around the oxygen. In the section, "What Does Water Really Look Like? The VSEPR Theory," later in this chapter, you see why I put the hydrogen atoms at a 90-degree angle to each other, but it really doesn't matter when writing electron-dot (or Lewis) formulas.

2. **Take all the valence electrons from all the atoms and throw them into an electron pot.**

   Each hydrogen atom has one electron, and the oxygen atom has six valence electrons (VIA family), so you have eight electrons in your electron pot. Those are the electrons you use when making your bonds and completing each atom's octet.

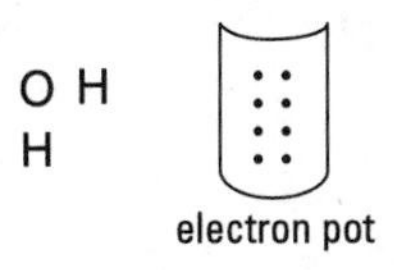

3. **Use the $N - A = S$ equation to figure the number of bonds in this molecule. In this equation,**

   N equals the sum of the number of valence electrons *n*eeded by each atom. N has only two possible values — 2 or 8. If the atom is hydrogen, it's 2; if it's anything else, it's 8.

   A equals the sum of the number of valence electrons *a*vailable for each atom. If you're doing the structure of an ion, you add one electron for every unit of negative charge if it's an anion or subtract one electron for every unit of positive charge if it's a cation. *A* is the number of valence electrons in your electron pot.

   S equals the number of electrons *s*hared in the molecule. And if you divide *S* by 2, you have the number of covalent bonds in the molecule.

   So in the case of water,

   $N = 8 + 2(2) = 12$ (8 valence electrons for the oxygen atom, plus 2 each for the two hydrogen atoms)

   $A = 6 + 2(1) = 8$ (6 valence electrons for the oxygen atom, plus 1 for each of the two hydrogen atoms)

   $S = 12 - 8 = 4$ (four electrons shared in water), and $S/2 = 4/2 = 2$ bonds

   You now know that there are two bonds (two shared pairs of electrons) in water.

4. **Distribute the electrons from your electron pot to account for the bonds.**

   You use four electrons from the eight in the pot, which leaves you with four to distribute later. There has to be at least one bond from your central atom to the atoms surrounding it.

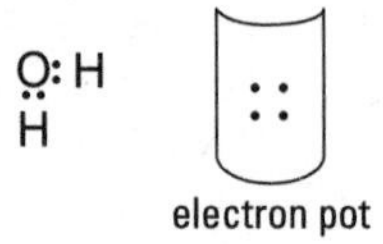

   O: H
   H

   electron pot

5. **Distribute the rest of the electrons (normally in pairs) so that each atom achieves its full octet of electrons.**

   Remember that hydrogen needs only two electrons to fill its valence energy level. In this case, each hydrogen atom has two electrons, but the oxygen atom has only four electrons, so the remaining four electrons are placed around the oxygen. This empties your electron pot. The completed electron-dot formula for water is shown in Figure 7-6.

:O: H
H

Figure 7-6: Electron-dot formula of H_2O.

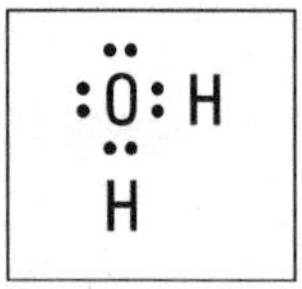

Notice that there are actually two types of electrons shown in this structural formula: *bonding electrons,* the electrons that are shared between two atoms, and *nonbonding electrons,* the electrons that are not being shared. The last four electrons (two electron pairs) that you put around oxygen are not being shared, so they're nonbonding electrons.

Writing the Lewis formula for water

If you want the Lewis formula for water, all you have to do is substitute a dash for every bonding pair of electrons. This structural formula is shown in Figure 7-7.

:O – H
|
H

Figure 7-7: The Lewis formula for H_2O.

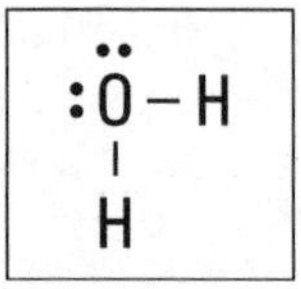

Writing the Lewis formula for C_2H_4O

Here's an example of a Lewis formula that's a little more complicated — C_2H_4O.

The compound has the following framework:

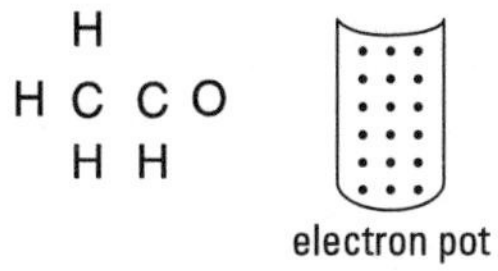

```
   H
H  C  C  O
   H  H
```

electron pot

Notice that it has not one but two central atoms — the two carbon atoms. You can put 18 valence electrons into the electron pot: four for each carbon atom, two for each hydrogen atom, and six for the oxygen atom.

Now apply the N – A = S equation:

> N = 2(8) + 4(2) + 8 = 32 (2 carbon atoms with 8 valence electrons each, plus 4 hydrogen atoms with 2 valence electrons each, plus an oxygen atom with 8 electrons)
>
> A = 2(4) + 4(1) + 6 = 18 (4 electrons for each of the two carbon atoms, plus 1 electron for each of the 4 hydrogen atoms, plus 6 valence electrons for the oxygen atom)
>
> S = 32–18 = 14, and S/2 = 14/2 = 7 bonds

Add single bonds between the carbon atoms and the hydrogen atom, between the two carbon atoms, and between the carbon atom and oxygen atom. That's six of your seven bonds.

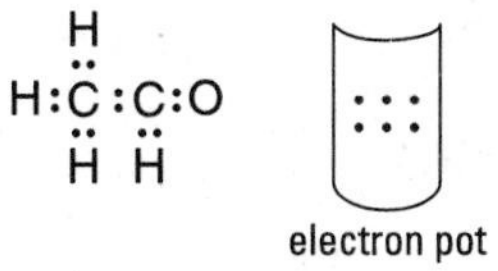

```
  H
  ··
H:C:C:O
  ·· ··
  H  H
```

electron pot

There's only one place that the seventh bond can go, and that's between the carbon atom and the oxygen atom. It can't be between a carbon atom and a hydrogen atom, because that would overfill hydrogen's valence energy level. And it can't be between the two carbon atoms, because that would give the carbon on the left ten electrons instead of eight. So there must be a double bond between the carbon atom and the oxygen atom. The four remaining electrons in the pot must be distributed around the oxygen atom, because all the other atoms have reached their octet. The electron-dot formula is shown in Figure 7-8.

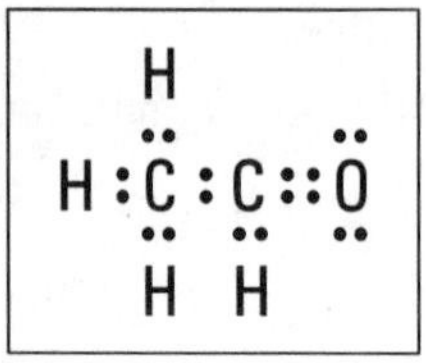

Figure 7-8: Electron-dot formula of C_2H_4O.

```
   H
   ..      ..
H :C : C ::O
   ..  ..  ..
   H   H
```

If you convert the bonding pairs to dashes, you have the Lewis formula of C_2H_4O, as shown in Figure 7-9.

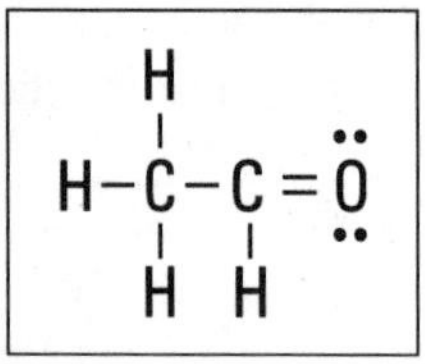

Figure 7-9: The Lewis formula for C_2H_4O.

```
   H
   |       ..
H–C – C = O
   |   |   ..
   H   H
```

I like the Lewis formula because it enables you to show a lot of information without having to write all those little dots. But it, too, is rather bulky. Sometimes chemists (who are, in general, a lazy lot) use *condensed structural formulas* to show bonding patterns. They may condense the Lewis formula by omitting the nonbonding electrons and grouping atoms together and/or by omitting certain dashes (covalent bonds). A couple of condensed formulas for C_2H_4O are shown in Figure 7-10.

Figure 7-10: Condensed structural formulas for C_2H_4O.

$CH_3 - CH = O$

CH_3CHO

Some Atoms Are More Attractive Than Others

When a chlorine atom covalently bonds to another chlorine atom, the shared electron pair is shared equally. The electron density that comprises the covalent bond is located halfway between the two atoms. Each atom attracts the two bonding electrons equally. But what happens when the two atoms involved in a bond aren't the same? The two positively charged nuclei have different attractive forces; they "pull" on the electron pair to different degrees. The end result is that the electron pair is shifted toward one atom. But the question is, "Which atom does the electron pair shift toward?" Electronegativities provide the answer.

Attracting electrons: Electronegativities

Electronegativity is the strength an atom has to attract a bonding pair of electrons to itself. The larger the value of the electronegativity, the greater the atom's strength to attract a bonding pair of electrons. Figure 7-11 shows the electronegativity values of the various elements below each element symbol on the periodic table. Notice that, with a few exceptions, the electronegativities increase, from left to right, in a period, and decrease, from top to bottom, in a family.

Electronegativities are useful because they give information about what will happen to the bonding pair of electrons when two atoms bond. For example, look at the Cl_2 molecule. Chlorine has an electronegativity value of 3.0, as shown in Figure 7-11. Each chlorine atom attracts the bonding electrons with a force of 3.0. Because there's an equal attraction, the bonding electron pair is shared equally between the two chlorine atoms and is located halfway between the two atoms. A bond in which the electron pair is equally shared is called a *nonpolar covalent bond.* You have a nonpolar covalent bond anytime the two atoms involved in the bond are the same or anytime the difference in the electronegativities of the atoms involved in the bond is very small.

Now consider hydrogen chloride (HCl). Hydrogen has an electronegativity of 2.1, and chlorine has an electronegativity of 3.0. The electron pair that is bonding HCl together shifts toward the chlorine atom because it has a larger electronegativity value. A bond in which the electron pair is shifted toward one atom is called a *polar covalent bond.* The atom that more strongly attracts the bonding electron pair is slightly more negative, while the other atom is slightly more positive. The larger the difference in the electronegativities, the more negative and positive the atoms become.

Now look at a case in which the two atoms have extremely different electronegativities — sodium chloride (NaCl). Sodium chloride is ionically bonded (see Chapter 6 for information on ionic bonds). An electron has transferred from sodium to chlorine. Sodium has an electronegativity of 1.0, and chlorine has an electronegativity of 3.0. That's an electronegativity difference of 2.0 (3.0 – 1.0), making the bond between the two atoms very, very polar. In fact, the electronegativity difference provides another way of predicting the kind of bond that will form between two elements.

Figure 7-11: Electro-negativities of the elements.

Electronegativities of the Elements

Decreasing →

Increasing →

1 H 2.1																
3 Li 1.0	4 Be 1.5											5 B 2.0	6 C 2.5	7 N 3.0	8 O 3.5	9 F 4.0
11 Na 0.9	12 Mg 1.2											13 Al 1.5	14 Si 1.8	15 P 2.1	16 S 2.5	17 Cl 3.0
19 K 0.8	20 Ca 1.0	21 Sc 1.3	22 Ti 1.5	23 V 1.6	24 Cr 1.6	25 Mn 1.5	26 Fe 1.8	27 Co 1.9	28 Ni 1.9	29 Cu 1.9	30 Zn 1.6	31 Ga 1.6	32 Ge 1.8	33 As 2.0	34 Se 2.4	35 Br 2.8
37 Rb 0.8	38 Sr 1.0	39 Y 1.2	40 Zr 1.4	41 Nb 1.6	42 Mo 1.8	43 Tc 1.9	44 Ru 2.2	45 Rh 2.2	46 Pd 2.2	47 Ag 1.9	48 Cd 1.7	49 In 1.7	50 Sn 1.8	51 Sb 1.9	52 Te 2.1	53 I 2.5
55 Cs 0.7	56 Ba 0.9	57 La 1.1	72 Hf 1.3	73 Ta 1.5	74 W 1.7	75 Re 1.9	76 Os 2.2	77 Ir 2.2	78 Pt 2.2	79 Au 2.4	80 Hg 1.9	81 Tl 1.8	82 Pb 1.9	83 Bi 1.9	84 Po 2.0	85 At 2.2
87 Fr 0.7	88 Ra 0.9	89 Ac 1.1														

Electronegativity Difference	*Type of Bond Formed*
0.0 to 0.2	nonpolar covalent
0.3 to 1.4	polar covalent
> 1.5	ionic

The presence of a polar covalent bond in a molecule can have some pretty dramatic effects on the properties of a molecule.

Polar covalent bonding

If the two atoms involved in the covalent bond are not the same, the bonding pair of electrons are pulled toward one atom, with that atom taking on a slight (partial) negative charge and the other atom taking on a partial positive charge. In most cases, the molecule has a positive end and a negative end, called a *dipole* (think of a magnet). Figure 7-12 shows a couple of examples of molecules in which dipoles have formed. (The little Greek symbol by the charges refers to a *partial* charge.)

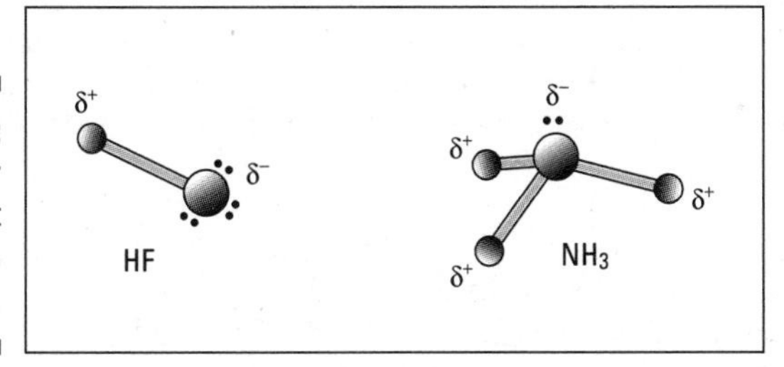

Figure 7-12: Polar covalent bonding in HF and NH_3.

In hydrogen fluoride (HF), the bonding electron pair is pulled much closer to the fluorine atom than to the hydrogen atom, so the fluorine end becomes partially negatively charged and the hydrogen end becomes partially positively charged. The same thing takes place in ammonia (NH_3); the nitrogen has a greater electronegativity than hydrogen, so the bonding pairs of electrons are more attracted to it than to the hydrogen atoms. The nitrogen atom takes on a partial negative charge, and the hydrogen atoms take on a partial positive charge.

The presence of a polar covalent bond explains why some substances act the way they do in a chemical reaction: Because this type of molecule has a positive end and a negative end, it can attract the part of another molecule with the opposite charge.

In addition, this type of molecule can act as a weak electrolyte because a polar covalent bond allows the substance to act as a conductor. So if a chemist wants a material to act as a good *insulator* (a device used to separate conductors), the chemist would look for a material with as weak a polar covalent bond as possible.

Water: A really strange molecule

Water (H_2O) has some very strange chemical and physical properties. It can exist in all three states of matter at the same time. Imagine that you're sitting in your hot tub (filled with *liquid* water) watching the steam *(gas)* rise from the surface as you enjoy a cold drink from a glass filled with ice *(solid)* cubes. Very few other chemical substances can exist in all these physical states in this close of a temperature range.

And those ice cubes are floating! In the solid state, the particles of matter are usually much closer together than they are in the liquid state. So if you put a solid into its corresponding liquid, it sinks. But this is not true of water. Its solid state is less dense than its liquid state, so it floats. Imagine what would happen if ice sank. In the winter, the lakes would freeze, and the ice would sink to the bottom, exposing more water. The extra exposed water would then freeze and sink, and so on, until the entire lake was frozen solid. This would destroy the aquatic life in the lake in no time. So instead, the ice floats and insulates the water underneath it, protecting aquatic life. And water's boiling point is unusually high. Other compounds similar in weight to water have a *much* lower boiling point.

Another unique property of water is its ability to dissolve a large variety of chemical substances. It dissolves salts and other ionic compounds, as well as polar covalent compounds such as alcohols and organic acids. In fact, water is sometimes called the universal solvent because it can dissolve so many things. It can also absorb a large amount of heat, which allows large bodies of water to help moderate the temperature on earth.

Water has many unusual properties because of its polar covalent bonds. Oxygen has a larger electronegativity than hydrogen, so the electron pairs are pulled in closer to the oxygen atom, giving it a partial negative charge. Subsequently, both of the hydrogen atoms take on a partial positive charge. The partial charges on the atoms created by the polar covalent bonds in water are shown in Figure 7-13.

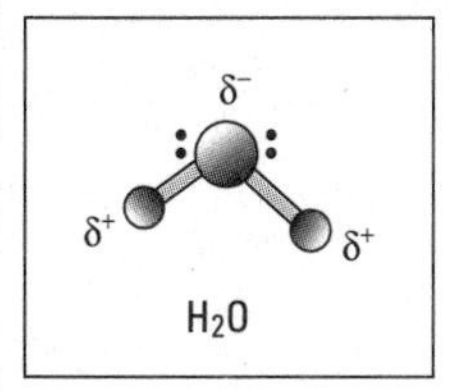

Figure 7-13: Polar covalent bonding in water.

Water is a dipole and acts like a magnet, with the oxygen end having a negative charge and the hydrogen end having a positive charge. These charged ends can attract other water molecules. The partially negatively charged oxygen atom of one water molecule can attract the partially positively charged hydrogen atom of another water molecule. This attraction between the molecules occurs frequently and is a type of *intermolecular force* (force between different molecules).

Intermolecular forces can be of three different types. The first type is called a *London force* or *dispersion force.* This very weak type of attraction generally occurs between nonpolar covalent molecules, such as nitrogen (N_2), hydrogen (H_2), or methane (CH_4). It results from the ebb and flow of the electron orbitals, giving a very weak and very brief charge separation around the bond.

The second type of intermolecular force is called a *dipole-dipole interaction.* This intermolecular force occurs when the positive end of one dipole molecule is attracted to the negative end of another dipole molecule. It's much stronger than a London force, but it's still pretty weak.

The third type of interaction is really just an extremely strong dipole-dipole interaction that occurs when a hydrogen atom is bonded to one of three extremely electronegative elements — O, N, or F. These three elements have a very strong attraction for the bonding pair of electrons, so the atoms involved in the bond take on a large amount of partial charge. This bond turns out to be highly polar — and the higher the polarity, the more effective the bond. When the O, N, or F on one molecule attracts the hydrogen of another molecule, the dipole-dipole interaction is very strong. This strong interaction (only about 5 percent of the strength of an ordinary covalent bond but still very strong for an intermolecular force) is called a *hydrogen bond.* The hydrogen bond is the type of interaction that's present in water (see Figure 7-14).

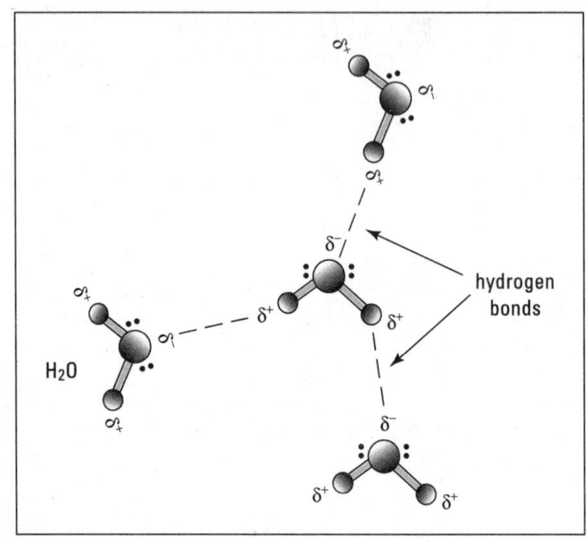

Figure 7-14: Hydrogen bonding in water.

Water molecules are stabilized by these hydrogen bonds, so breaking up (separating) the molecules is very hard. The hydrogen bonds account for water's high boiling point and ability to absorb heat. When water freezes, the hydrogen bonds lock water into an open framework that includes a lot of empty space. In liquid water, the molecules can get a little closer to each other, but when the solid forms, the hydrogen bonds result in a structure that contains large holes. The holes increase the volume and decrease the density. This process explains why the density of ice is less than that of liquid water (the reason ice floats). The structure of ice is shown in Figure 7-15, with the hydrogen bond indicated by dotted lines.

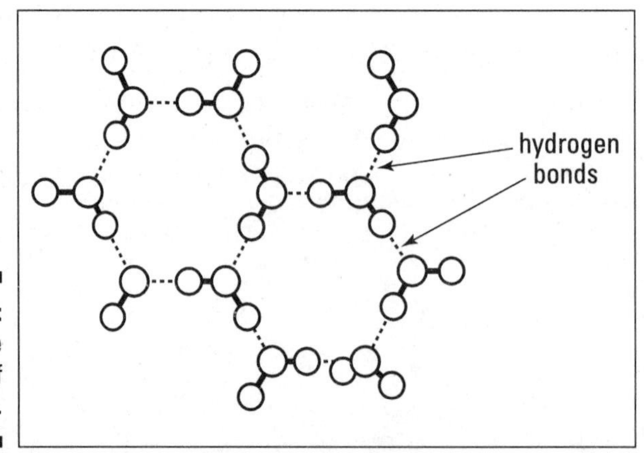

Figure 7-15: The structure of ice.

What Does Water Really Look Like? The VSEPR Theory

The *molecular geometry* of a molecule, how the atoms are arranged in three-dimensional space, is important for chemists to know because it often explains why certain reactions will or won't occur. In the area of medicine, for example, the molecular geometry of a drug may lead to side reactions. Molecular geometry also explains why water is a dipole (a molecule with a positive end and a negative end, like a magnet) and carbon dioxide is not.

The *VSEPR (Valence Shell Electron-Pair Repulsion) theory* allows chemists to predict the molecular geometry of molecules. The VSEPR theory assumes that the electron pairs around an atom, whether they're bonding (shared between two atoms) or nonbonding (not shared), will try to get as far apart from each other in space to minimize the repulsion between themselves. It's like going to a fancy party and seeing someone else wearing the exact same outfit. You're gonna try to stay as far away from that person as possible!

Electron-pair geometry is the arrangement of the electron pairs, bonding and nonbonding, around a central atom. After you determine the electron-pair geometry, you can imagine the nonbonding electrons being invisible and see what's left. What's left is what I call the *molecular geometry,* or *shape,* the arrangement of the other atoms around a central atom.

To determine the molecular geometry, or shape, of a molecule using the VSEPR theory, follow these steps:

1. **Determine the Lewis formula (see "Covalent Bond Basics," earlier in the chapter) of the molecule.**
2. **Determine the total number of electron pairs around the central atom.**
3. **Using Table 7-2, determine the electron-pair geometry.**

   (Table 7-2 relates the number of bonding and nonbonding electron pairs to the electron-pair geometry and molecular shape.)
4. **Imagine that the nonbonding electron pairs are invisible, and use Table 7-2 to determine the molecular shape.**

Table 7-2 Predicting Molecular Shape with the VSEPR Theory

Total Number of Electron Pairs	*Number of Bonding Pairs*	*Electron-pair Geometry*	*Molecular Geometry*
2	2	linear	linear
3	3	trigonal planar	trigonal planar
3	2	trigonal planar	bent, V-shaped
3	1	trigonal planar	linear
4	4	tetrahedral	tetrahedral
4	3	tetrahedral	trigonal pyramidal
4	2	tetrahedral	bent, V- shaped
5	5	trigonal bipyramidal	trigonal bipyramidal
5	4	trigonal bipyramidal	Seesaw
5	3	trigonal bipyramidal	T-shaped
5	2	trigonal bipyramidal	linear
6	6	octahedral	octahedral
6	5	octahedral	square pyramidal
6	4	octahedral	square planar

Even though you normally don't have to worry about more than four electron pairs around the central atom (octet rule), I put some of the less common exceptions to the octet rule in Table 7-2. Figure 7-16 shows some of the more common shapes mentioned in the table.

To determine the shapes of water (H_2O) and ammonia (NH_3), the first thing you have to do is determine the Lewis formula for each compound. Follow the rules outlined in the section, "Structural formula: Add the bonding pattern" (the N – A = S rules), and write the Lewis formulas as shown in Figure 7-17.

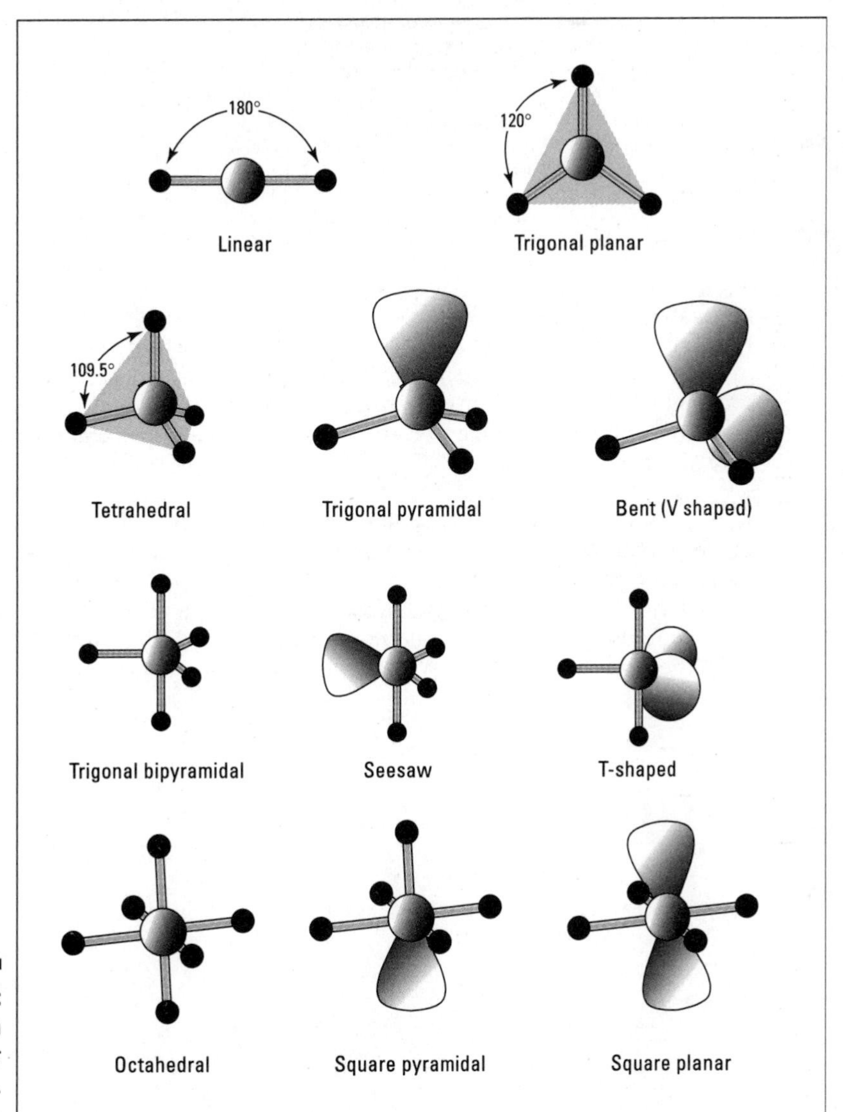

Figure 7-16: Common molecular shapes.

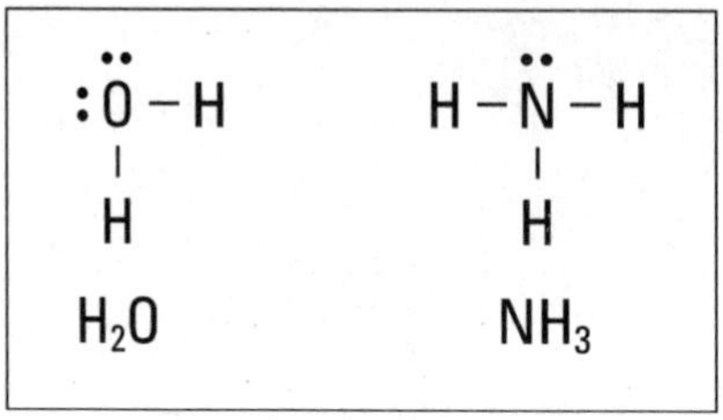

$$\begin{array}{cc} :\ddot{O}-H & H-\ddot{N}-H \\ | & | \\ H & H \\ H_2O & NH_3 \end{array}$$

Figure 7-17: Lewis formulas for H_2O and NH_3.

For water, there are four electron pairs around the oxygen atom, so the electron-pair geometry is tetrahedral. Only two of these four electron pairs are involved in bonding, so the molecular shape is bent or V-shaped. Because the molecular shape for water is V-shaped, I always show water with the hydrogen atoms at about a 90-degree angle to each other — it's a good approximation of the actual shape.

Ammonia also has four electron pairs around the nitrogen central atom, so its electron-pair geometry is tetrahedral, as well. Only one of the four electron pairs is nonbonding, however, so its molecular shape is trigonal pyramidal. This shape is like a three-legged milk stool, with the nitrogen being the seat — the lone pair of nonbonding electrons would then stick straight up from the seat. You'd get a surprise if you sat on an ammonia stool!

Chapter 8

Chemical Cooking: Chemical Reactions

In This Chapter

- Differentiating between reactants and products
- Finding out how reactions occur
- Taking a look at the different types of reactions
- Understanding how to balance reactions
- Figuring out chemical equilibrium
- Checking out speeds of reaction

Chemists do a lot of things: They measure the physical properties of substances; they analyze mixtures to find out what they're composed of; and they make new substances. The process of making chemical compounds is called *synthesis.* Synthesis depends on chemical reactions. I always thought that it'd be neat to be a synthetic organic chemist and work on the creation of new and potentially important compounds. I can just imagine the thrill of working for months, or even years, and finally ending up with a little pile of "stuff" that nobody in the world has ever seen. Hey, I am a nerd, after all!

In this chapter, I discuss chemical reactions — how they occur and how to write a balanced chemical equation. I also tell you about chemical equilibrium and explain why chemists can't get the amount of product out of a reaction that they thought they could. And finally, I discuss the speed of reaction and why you shouldn't leave that turkey sitting out on the table after finishing your Thanksgiving feast.

What You Have and What You'll Get: Reactants and Products

In a chemical reaction, substances (elements and/or compounds) are changed into other substances (compounds and/or elements). You can't change one element into another in a chemical reaction — that happens in nuclear reactions, as I describe in Chapter 5. Instead, you create a new substance with chemical reactions.

A number of clues show that a chemical reaction has taken place — something new is visibly produced, a gas is created, heat is given off or taken in, and so on. The chemical substances that are eventually changed are called the *reactants,* and the new substances that are formed are called the *products. Chemical equations* show the reactants and products, as well as other factors such as energy changes, catalysts, and so on. With these equations, an arrow is used to indicate that a chemical reaction has taken place. In general terms, a chemical reaction follows this format:

$$\text{Reactants} \rightarrow \text{Products}$$

For example, take a look at the reaction that occurs when you light your natural gas range in order to fry your breakfast eggs. Methane (natural gas) reacts with the oxygen in the atmosphere to produce carbon dioxide and water vapor. (If your burner isn't properly adjusted to give that nice blue flame, you may also get a significant amount of carbon monoxide along with carbon dioxide.) The chemical equation that represents this reaction is written like this:

$$CH_4(g) + 2\ O_2(g) \rightarrow CO_2(g) + 2\ H_2O(g)$$

You can read the equation like this: One molecule of methane gas, $CH_4(g)$, reacts with two molecules of oxygen gas, $O_2(g)$, to form one molecule of carbon dioxide gas, $CO_2(g)$, and two molecules of water vapor, $H_2O(g)$. The *2* in front of the oxygen gas and the *2* in front of the water vapor are called the reaction *coefficients.* They indicate the number of each chemical species that reacts or is formed. I show you how to figure out the value of the coefficients in the section "Balancing Chemical Reactions," later in the chapter.

Methane and oxygen (oxygen is a diatomic — two-atom — element) are the reactants, while carbon dioxide and water are the products. All the reactants and products are gases (indicated by the *g*'s in parentheses).

In this reaction, all reactants and products are invisible. The heat being evolved is the clue that tells you a reaction is taking place. By the way, this is a good example of an *exothermic* reaction, a reaction in which heat is given off. A lot of reactions are exothermic. Some reactions, however, absorb energy rather than release it. These reactions are called *endothermic* reactions. Cooking involves a lot of endothermic reactions — frying those eggs, for example. You can't just break the shells and let the eggs lie on the pan and then expect the myriad chemical reactions to take place without heating the pan (except when you're outside in Texas during August; there, the sun will heat the pan just fine).

Thinking about cooking those eggs brings to mind another issue about exothermic reactions. You have to ignite the methane coming out of the burners with a match, lighter, pilot light, or built-in electric igniter. In other words, you have to put in a little energy to get the reaction going. The energy you have to supply to get a reaction going is called the *activation energy* of the reaction. (In the next section, I show you that there's also an activation energy associated with endothermic reactions, but it isn't nearly as obvious.)

But what really happens at the molecular level when the methane and oxygen react? Divert thine eyes to the very next section to find out.

How Do Reactions Occur? Collision Theory

In order for a chemical reaction to take place, the reactants must collide. It's like playing pool. In order to drop the 8-ball into the corner pocket, you must hit it with the cue ball. This collision transfers *kinetic energy* (energy of motion) from one ball to the other, sending the second ball (hopefully) toward the pocket. The collision between the molecules provides the energy needed to break the necessary bonds so that new bonds can be formed.

But wait a minute. When you play pool, not every shot you make causes a ball to go into the pocket. Sometimes you don't hit the ball hard enough, and you don't transfer enough energy to get the ball to the pocket. This is also true with molecular collisions and reactions. Sometimes, even if there is a collision, not enough kinetic energy is available to be transferred — the molecules aren't moving fast enough. You can help the situation somewhat by heating the mixture of reactants. The temperature is a measure of the average kinetic energy of the molecules; raising the temperature increases the kinetic energy available to break bonds during collisions.

Sometimes, even if you hit the ball hard enough, it doesn't go into the pocket because you didn't hit it in the right spot. The same is true during a molecular collision. The molecules must collide in the right orientation, or hit at the right spot, in order for the reaction to occur.

Here's an example: Suppose you have an equation showing molecule *A-B* reacting with *C* to form *C-A* and *B,* like this:

A-B + C → C-A + B

The way this equation is written, the reaction requires that reactant *C* collide with *A-B* on the *A* end of the molecule. (You know this because the product side shows *C* hooked up with *A* — *C-A.*) If it hits the *B* end, nothing will happen. The *A* end of this hypothetical molecule is called the *reactive site,* the place on the molecule that the collision must take place in order for the reaction to occur. If *C* collides at the *A* end of the molecule, then there's a chance that enough energy can be transferred to break the *A-B* bond. After the *A-B* bond is broken, the *C-A* bond can be formed. The equation for this reaction process can be shown in this way (I show the breaking of the *AB* bond and the forming of the *CA* bond as "squiggly" bonds):

C~A~B → C-A + B

So in order for this reaction to occur, there must be a collision between *C* and *A-B* at the reactive site. The collision between *C* and *A-B* has to transfer enough energy to break the *A-B* bond, allowing the *C-A* bond to form.

Energy is required to break a bond between atoms.

Note that this example is a simple one. I've assumed that only one collision is needed, making this a one-step reaction. Many reactions are one-step, but many others require several steps in going from reactants to final products. In the process, several compounds may be formed that react with each other to give the final products. These compounds are called *intermediates.* They're shown in the reaction *mechanism,* the series of steps that the reaction goes through in going from reactants to products. But in this chapter, I keep it simple and pretty much limit my discussion to one-step reactions.

An exothermic example

Imagine that the hypothetical reaction A-B + C → C-A + B is exothermic — a reaction in which heat is given off (released) when going from reactants to products. The reactants start off at a higher energy state than the products, so energy is released in going from reactants to products. Figure 8-1 shows an energy diagram of this reaction.

In Figure 8-1, E_a is the activation energy for the reaction — the energy that you have to put in to get the reaction going. I show the collision of *C* and *A-B* with the breaking of the *A-B* bond and the forming of the *C-A* bond at the top of an activation energy hill. This grouping of reactants at the top of the activation energy hill is sometimes called the *transition state* of the reaction. As I show in Figure 8-1, the difference in the energy level of the reactants and the energy level of the products is the amount of energy (heat) that is released in the reaction.

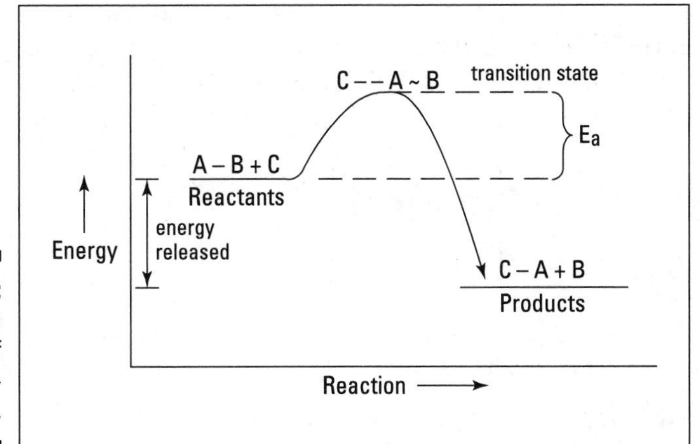

Figure 8-1: Exothermic reaction of A-B + C → C-A + B.

An endothermic example

Suppose that the hypothetical reaction A-B + C → C-A + B is endothermic — a reaction in which heat is absorbed in going from reactants to products — so the reactants are at a lower energy state than the products. Figure 8-2 shows an energy diagram of this reaction.

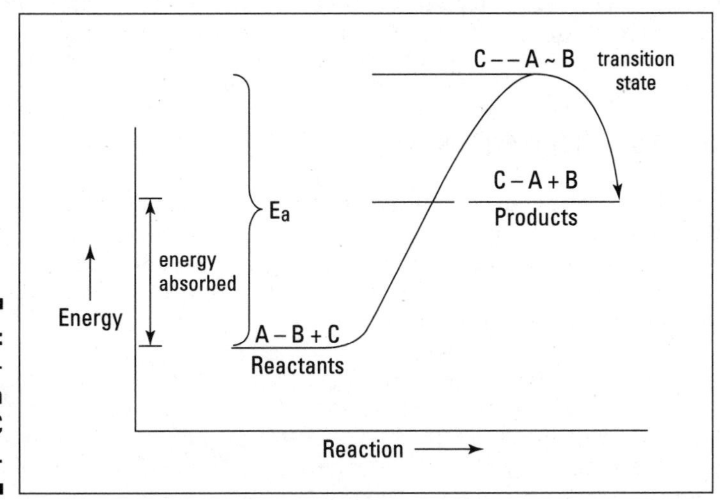

Figure 8-2: Endothermic reaction of A-B + C → C-A + B.

Just as with the exothermic-reaction energy diagram shown in Figure 8-1, this diagram shows that an activation energy is associated with the reaction (represented by E_a). In going from reactants to products, you have to put in more energy initially to get the reaction started, and then you get that energy back out as the reaction proceeds. Notice that the transition state appears at the top of the activation energy hill — just like in the exothermic-reaction energy diagram. The difference is that, in going from reactants to products, energy (heat) must be absorbed in the endothermic example.

What Kind of Reaction Do You Think I Am?

Several general types of chemical reactions can occur based on what happens when going from reactants to products. The more common reactions are

- Combination
- Decomposition
- Single displacement
- Double displacement
- Combustion
- Redox

Combination reactions

In *combination reactions,* two or more reactants form one product. The reaction of sodium and chlorine to form sodium chloride,

$$2\ Na(s) + Cl_2(g) \rightarrow 2\ NaCl(s)$$

and the burning of coal (carbon) to give carbon dioxide,

$$C(s) + O_2(g) \rightarrow CO_2(g)$$

are examples of combination reactions.

Note that, depending on conditions or the relative amounts of the reactants, more than one product can be formed in a combination reaction. Take the burning of coal, for example. If an excess of oxygen is present, the product is carbon dioxide. But if a limited amount of oxygen is available, the product is carbon monoxide:

$$2\,C(s) + O_2(g) \rightarrow 2\,CO(g) \qquad \text{(limited oxygen)}$$

Decomposition reactions

Decomposition reactions are really the opposite of combination reactions. In *decomposition reactions,* a single compound breaks down into two or more simpler substances (elements and/or compounds). The decomposition of water into hydrogen and oxygen gases,

$$2\,H_2O(l) \rightarrow 2\,H_2(g) + O_2(g)$$

and the decomposition of hydrogen peroxide to form oxygen gas and water,

$$2\,H_2O_2(l) \rightarrow 2\,H_2O(l) + O_2(g)$$

are examples of decomposition reactions.

Single displacement reactions

In *single displacement reactions,* a more active element displaces (kicks out) another less active element from a compound. For example, if you put a piece of zinc metal into a copper(II) sulfate solution (by the way, Chapter 6 explains why copper(II) sulfate is named the way it is — in case you're wondering), the zinc displaces the copper, as shown in this equation:

$$Zn(s) + CuSO_4(aq) \rightarrow ZnSO_4(aq) + Cu(s)$$

The notation *(aq)* indicates that the compound is dissolved in water — in an *aq*ueous solution. Because zinc replaces copper in this case, it's said to be more active. If you place a piece of copper in a zinc sulfate solution, nothing will happen. Table 8-1 shows the activity series of some common metals. Notice that because zinc is more active in the table, it will replace copper, just as the preceding equation shows.

Table 8-1 The Activity Series of Some Common Metals

Activity	*Metal*
Most active	Alkali and alkaline earth metals
	Al
	Zn
	Cr
	Fe
	Ni
	Sn
	Pb
	Cu
	Ag
Least Active	Au

Take another look at the reaction between zinc metal and copper(II) sulfate solution:

$$Zn(s) + CuSO_4(aq) \rightarrow ZnSO_4(aq) + Cu(s)$$

I've written this reaction as a molecular equation, showing all species in the neutral form. However, these reactions normally occur in an aqueous (water) solution. When the ionically-bonded $CuSO_4$ is dissolved in water, it breaks apart into *ions* (atoms or groups of atoms that have an electrical charge due to the loss or gain of electrons). The copper ion has a +2 charge because it lost two electrons. It's a *cation,* a positively charged ion. The sulfate ion has a -2 charge because it has two extra electrons. It's an *anion,* a negatively charged ion. (Check out Chapter 6 for a more complete discussion of ionic bonding.)

$$Zn(s) + Cu^{2+} + SO_4^{2-} \rightarrow Zn^{2+} + SO_4^{2-} + Cu(s)$$

Equations written in this form, in which the ions are shown separately, are called *ionic equations* (because they show the reaction and production of ions). Notice that the sulfate ion, SO_4^{2-}, hasn't changed in the reaction. Ions that don't change during the reaction and are found on both sides of the equation in an identical form are called spectator ions. Chemists (a lazy, lazy lot, they are) often omit the spectator ions and write the equation showing only those chemical substances that are changed during the reaction. This is called the net-ionic equation:

$$Zn(s) + Cu^{2+} \rightarrow Zn^{2+} + Cu(s)$$

Double displacement reactions

In single displacement reactions, only one chemical species is displaced. In *double displacement reactions,* or *metathesis reactions,* two species (normally ions) are displaced. Most of the time, reactions of this type occur in a solution, and either an insoluble solid (precipitation reactions) or water (neutralization reactions) will be formed.

Precipitation reactions

If you mix a solution of potassium chloride and a solution of silver nitrate, a white insoluble solid is formed in the resulting solution. The formation of an insoluble solid in a solution is called *precipitation.* Here are the molecular, ionic, and net-ionic equations for this double-displacement reaction:

$$KCl(aq) + AgNO_3(aq) \rightarrow AgCl(s) + KNO_3(aq)$$

$$K^+ + Cl^- + Ag^+ + NO_3^- \rightarrow AgCl(s) + K^+ + NO_3^-$$

$$Cl^- + Ag^+ \rightarrow AgCl(s)$$

The white insoluble solid that's formed is silver chloride. You can drop out the potassium cation and nitrate anion spectator ions, because they don't change during the reaction and are found on both sides of the equation in an identical form. (If you're totally confused about all those plus and minus symbols in the equations, or don't know what a cation or an anion is, just flip to Chapter 6. It tells all you need to know about this stuff.)

In order to write these equations, you have to know something about the solubility of ionic compounds. Don't fret. Here you go: If a compound is soluble, it will remain in its free ion form, but if it's insoluble, it will precipitate (form a solid). Table 8-2 gives the solubilities of selected ionic compounds.

Table 8-2 Solubilities of Selected Ionic Compounds

Water Soluble	***Water Insoluble***
All chlorides, bromides, iodides	*except those of* Ag^+, Pb^{2+}, Hg_2^{2+}
All compound of NH_4^+	Oxides
All compounds of alkali metals	Sulfides
All acetates	most phosphates
All nitrates	most hydroxides
All chlorates	
All sulfates	*except* $PbSO_4$, $BaSO_4$ and $SrSO_4$

To use Table 8-2, take the cation of one reactant and combine it with the anion of the other reactant, and vice versa (keeping the neutrality of the compounds). This allows you to predict the possible products of the reaction. Then look up the solubilities of the possible products in the table. If the compound is insoluble, it will precipitate. If it is soluble, it will remain in solution.

Neutralization reactions

The other type of double-displacement reaction is the reaction between an acid and a base. This double-displacement reaction, called a *neutralization reaction,* forms water. Take a look at the mixing solutions of sulfuric acid (auto battery acid, H_2SO_4) and sodium hydroxide (lye, NaOH). Here are the molecular, ionic, and net-ionic equations for this reaction:

$$H_2SO_4(aq) + 2\ NaOH(aq) \rightarrow Na_2SO_4(aq) + 2\ H_2O(l)$$

$$2\ H^+ + SO_4^{2-} + 2\ Na^+ + 2\ OH^- \rightarrow 2\ Na^+ + SO_4^{2-} + 2\ H_2O(l)$$

$$2\ H^+ + 2\ OH^- \rightarrow 2\ H_2O(l) \text{ or } H^+ + OH^- \rightarrow H_2O(l)$$

To go from the ionic equation to the net-ionic equation, the spectator ions (those that don't really react and that appear in an unchanged form on both sides on the arrow) are dropped out. Then the coefficients in front of the reactants and products are reduced down to the lowest common denominator.

You can find more about acid-base reactions in Chapter 12.

Combustion reactions

Combustion reactions occur when a compound, usually one containing carbon, combines with the oxygen gas in the air. This process is commonly called *burning.* Heat is the most-useful product of most combustion reactions.

Here's the equation that represents the burning of propane:

$$C_3H_8(g) + 5\ O_2(g) \rightarrow 3\ CO_2(g) + 4\ H_2O(l)$$

Propane belongs to a class of compounds called *hydrocarbons,* compounds composed only of carbon and hydrogen. The product of this reaction is heat. You don't burn propane in your gas grill to add carbon dioxide to the atmosphere — you want the heat for cooking your steaks.

Combustion reactions are also a type of redox reaction.

Redox reactions

Redox reactions, or *reduction-oxidation reactions,* are reactions in which electrons are exchanged:

$2\ Na(s) + Cl_2(g) \rightarrow 2\ NaCl(s)$

$C(s) + O_2(g) \rightarrow CO_2(g)$

$Zn(s) + CuSO_4(aq) \rightarrow ZnSO_4(aq) + Cu(s)$

The preceding reactions are examples of other types of reactions (such as combination, combustion, and single-replacement reactions), but they're all redox reactions. They all involve the transfer of electrons from one chemical species to another. Redox reactions are involved in combustion, rusting, photosynthesis, respiration, batteries, and more. I talk about redox reactions in some detail in Chapter 9.

Balancing Chemical Reactions

If you carry out a chemical reaction and carefully sum up the masses of all the reactants, and then compare the sum to the sum of the masses of all the products, you see that they're the same. In fact, a law in chemistry, the *Law of Conservation of Mass,* states, "In an ordinary chemical reaction, matter is neither created nor destroyed." This means that you have neither gained nor lost any atoms during the reaction. They may be combined differently, but they're still there.

A chemical equation represents the reaction. That chemical equation is used to calculate how much of each element is needed and how much of each element will be produced. And that chemical equation needs to obey the Law of Conservation of Mass.

You need to have the same number of each kind of element on both sides of the equation. The equation should *balance.* In this section, I show you how to balance chemical equations.

Smell that ammonia

My favorite reaction is called the *Haber process,* a method for preparing ammonia (NH_3) by reacting nitrogen gas with hydrogen gas:

$N_2(g) + H_2(g) \rightarrow NH_3(g)$

This equation shows you what happens in the reaction, but it doesn't show you how much of each element you need to produce the ammonia. To find out how much of each element you need, you have to balance the equation — make sure that the number of atoms on the left side of the equation equals the number of atoms on the right.

You know the reactants and the product for this reaction, and you can't change them. You can't change the compounds, and you can't change the subscripts, because that would change the compounds. So the only thing you can do to balance the equation is add *coefficients,* whole numbers in front of the compounds or elements in the equation. Coefficients tell you how many atoms or molecules you have.

For example, if you write *2 H_2O,* it means you have two water molecules:

$$2H_2O = \begin{matrix} H_2O \\ + \\ H_2O \end{matrix}$$

Each water molecule is composed of two hydrogen atoms and one oxygen atom. So with *2 H_2O,* you have a total of 4 hydrogen atoms and 2 oxygen atoms:

$$2H_2O = \begin{matrix} H_2O = 2H + 1O \\ + \qquad\quad + \\ H_2O = 2H + 1O \\ \hline 4H + 2O \end{matrix}$$

In this chapter, I show you how to balance equations by using a method called *balancing by inspection,* or as I call it, "fiddling with coefficients." You take each atom in turn and balance it by adding appropriate coefficients to one side or the other.

With that in mind, take another look at the equation for preparing ammonia:

$$N_2(g) + H_2(g) \rightarrow NH_3(g)$$

In most cases, it's a good idea to wait until the end to balance hydrogen atoms and oxygen atoms; balance the other atoms first.

So in this example, you need to balance the nitrogen atoms first. You have 2 nitrogen atoms on the left side of the arrow (reactant side) and only 1 nitrogen atom on the right side (product side). In order to balance the nitrogen atoms, use a coefficient of 2 in front of the ammonia on the right.

$$N_2(g) + H_2(g) \rightarrow 2\ NH_3(g)$$

Now you have 2 nitrogen atoms on the left and 2 nitrogen atoms on the right.

Next, tackle the hydrogen atoms. You have 2 hydrogen atoms on the left and 6 hydrogen atoms on the right (2 NH_3 molecules, each with 3 hydrogen atoms, for a total of 6 hydrogen atoms). So put a 3 in front of the H_2 on the left, giving you:

$$N_2(g) + 3\ H_2(g) \rightarrow 2\ NH_3(g)$$

That should do it. Do a check to be sure: You have 2 nitrogen atoms on the left and 2 nitrogen atoms on the right. You have 6 hydrogen atoms on the left ($3 \times 2 = 6$) and 6 hydrogen atoms on the right ($2 \times 3 = 6$). The equation is balanced. You can read the equation this way: 1 nitrogen molecule reacts with 3 hydrogen molecules to yield 2 ammonia molecules.

Here's a tidbit for you: This equation would have also balanced with coefficients of 2, 6, and 4 instead of 1, 3, and 2. In fact, any multiple of 1, 3, and 2 would have balanced the equation, but chemists have agreed to always show the lowest whole-number ratio (see the discussion on empirical formulas in Chapter 7 for details).

Flick that bic

Take a look at an equation showing the burning of butane, a hydrocarbon, with excess oxygen available. (This is the reaction that takes place when you light a butane lighter.) The unbalanced reaction is

$$C_4H_{10}(g) + O_2(g) \rightarrow CO_2(g) + H_2O(g)$$

Because it's always a good idea to wait until the end to balance hydrogen atoms and oxygen atoms, balance the carbon atoms first. You have 4 carbon atoms on the left and one carbon atom on the right, so add a coefficient of 4 in front of the carbon dioxide:

$$C_4H_{10}(g) + O_2(g) \rightarrow 4\ CO_2(g) + H_2O(g)$$

Balance the hydrogen atoms next. You have 10 hydrogen atoms on the left and 2 hydrogen atoms on the right, so use a coefficient of 5 in front of the water on the right:

$$C_4H_{10}(g) + O_2(g) \rightarrow 4\ CO_2(g) + 5\ H_2O(g)$$

Now work on balancing the oxygen atoms. You have 2 oxygen atoms on the left and a total of 13 oxygen atoms on the right $[(4 \times 2) + (5 \times 1) = 13]$. What can you multiply 2 with in order for it to equal 13? How about 6.5?

$$C_4H_{10}(g) + 6.5\ O_2(g) \rightarrow 4\ CO_2(g) + 5\ H_2O(g)$$

But you're not done. You want the lowest *whole-number* ratio of coefficients. You'll have to multiply the entire equation by 2 in order to generate whole numbers:

$$[C_4H_{10}(g) + 6.5\ O_2(g) \rightarrow 4\ CO_2(g) + 5\ H_2O(g)] \times 2$$

Multiply every coefficient by 2 (don't touch the subscripts!) to get

$$2\ C_4H_{10}(g) + 13\ O_2(g) \rightarrow 8\ CO_2(g) + 10\ H_2O(g)$$

If you check the atom count on both sides of the equation, you find that the equation is balanced, and the coefficients are in the lowest whole-number ratio.

After balancing an equation, make sure that the same number of each atom is on both sides and that the coefficients are in the lowest whole-number ratio.

Most simple reactions can be balanced in this fashion. But one class of reactions is so complex that this method doesn't work well when applied to them. They're redox reactions. A special method is used for balancing these equations, and I show it to you in Chapter 9.

Chemical Equilibrium

My favorite reaction is the Haber process, the synthesis of ammonia from nitrogen and hydrogen gases. After balancing the reaction (see the section "Smell that ammonia," earlier in this chapter), you end up with

$$N_2(g) + 3\ H_2(g) \rightarrow 2\ NH_3(g)$$

Written this way, the reaction says that hydrogen and nitrogen react to form ammonia — and this keeps on happening until you use up one or both of the reactants. But this isn't quite true. (Yep. It's hair-splitting time.)

If this reaction occurs in a closed container (which it has to, with everything being gases), then the nitrogen and hydrogen react and ammonia *is* formed — *but* some of the ammonia soon starts to decompose to nitrogen and hydrogen, like this:

$$2\ NH_3(g) \rightarrow N_2(g) + 3\ H_2(g)$$

In the container, then, you actually have *two* exactly opposite reactions occurring — nitrogen and hydrogen combine to give ammonia, and ammonia decomposes to give nitrogen and hydrogen.

Instead of showing the two separate reactions, you can show one reaction and use a double arrow like this:

$N_2(g) + 3\ H_2(g) \leftrightarrow 2\ NH_3(g)$

You put the nitrogen and hydrogen on the left because that's what you initially put into the reaction container.

Now these two reactions occur at different speeds, but sooner or later, the two speeds become the same, and the relative amounts of nitrogen, hydrogen, and ammonia become constant. This is an example of a chemical equilibrium. A *dynamic chemical equilibrium* is established when two exactly opposite chemical reactions are occurring at the same place, at the same time, with the same rates (speed) of reaction. I call this example a *dynamic* chemical equilibrium, because when the reactions reach equilibrium, things don't just stop. At any given time, you have nitrogen and hydrogen reacting to form ammonia, and ammonia decomposing to form nitrogen and hydrogen. When the system reaches equilibrium, the amounts of all chemical species become *constant* but not necessarily the same.

Here's an example to help you understand what I mean by this dynamic stuff: I was raised on a farm in North Carolina, and my mother, Grace, *loved* small dogs. Sometimes we'd have close to a dozen dogs running around the house. When Mom opened the door to let them outside, they'd start running out. But some would change their minds after they got outside and would then start running back into the house. They'd then get caught up in the excitement of the other dogs and start running back outside again. There'd be a never-ending cycle of dogs running in and out of the house. Sometimes there'd only be two or three in the house, with the rest outside, or vice versa. The number of dogs inside and outside would be *constant* but not the same. And at any given point, there'd be dogs running out of the house and dogs running into the house. It was a dynamic equilibrium (and a noisy one).

Sometimes there's a lot of product (chemical species on the right-hand side of the double arrow) when the reaction reaches equilibrium, and sometimes there's very little. You can tell the relative amounts of reactants and products at equilibrium if you know the equilibrium constant for the reaction.

Look at a hypothetical equilibrium reaction:

$aA + bB \leftrightarrow cC + dD$

The capital letters stand for the chemical species, and the small letters represent the coefficients in the balanced chemical equation. The *equilibrium constant* (represented as K_{eq}) is mathematically defined as

$$K_{eq} = \frac{[C]^c[D]^d}{[A]^a[B]^b}$$

The numerator contains the product of the two chemical species on the right-hand side of the equation, with each chemical species raised to the power of its coefficient in the balanced chemical equation. The denominator is the same, but you use the chemical species on the left-hand side of the equation. (It's not important right now, but those brackets stand for something called the molar concentration. You can find out what that is in Chapter 11.) Note that sometimes chemists use the K_c notation instead of the K_{eq} form.

The numerical value of the equilibrium constant gives you a clue as to the relative amounts of products and reactants.

The larger the value of the equilibrium constant (K_{eq}), the more products are present at equilibrium. If, for example, you have a reaction that has an equilibrium constant of 0.001 at room temperature and 0.1 at 100 degrees Celsius, you can say that you will have much more product at the higher temperature.

Now I happen to know that the K_{eq} for the *Haber process* (the ammonia synthesis) is 3.5×10^8 at room temperature. This large value indicates that, at equilibrium, there's a lot of ammonia produced from the nitrogen and hydrogen, but there's still hydrogen and nitrogen left at equilibrium. If you're, say, an industrial chemist making ammonia, you want as much of the reactants as possible to be converted to product. You'd like the reaction to go to completion (meaning you'd like the reactants to keep creating the product until they're all used up), but you know that it's an equilibrium reaction, and you can't change that. But it would be nice if you could, in some way, manipulate the system to get a little bit more product formed. There is such a way — through Le Chatelier's Principle.

Le Chatelier's Principle

A French chemist, Henri Le Chatelier, discovered that if you apply a change of condition (called *stress*) to a chemical system that's at equilibrium, the reaction will return to equilibrium by shifting in such a way as to counteract the change (the stress). This is called *Le Chatelier's Principle.*

You can stress an equilibrium system in three ways:

- Change the concentration of a reactant or product.
- Change the temperature.
- Change the pressure on a system that contains gases.

Now if you're a chemist who's looking for a way to make as much ammonia (money) as possible for a chemical company, you can use Le Chatelier's Principle to help you along. In this section, I show you how.

But first, I want to show you a quick, useful analogy. A reaction at equilibrium is like one of my favorite pieces of playground equipment, a teeter-totter. Everything is well balanced, as shown in Figure 8-3.

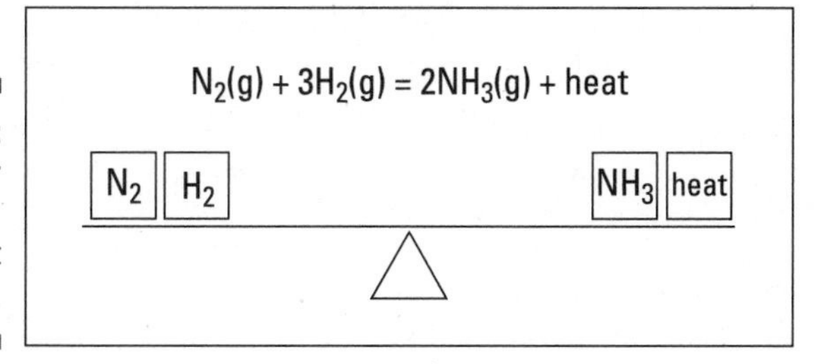

Figure 8-3: The Haber ammonia system at equilibrium.

The Haber process, the synthesis of ammonia from nitrogen and hydrogen gases, is exothermic: It gives off heat. I show that heat on the right-hand side of the teeter-totter in the figure.

Changing the concentration

Suppose that you have the ammonia system at equilibrium (see Figure 8-3, as well as the section, "Chemical Equilibrium," earlier in this chapter), and you then put in some more nitrogen gas. Figure 8-4 shows what happens to the teeter-totter when you add more nitrogen gas.

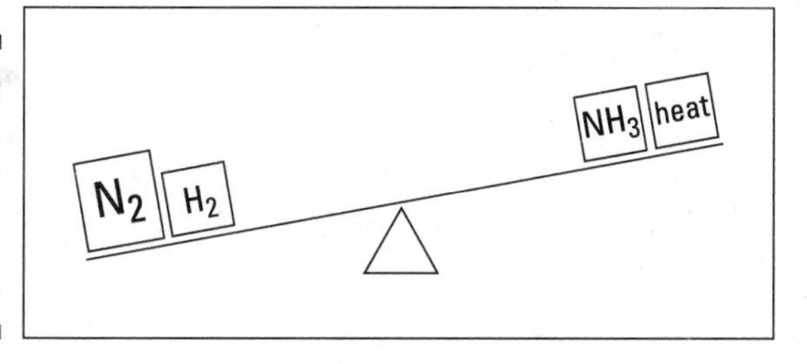

Figure 8-4: Increasing the concentration of a reactant.

In order to reestablish the balance (equilibrium), weight has to be shifted from the left to the right, using up some nitrogen and hydrogen, and forming more ammonia and heat. Figure 8-5 shows this shifting of weight.

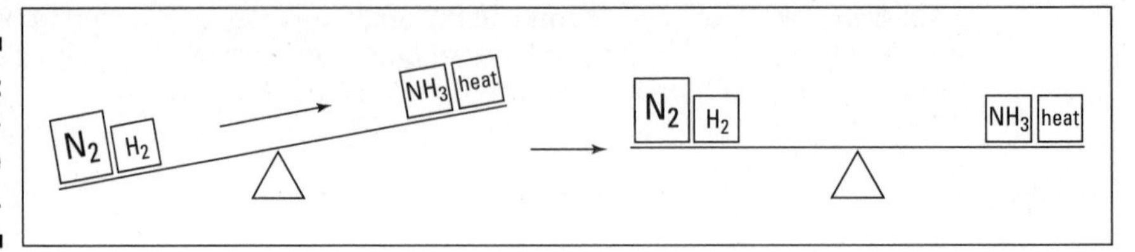

Figure 8-5: Reestablishing the equilibrium.

The equilibrium has been reestablished. There is less hydrogen and more nitrogen, ammonia, and heat than you had before you added the additional nitrogen. The same thing would happen if you had a way of removing ammonia as it was formed. The right-hand side of the teeter-totter would again be lighter, and weight would be shifted to the right in order to reestablish the equilibrium. Again, more ammonia would be formed. In general, if you add more of a reactant or product, the reaction will shift to the other side to use it up. If you remove some reactant or product, the reaction shifts to that side in order to replace it.

Changing the temperature

Suppose that you heat the reaction mixture. You know that the reaction is exothermic — heat is given off, showing up on the right-hand side of the equation. So if you heat the reaction mixture, the right side of the teeter-totter gets heavier, and weight must be shifted to the left in order to reestablish the equilibrium. This weight shift uses up ammonia and produces more nitrogen and hydrogen. And as the reaction shifts, the amount of heat also decreases, lowering the temperature of the reaction mixture. Figure 8-6 shows this shift in weight.

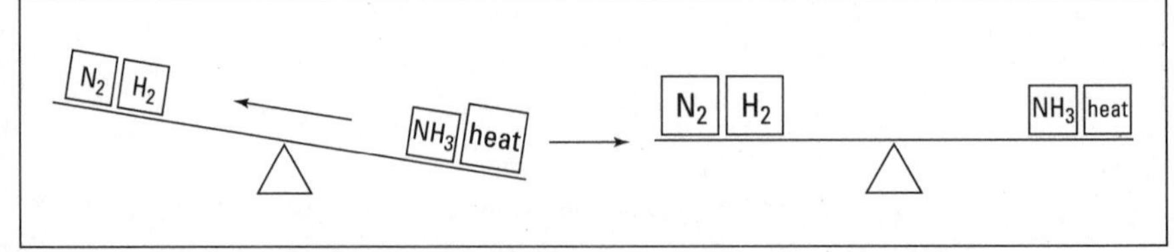

Figure 8-6: Increasing the temperature on an exothermic reaction and reestablishing the equilibrium.

That's not what you want! You want more ammonia, not more nitrogen and hydrogen. So you have to cool the reaction mixture, using up heat, and then the equilibrium shifts to the right in order to replace it. This process helps you make more ammonia and more profit. That's more like it.

In general, heating a reaction causes it to shift to the endothermic side. (If you have an exothermic reaction where heat is produced on the right side, then the left side is the endothermic side.) Cooling a reaction mixture causes the equilibrium to shift to the exothermic side.

Changing the pressure

Changing the pressure only affects the equilibrium if there are reactants and/or products that are gases. In the Haber process, all species are gases, so there is a pressure effect.

My teeter-totter analogy of equilibrium systems begins to break down when I explain pressure effects, so I have to take another approach. Think about the sealed container where your ammonia reaction is occurring. (The reaction has to occur in a sealed container with everything being gases.) You have nitrogen, hydrogen, and ammonia gases inside. There is pressure in the sealed container, and that pressure is due to the gas molecules hitting the inside walls of the container.

Now suppose that the system is at equilibrium, and you want to increase the pressure. You can do so by making the container smaller (with a piston type of arrangement) or by putting in an unreactive gas, such as neon. You get more collisions on the inside walls of the container, and, therefore, you have more pressure. Increasing the pressure stresses the equilibrium; in order to remove that stress and reestablish the equilibrium, the pressure must be reduced.

Take another look at the Haber reaction and see if there may be some clues as to how this may happen.

$$N_2(g) + 3\ H_2(g) \leftrightarrow 2\ NH_3(g)$$

Every time the forward (left to right) reaction takes place, four molecules of gas (one nitrogen and three hydrogen) form two molecules of ammonia gas. This reaction reduces the number of molecules of gas in the container. The reverse reaction (right to left) takes two ammonia gas molecules and makes four gas molecules (nitrogen and hydrogen). This reaction increases the number of gas molecules in the container.

The equilibrium has been stressed by an increase in pressure; reducing the pressure will relieve the stress. Reducing the number of gas molecules in the container will reduce the pressure (fewer collisions on the inside walls of the container), so the forward (left to right) reaction is favored because four gas molecules are consumed and only two are formed. As a result of the forward reaction, more ammonia is produced!

In general, increasing the pressure on an equilibrium mixture causes the reaction to shift to the side containing the *fewest* number of gas molecules.

Reacting Fast and Reacting Slow: Chemical Kinetics

Say that you're a chemist who wants to make as much ammonia as possible from a given amount of hydrogen and nitrogen. Manipulating the equilibrium (see the preceding section) isn't your total solution. You want to produce as much as possible, *as fast as possible*. So there's something else you must consider — the kinetics of the reaction.

Kinetics is the study of the speed of a reaction. Some reactions are fast; others are slow. Sometimes chemists want to speed the slow ones up and slow the fast ones down. There are several factors that affect the speed of a reaction:

- Nature of the reactants
- Particle size of the reactants
- Concentration of the reactants
- Pressure of gaseous reactants
- Temperature
- Catalysts

Nature of the reactants

In order for a reaction to occur, there must be a collision between the reactants at the reactive site of the molecule (see "How Do Reactions Occur? Collision Theory," earlier in this chapter). The larger and more complex the reactant molecules, the less chance there is of a collision at the reactive site. Sometimes, in very complex molecules, the reactive site is totally blocked off by other parts of the molecule, so no reaction occurs. There may be a lot of collisions, but only the ones that occur at the reactive site have any chance of leading to chemical reaction.

In general, the reaction rate is slower when the reactants are large and complex molecules.

Particle size of the reactants

Reaction depends on collisions. The more surface area on which collisions can occur, the faster the reaction. You can hold a burning match to a large chunk of coal and nothing will happen. But if you take that same piece of coal, grind it up very, very fine, throw it up into the air, and strike a match, you'll get an explosion because of the increased surface area of the coal.

Concentration of the reactants

Increasing the number of collisions speeds up the reaction rate. The more reactant molecules there are colliding, the faster the reaction will be. For example, a wood splint burns okay in air (20 percent oxygen), but it burns *much* faster in pure oxygen.

In most simple cases, increasing the concentration of the reactants increases the speed of the reaction. However, if the reaction is complex and has a complex *mechanism* (series of steps in the reaction), this may not be the case. In fact, determining the concentration effect on the rate of reaction can give you clues as to which reactant is involved in the rate-determining step of the mechanism. (This information can then be used to help figure out the reaction mechanism.) You can do this by running the reaction at several different concentrations and observing the effect on the rate of reaction. If, for example, changing the concentration of one reactant has no effect on the rate of reaction, then you know that reactant is not involved in the slowest step (the rate-determining step) in the mechanism.

Pressure of gaseous reactants

The pressure of gaseous reactants has basically the same effect as concentration. The higher the reactant pressure, the faster the reaction rate. This is due to (you guessed it!) the increased number of collisions. But if there's a complex mechanism involved, changing the pressure may not have the expected result.

Temperature

Okay, why did mom tell you to put that turkey in the refrigerator after Thanksgiving dinner? Because it would've spoiled if you didn't. And what is spoilage? It's increased bacterial growth. So when you put the turkey in the refrigerator, the cold temperature inside the fridge slowed down the rate of bacterial growth.

Bacterial growth is simply a *biochemical reaction,* a chemical reaction involving living organisms. In most cases, increasing the temperature causes the reaction rate to increase. In organic chemistry, there's a general rule that says increasing the temperature 10 degrees Celsius will cause the reaction rate to double.

But why is this true? Part of the answer is (you guessed it!) an increased number of collisions. Increasing the temperature causes the molecules to move faster, so there's an increased chance of them colliding with each other and reacting. But this is only part of the story. Increasing the temperature also increases the average kinetic energy of the molecules. Look at Figure 8-7 for an example of how increasing the temperature affects the kinetic energy of the reactants and increases the reaction rate.

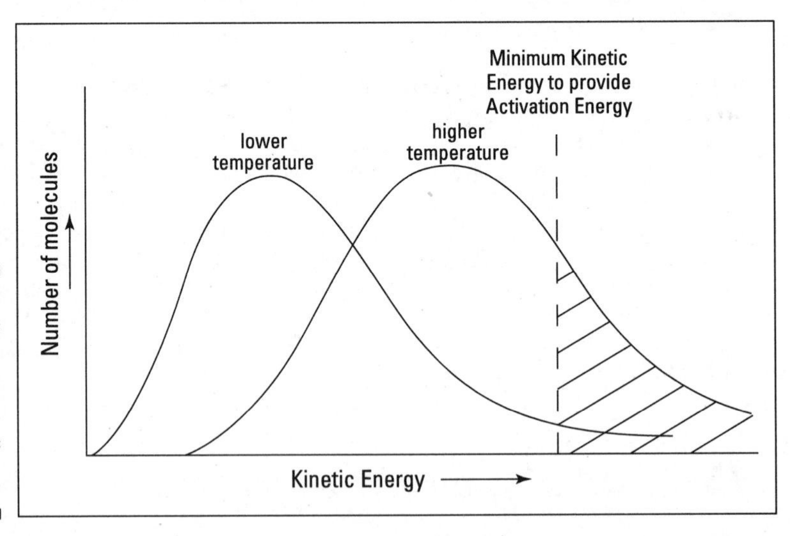

Figure 8-7: The effect of temperature on the kinetic energy of reactants.

At a given temperature, not all the molecules are moving with the same kinetic energy. A small number of molecules are moving very slow (low kinetic energy), while a few are moving very fast (high kinetic energy). A vast majority of the molecules are somewhere in between these two extremes.

In fact, temperature is a measure of the average kinetic energy of the molecules. As you can see in Figure 8-7, increasing the temperature increases the average kinetic energy of the reactants, essentially shifting the curve to the right toward higher kinetic energies. But also notice that I've marked the minimum amount of kinetic energy needed by the reactants to provide the activation energy (the energy required to get a reaction going) during collision. The reactants have to collide at the reactive site, but they also have to transfer *enough* energy to break bonds so that new bonds can be formed. If the reactants don't have enough energy, a reaction won't occur even if the reactants do collide at the reactive site.

Notice that at the lower temperature, very few of the reactant molecules have the minimum amount of kinetic energy needed to provide the activation energy. At the higher temperature, many more molecules possess the minimum amount of kinetic energy needed, which means a lot more collisions will be energetic enough to lead to reaction.

Increasing the temperature not only increases the number of collisions but also increases the number of collisions that are effective — that transfer enough energy to cause a reaction to take place.

Catalysts

Catalysts are substances that increase the reaction rate without themselves being changed at the end of the reaction. They increase the reaction rate by lowering the activation energy for the reaction.

Look at Figure 8-1, for example. If the activation energy hill were lower, it'd be easier for the reaction to occur and the reaction rate would be faster. You can see the same thing in Figure 8-7. If you shift to the left that dotted line representing the minimum amount of kinetic energy needed to provide the activation energy, then many more molecules will have the minimum energy needed, and the reaction will be faster.

Catalysts lower the activation energy of a reaction in one of two ways:

- Providing a surface and orientation
- Providing an alternative mechanism (series of steps for the reaction to go through) with a lower activation energy

Surface and orientation — heterogeneous catalysis

In the section "How Do Reactions Occur? Collision Theory," I describe how molecules react, using this generalized example:

C-A-B → C-A + B

Reactant *C* must hit the reactive site on the *A* end of molecule *A-B* in order to break the *A-B* bond and form the *C-A* bond shown in the equation. The probability of the collision occurring in the proper orientation is pretty much driven by chance. The reactants are moving around, running into each other, and sooner or later the collision may occur at the reactive site. But what would happen if you could tie the *A-B* molecule down with the *A* end exposed? It'd be much easier and more probable for *C* to hit *A* with this scenario.

This is what a heterogeneous catalyst accomplishes: It ties one molecule to a surface while providing proper orientation to make the reaction easier. The process of heterogeneous catalysis is shown in Figure 8-8.

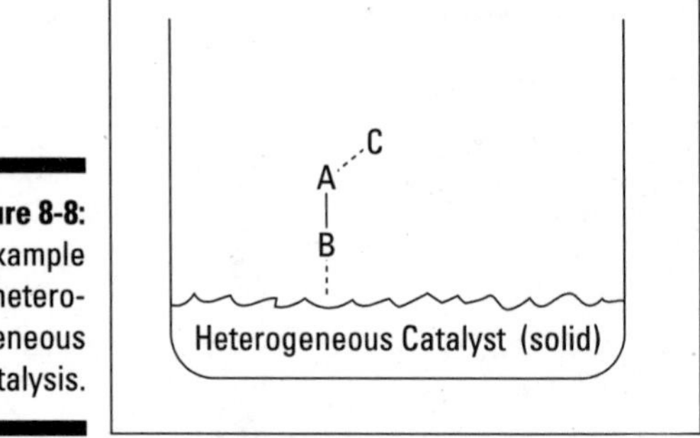

Figure 8-8: An example of heterogeneous catalysis.

The catalyst is called a *heterogeneous* catalyst because it's in a different phase than the reactants. This catalyst is commonly a finely divided solid metal or metal oxide, while the reactants are gases or in solution. This heterogeneous catalyst tends to attract one part of a reactant molecule due to rather complex interactions that are not fully understood. After the reaction takes place, the forces that bound the *B* part of the molecule to the surface of the catalyst are no longer there. So *B* can drift off, and the catalyst can be ready to do it again.

Most of us sit very close to a heterogeneous catalyst every day — the catalytic converter in our automobile. It contains finely divided platinum and/or palladium metal and speeds up the reaction that causes harmful gases from gasoline (such as carbon monoxide and unburned hydrocarbons) to decompose into mostly harmless products (such as water and carbon dioxide).

Alternative mechanism — homogeneous catalysis

The second type of catalyst is a *homogeneous catalyst* — one that's in the same phase as the reactants. It provides an alternative mechanism, or reaction pathway, that has a lower activation energy than the original reaction.

For an example, check out the decomposition reaction of hydrogen peroxide:

$$2\ H_2O_2(l) \rightarrow 2\ H_2O(l) + O_2(g)$$

This is a slow reaction, especially if it's kept cool in a dark bottle. It may take years for that bottle of hydrogen peroxide in your medicine cabinet to decompose. But if you put a little bit of a solution containing the ferric ion in the bottle, the reaction will be much faster, even though it will be a two-step mechanism instead of a one-step mechanism:

(Step 1) $2\ Fe^{3+} + H_2O_2(l) \rightarrow 2\ Fe^{2+} + O_2(g) + 2\ H^+$

(Step 2) $2\ Fe^{2+} + H_2O_2(l) + 2\ H^+ \rightarrow 2\ Fe^{3+} + 2\ H_2O(l)$

If you add the two preceding reactions together and cancel the species that are identical on both sides, you get the original, uncatalyzed reaction:

$\mathbf{2\ Fe^{3+}} + H_2O_2(l) + \mathbf{2\ Fe^{2+}} + H_2O_2(l) + \mathbf{2\ H^+} \rightarrow \mathbf{2\ Fe^{2+}} + O_2(g) + \mathbf{2\ H^+} + \mathbf{2\ Fe^{3+}} +$ $2\ H_2O(l)$ (species to be cancelled are bolded)

$$2\ H_2O_2(l) \rightarrow 2\ H_2O(l) + O_2(g)$$

The ferric ion catalyst was changed in the first step and then changed back in the second step. This two-step catalyzed pathway has a lower activation energy and is faster.

Chapter 9

Electrochemistry: Batteries to Teapots

In This Chapter

- Finding out about redox reactions
- Figuring out how to balance redox equations
- Taking a look at electrochemical cells
- Checking out electroplating
- Discovering how burning fuels and burning foods are similar

Many of the things we deal with in real life are related either directly or indirectly to electrochemical reactions. Think of all the things around you that contain batteries — flashlights, watches, automobiles, calculators, PDAs, pacemakers, cell phones, toys, garage door openers, and so on.

Do you drink from an aluminum can? The aluminum was extracted by an electrochemical reaction. Do you have a car with a chrome bumper? That chrome is electroplated onto the bumper, just like the silver on Grandmother Grace's tea service or the gold on that five-dollar gold chain. Do you watch television, use electric lights or an electric blender, or have a desktop computer? There's a good chance that the electricity you use for these things is generated from the combustion of some fossil fuel. Combustion is a redox reaction. So are respiration, photosynthesis, and many other biochemical processes that we depend upon for life. Electrochemical and redox reactions surround us.

In this chapter, I explain redox reactions, go through the balancing of this type of equation, and then show you some applications of redox reactions in an area of chemistry called electrochemistry.

There Go Those Pesky Electrons: Redox Reactions

Redox reactions — reactions in which there's a simultaneous transfer of electrons from one chemical species to another — are really composed of two different reactions: *oxidation* (a loss of electrons) and *reduction* (a gain of electrons). These reactions are coupled, as the electrons that are lost in the oxidation reaction are the same electrons that are gained in the reduction reaction. In fact, these two reactions (reduction and oxidation) are commonly called *half-reactions,* because it takes these two halves to make a whole reaction, and the overall reaction is called a *redox* (*red*uction/*ox*idation) reaction. In Chapter 8, I describe a redox reaction that occurs between zinc metal and the cupric (Cu^{2+}) ion. The zinc metal loses electrons and the cupric ion gains them.

Now where did I put those electrons? Oxidation

There are three definitions you can use for oxidation:

- The loss of electrons
- The gain of oxygen
- The loss of hydrogen

Because I typically deal with electrochemical cells, I normally use the definition that describes the loss of the electrons. The other definitions are useful in processes such as combustion and photosynthesis.

Loss of electrons

One way to define oxidation is with the reaction in which a chemical substance loses electrons in going from reactant to product. For example, when sodium metal reacts with chlorine gas to form sodium chloride (NaCl), the sodium metal loses an electron, which is then gained by chlorine. The following equation shows sodium losing the electron:

$$Na(s) \rightarrow Na^{+} + e^{-}$$

When it loses the electron, chemists say that the sodium metal has been oxidized to the sodium cation. (A *cation* is an ion with a positive charge due to the loss of electrons — see Chapter 6.)

Reactions of this type are quite common in *electrochemical reactions,* reactions that produce or use electricity. (For more info about electrochemical reactions, flip to the section, "Power On the Go: Electrochemical Cells," later in this chapter.)

Gain of oxygen

Sometimes, in certain oxidation reactions, it's obvious that oxygen has been gained in going from reactant to product. Reactions where the gain of oxygen is more obvious than the gain of electrons include combustion reactions (*burning*) and the *rusting* of iron. Here are two examples:

$$C(s) + O_2(g) \rightarrow CO_2(g) \quad \text{(burning of coal)}$$

$$2\,Fe(s) + 3\,O_2(g) \rightarrow 2\,Fe_2O_3(s) \quad \text{(rusting of iron)}$$

In these cases, chemists say that the carbon and the iron metal have been oxidized to carbon dioxide and rust, respectively.

Loss of hydrogen

In other reactions, oxidation can best be seen as the loss of hydrogen. Methyl alcohol (wood alcohol) can be oxidized to formaldehyde:

$$CH_3OH(l) \rightarrow CH_2O(l) + H_2(g)$$

In going from methanol to formaldehyde, the compound went from having four hydrogen atoms to having two hydrogen atoms.

Look what I found! Reduction

Like oxidation, there are three definitions you can use to describe reduction:

- The gain of electrons
- The loss of oxygen
- The gain of hydrogen

Gain of electrons

Reduction is often seen as the gain of electrons. In the process of electroplating silver onto a teapot (see the section, "Five Dollars for a Gold Chain? Electroplating," later in this chapter), for example, the silver cation is reduced to silver metal by the gain of an electron. The following equation shows the silver cation gaining the electron:

$$Ag^+ + e^- \rightarrow Ag$$

When it gains the electron, chemists say that the silver cation has been reduced to silver metal.

Loss of oxygen

In other reactions, it's easier to see reduction as the loss of oxygen in going from reactant to product. For example, iron ore (primarily rust, Fe_2O_3) is reduced to iron metal in a blast furnace by a reaction with carbon monoxide:

$$Fe_2O_3(s) + 3\ CO(g) \rightarrow 2\ Fe(s) + 3\ CO_2(g)$$

The iron has lost oxygen, so chemists say that the iron ion has been reduced to iron metal.

Gain of hydrogen

In certain cases, a reduction can also be described as the gain of hydrogen atoms in going from reactant to product. For example, carbon monoxide and hydrogen gas can be reduced to methyl alcohol:

$$CO(g) + 2\ H_2(g) \rightarrow CH_3OH(l)$$

In this reduction process, the CO has gained the hydrogen atoms.

One's loss is the other's gain

Neither oxidation nor reduction can take place without the other. When those electrons are lost, something has to gain them.

Consider, for example, the *net-ionic equation* (the equation showing just the chemical substances that are changed during a reaction — see Chapter 8) for a reaction with zinc metal and an aqueous copper(II) sulfate solution:

$$Zn(s) + Cu^{2+} \rightarrow Zn^{2+} + Cu$$

This overall reaction is really composed of two half-reactions:

$Zn(s) \rightarrow Zn^{2+} + 2e^-$ (oxidation half-reaction — the loss of electrons)

$Cu^{2+} + 2e^- \rightarrow Cu(s)$ (reduction half-reaction — the gain of electrons)

To help yourself remember which is oxidation and which is reduction in terms of electrons, memorize the phrase "LEO goes GER" (*L*ose *E*lectrons *O*xidation; *G*ain *E*lectrons *R*eduction).

Zinc loses two electrons; the copper(II) cation gains those same two electrons. Zn is being oxidized. But without Cu^{2+} present, nothing will happen. That copper cation is the *oxidizing agent.* It's a necessary agent for the oxidation process to proceed. The oxidizing agent accepts the electrons from the chemical species that is being oxidized.

Cu^{2+} is reduced as it gains electrons. The species that furnishes the electrons is called the *reducing agent.* In this case, the reducing agent is zinc metal.

The oxidizing agent is the species that's being reduced, and the reducing agent is the species that's being oxidized. Both the oxidizing and reducing agents are on the left (reactant) side of the redox equation.

Playing the numbers: Oxidation numbers, that is

Oxidation numbers are bookkeeping numbers. They allow chemists to do things such as balance redox equations. Oxidation numbers are positive or negative numbers, but don't confuse them with charges on ions or valences. Oxidation numbers are assigned to elements using these rules:

- **Rule 1:** The oxidation number of an element in its free (uncombined) state is zero (for example, Al(s) or Zn(s)). This is also true for elements found in nature as *diatomic* (two-atom) elements (H_2, O_2, N_2, F_2, Cl_2, Br_2, or I_2) and for sulfur, found as S_8.
- **Rule 2:** The oxidation number of a *monatomic* (one-atom) ion is the same as the charge on the ion (for example, $Na^+ = +1$, $S^{2-} = -2$).
- **Rule 3:** The sum of all oxidation numbers in a neutral compound is zero. The sum of all oxidation numbers in a *polyatomic* (many-atom) ion is equal to the charge on the ion. This rule often allows chemists to calculate the oxidation number of an atom that may have multiple oxidation states, if the other atoms in the ion have known oxidation numbers. (See Chapter 6 for examples of atoms with multiple oxidation states.)
- **Rule 4:** The oxidation number of an alkali metal (IA family) in a compound is +1; the oxidation number of an alkaline earth metal (IIA family) in a compound is +2.
- **Rule 5:** The oxidation number of oxygen in a compound is usually –2. If, however, the oxygen is in a class of compounds called *peroxides* (for example, hydrogen peroxide, or H_2O_2), then the oxygen has an oxidation number of –1. If the oxygen is bonded to fluorine, the number is +1.
- **Rule 6:** The oxidation state of hydrogen in a compound is usually +1. If the hydrogen is part of a *binary metal hydride* (compound of hydrogen and some metal), then the oxidation state of hydrogen is –1.
- **Rule 7:** The oxidation number of fluorine is always –1. Chlorine, bromine, and iodine usually have an oxidation number of –1, unless they're in combination with an oxygen or fluorine. (For example, in ClO^-, the oxidation number of oxygen is –2 and the oxidation number of chlorine is +1; remember that the sum of all the oxidation numbers in ClO^- have to equal –1.)

These rules give you another way to define oxidation and reduction — in terms of oxidation numbers. For example, consider this reaction, which shows oxidation by the loss of electrons:

$$Zn(s) \rightarrow Zn^{2+} + 2e^-$$

Notice that the zinc metal (the reactant) has an oxidation number of zero (rule 1), and the zinc cation (the product) has an oxidation number of +2 (rule 2). In general, you can say that a substance is oxidized when there's an *increase* in its oxidation number.

Reduction works the same way. Consider this reaction:

$$Cu^{2+} + 2e^- \rightarrow Cu(s)$$

The copper is going from an oxidation number of +2 to zero. A substance is reduced if there's a *decrease* in its oxidation number.

Balancing redox equations

Redox equations are often so complex that the inspection method (the fiddling with coefficients method) of balancing chemical equations doesn't work well with them (see Chapter 8 for a discussion of this balancing method). So chemists have developed two different methods of balancing redox equations. One method is called the *oxidation number method.* It's based on the changes in oxidation numbers that take place during the reaction. Personally, I don't think this method works nearly as well as the second method, the *ion-electron* (half-reaction) method, because it's sometimes difficult to determine the exact change in the numerical value of the oxidation numbers. So I'm just going to show you the second method.

Here's an overview of the ion-electron method: The unbalanced redox equation is converted to the ionic equation and then broken down into two half-reactions — oxidation and reduction. Each of these half-reactions is balanced separately and then combined to give the balanced ionic equation. Finally, the spectator ions are put into the balanced ionic equation, converting the reaction back to the molecular form. (Buzzword-o-rama, eh? For a discussion of molecular, ionic, and net-ionic equations, see Chapter 8.) It's important to follow the steps precisely and in the order listed. Otherwise, you may not be successful in balancing redox equations.

Now how about an example? I'm going to show you how to balance this redox equation with the ion-electron method:

$$Cu(s) + HNO_3(aq) \rightarrow Cu(NO_3)_2(aq) + NO(g) + H_2O(l)$$

Follow these steps:

1. **Convert the unbalanced redox reaction to the ionic form.**

   In this reaction, you show the nitric acid in the ionic form, because it's a strong acid (for a discussion of strong acids, see Chapter 12). Copper(II) nitrate is soluble (indicated by *(aq)*), so it's shown in its ionic form (see Chapter 8). Because NO(g) and water are molecular compounds, they remain shown in the molecular form:

   $Cu(s) + H^+ + NO_3^- \rightarrow Cu^{2+} + 2\ NO_3^- + NO(g) + H_2O(l)$

2. **If necessary, assign oxidation numbers and then write two half-reactions (oxidation and reduction) showing the chemical species that have had their oxidation numbers changed.**

   In some cases, it's easy to tell what has been oxidized and reduced; but in other cases, it isn't as easy. Start by going through the example reaction and assigning oxidation numbers. You can then use the chemical species that have had their oxidation numbers changed to write your unbalanced half-reactions:

   $Cu(s) + H^+ + NO_3^- \rightarrow Cu^{2+} + 2\ NO_3^- + NO(g) + H_2O(l)$

   0 +1 +5(-2)3 +2 +5(-2)3 +2 -2 (+1)2 -2

   Look closely. Copper changed its oxidation number (from 0 to 2) and so has nitrogen (from –2 to +2). Your unbalanced half-reactions are

   $Cu(s) \rightarrow Cu^{2+}$

   $NO_3^- \rightarrow NO$

3. **Balance all atoms, with the exception of oxygen and hydrogen.**

   It's a good idea to wait until the end to balance hydrogen and oxygen atoms, so always balance the other atoms first. You can balance them by inspection — fiddling with the coefficients. (You can't change subscripts; you can only add coefficients.) However, in this particular case, both the copper and nitrogen atoms already balance, with one each on both sides:

   $Cu(s) \rightarrow Cu^{2+}$

   $NO_3^- \rightarrow NO$

4. **Balance the oxygen atoms.**

   How you balance these atoms depends on whether you're dealing with acid or basic solutions:

   - In acid solutions, take the number of oxygen atoms needed and add that same number of water molecules to the side that needs oxygen.
   - In basic solutions, add 2 OH^- to the side that needs oxygen for every oxygen atom that is needed. Then, to the other side of the equation, add half as many water molecules as OH^- anions used.

An acidic solution will have some acid or H^+ shown; a basic solution will have an OH^- present. The example equation is in acidic conditions (nitric acid, HNO_3, which, in ionic form, is $H^+ + NO_3^-$). There's nothing to do on the half-reaction involving the copper, because there are no oxygen atoms present. But you do need to balance the oxygen atoms in the second half-reaction:

$Cu(s) \rightarrow Cu^{2+}$

$NO_3^- \rightarrow NO + 2\ H_2O$

5. **Balance the hydrogen atoms.**

   Again, how you balance these atoms depends on whether you're dealing with acid or basic solutions:

   - In acid solutions, take the number of hydrogen atoms needed and add that same number of H^+ to the side that needs hydrogen.
   - In basic solutions, add one water molecule to the side that needs hydrogen for every hydrogen atom that's needed. Then, to the other side of the equation, add as many OH^- anions as water molecules used.

   The example equation is in acidic conditions. You need to balance the hydrogen atoms in the second half-reaction:

   $Cu(s) \rightarrow Cu^{2+}$

   $4\ H^+ + NO_3^- \rightarrow NO + 2\ H_2O$

6. **Balance the ionic charge on each half-reaction by adding electrons.**

   $Cu(s) \rightarrow Cu^{2+} + 2\ e^-$ (oxidation)

   $3\ e^- + 4\ H^+ + NO_3^- \rightarrow NO + 2\ H_2O$ (reduction)

   The electrons should end up on opposite sides of the equation in the two half-reactions. Remember that you're using ionic charge, not oxidation numbers.

7. **Balance electron loss with electron gain between the two half-reactions.**

   The electrons that are lost in the oxidation half-reaction are the same electrons that are gained in the reduction half-reaction. The number of electrons lost and gained must be the same. But Step 6 shows a loss of 2 electrons and a gain of 3. So you must adjust the numbers using appropriate multipliers for both half-reactions. In this case, you have to find the lowest common denominator between 2 and 3. It's 6, so multiply the first half-reaction by 3 and the second half-reaction by 2.

   $3 \times [Cu(s) \rightarrow Cu^{2+} + 2\ e^-] = 3\ Cu(s) \rightarrow 3\ Cu^{2+} + 6\ e^-$

   $2 \times [3\ e^- + 4\ H^+ + NO_3^- \rightarrow NO + 2\ H_2O] = 6\ e^- + 8\ H^+ + 2\ NO_3^- \rightarrow 2\ NO + 4\ H_2O$

8. **Add the two half-reactions together and cancel anything common to both sides. The electrons should always cancel (the number of electrons should be the same on both sides).**

   $3\,Cu + \cancel{6e^-} + 8\,H^+ + 2\,NO_3^- \rightarrow 3\,Cu^{2+} + \cancel{6e^-} + 2\,NO + 4\,H_2O$

9. **Convert the equation back to the molecular form by adding the spectator ions.**

   If it's necessary to add spectator ions to one side of the equation, add the same number to the other side of the equation. For example, there are 8 H^+ on the left side of the equation. In the original equation, the H^+ was in the molecular form of HNO_3. You need to add the NO_3^- spectator ions back to it. You already have 2 on the left, so you simply add 6 more. You then add 6 NO_3^- to the right-hand side to keep things balanced. Those are the spectator ions that you need for the Cu^{2+} cation to convert it back to the molecular form that you want.

   $3\,Cu(s) + 8\,HNO_3(aq) \rightarrow 3\,Cu(NO_3)_2(aq) + 2\,NO(g) + 4\,H_2O(l)$

10. **Check to make sure that all the atoms are balanced, all the charges are balanced (if working with an ionic equation at the beginning), and all the coefficients are in the lowest whole-number ratio.**

That's how it's done. Reactions that take place in base are just as easy, as long as you follow the rules.

Power On the Go: Electrochemical Cells

In the section, "One's loss is the other's gain," I discuss a reaction in which I put a piece of zinc metal into a copper(II) sulfate solution. The copper metal begins spontaneously plating out on the surface of the zinc. The equation for this reaction is

$$Zn(s) + Cu^{2+} \rightarrow Zn^{2+} + Cu$$

This is an example of *direct* electron transfer. Zinc gives up two electrons (becomes oxidized) to the Cu^{2+} ion that accepts the electrons (reducing it to copper metal). In Chapter 8, I show you that nothing happens if you place a piece of copper metal into a solution containing Zn^{2+}, because zinc gives up electrons more easily than copper. I also show you the activity series of metals that allows you to predict whether or not a displacement (redox) reaction will take place.

Now this is a useful reaction if you want to plate out copper onto zinc. However, not many of us have a burning desire to do this! But if you were able to separate those two half-reactions so that when the zinc is oxidized, the electrons it releases are forced to travel through a wire to get to the Cu^{2+}, you'd have something useful. You'd have a *galvanic* or *voltaic cell,* a redox reaction that produces electricity. In this section, I show you how that Zn/Cu^{2+} reaction may be separated out so that you have an *indirect* electron transfer and can produce some useable electricity.

Galvanic cells are commonly called batteries, but sometimes this name is somewhat incorrect. A battery is composed of two or more cells connected together. You put a battery in your car, but you put a cell into your flashlight.

Nice cell there, Daniell

Take a look at Figure 9-1, which shows a Daniell cell that uses the Zn/Cu^{2+} reaction to produce electricity. (This cell is named after John Frederic Daniell, the British chemist who invented it in 1836.)

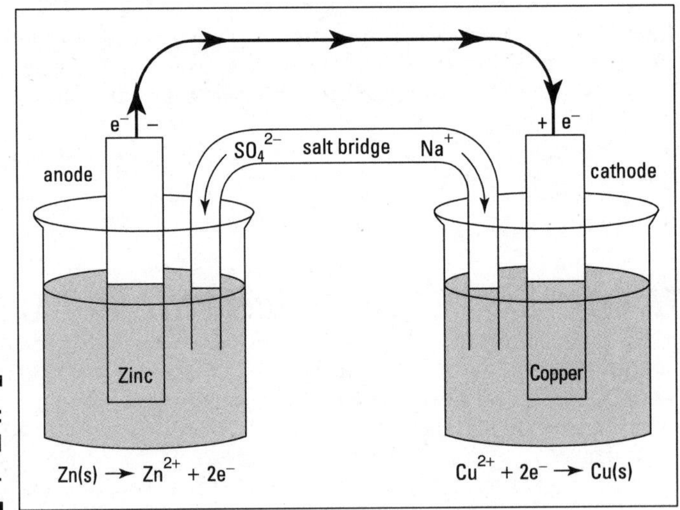

Figure 9-1: A Daniell cell.

In the Daniell cell, a piece of zinc metal is placed in a solution of zinc sulfate in one container, and a piece of copper metal is placed in a solution of copper(II) sulfate in another container. These strips of metal are called the cell's *electrodes.* They act as a terminal, or a holding place, for electrons. A wire connects the electrodes, but nothing happens until you put a salt bridge between the two containers. The *salt bridge,* normally a U-shaped hollow tube filled with a concentrated salt solution, provides a way for ions to move from one container to the other to keep the solutions electrically neutral. It's like running only one wire up to a ceiling light; the light won't work unless you put in a second wire to complete the circuit.

With the salt bridge in place, electrons can start to flow. It's the same basic redox reaction as the one I show you at the beginning of this section. Zinc is being oxidized, releasing electrons that flow through the wire to the copper electrode, where they're available for the Cu^{2+} ions to use in forming copper metal. Copper ions from the copper(II) sulfate solution are being plated out on the copper electrode, while the zinc electrode is being consumed. The cations in the salt bridge migrate to the container containing the copper electrode to replace the copper ions being consumed, while the anions in the salt bridge migrate toward the zinc side, where they keep the solution containing the newly formed Zn^{2+} cations electrically neutral.

The zinc electrode is called the *anode,* the electrode at which oxidation takes place, and is labeled with a "–" sign. The copper electrode is called the *cathode,* the electrode at which reduction takes place, and is labeled with a "+" sign.

This cell will produce a little over one volt. You can get just a little more voltage if you make the solutions that the electrodes are in very concentrated. But what can you do if you want, for example, two volts? You have a couple of choices. You can hook two of these cells up together and produce two volts, or you can choose two different metals from the activity series chart in Chapter 8 that are farther apart than zinc and copper. The farther apart the metals are on the activity series, the more voltage the cell will produce.

Let the light shine: Flashlight cells

The common flashlight cell (see Figure 9-2), a dry cell (it's not in a solution like a Daniell cell), is contained in a zinc housing that acts as the anode. The other electrode, the cathode, is a graphite rod in the middle of the cell. A layer of manganese oxide and carbon black (one of the many forms of carbon) surrounds the graphite rod, and a thick paste of ammonium chloride and zinc chloride serves as the electrolyte. The cell reactions are

$Zn(s) \rightarrow Zn^{2+} + 2\ e^-$ (anode reaction/oxidation)

$2\ MnO_2(s) + 2\ NH_4^+ + 2\ e^- \rightarrow Mn_2O_3(s) + 2\ NH_3(aq) + H_2O(l)$ (cathode reaction/reduction)

Note that the case is actually one of the electrodes; it's being used up in the reaction. If there's a thin spot in the case, a hole could form, and the cell could leak the corrosive contents. In addition, the ammonium chloride tends to corrode the metal case, again allowing for the possibility of leakage.

In the alkaline dry cell (alkaline battery), the acidic ammonium chloride of the regular dry cell is replaced by basic (alkaline) potassium hydroxide. With this chemical, corrosion of the zinc case is greatly reduced.

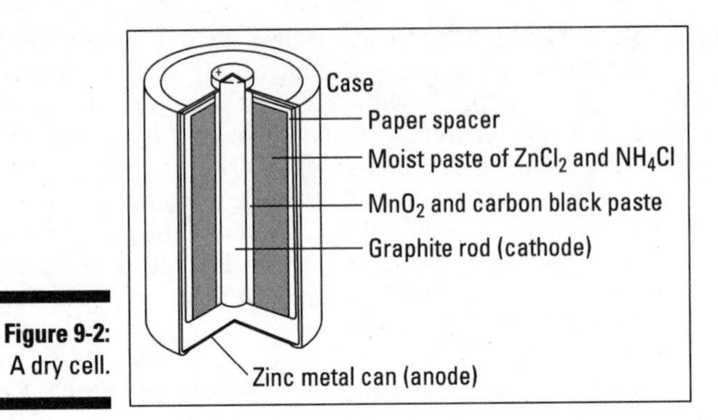

Figure 9-2: A dry cell.

Another cell with the same basic construction is the small mercury battery commonly used in watches, pacemakers, and so on. With this battery, the anode is zinc, as in the regular dry cell, but the cathode is steel. Mercury(II) oxide (HgO) and some alkaline paste form the electrolyte. You should dispose of this type of battery carefully, to keep the mercury from being released into the environment.

All these galvanic cells produce electricity until they run out of a reactant. Then they must be discarded. However, there are cells (batteries) that can be recharged, as the redox reaction can be reversed to regenerate the original reactants. Nickel-cadmium (Ni-Cad) and lithium batteries fall into this category. The most familiar type of rechargeable battery is probably the automobile battery.

Gentlemen, start your engines: Automobile batteries

The ordinary automobile battery, or lead storage battery, consists of six cells connected in series (see Figure 9-3). The anode of each cell is lead, while the cathode is lead dioxide (PbO_2). The electrodes are immersed in a sulfuric acid (H_2SO_4) solution. When you start your car, the following cell reactions take place:

$Pb(s) + H_2SO_4(aq) \rightarrow PbSO_4(s) + 2\,H^+ + 2\,e^-$ (anode)

$2\,e^- + 2\,H^+ + PbO_2(s) + H_2SO_4(aq) \rightarrow PbSO_4(s) + 2\,H_2O(l)$ (cathode)

$Pb(s) + PbO_2(s) + 2\,H_2SO_4(aq) \rightarrow 2\,PbSO_4 + 2\,H_2O(l)$ (overall reaction)

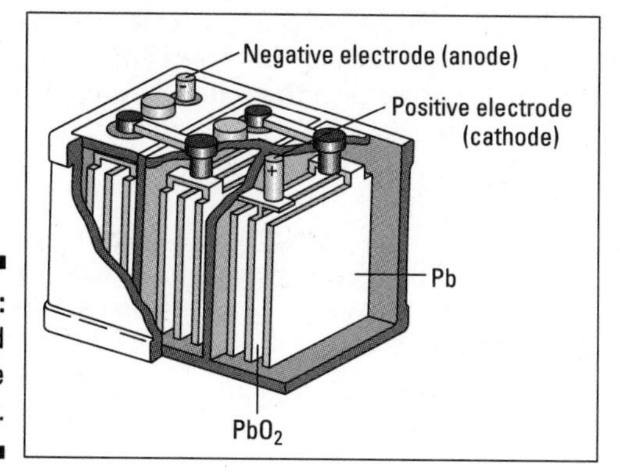

Figure 9-3: The lead storage battery.

When this reaction takes place, both electrodes become coated with solid lead (II) sulfate, and the sulfuric acid is used up.

After the automobile has been started, the alternator or generator takes over the job of producing electricity (for spark plugs, lights, and so on) and also recharges the battery. The alternator reverses both the flow of electrons into the battery and the original redox reactions, and regenerates the lead and lead dioxide:

$$2\ PbSO_4(s) + 2\ H_2O(l) \rightarrow Pb(s) + PbO_2(s) + 2\ H_2SO_4(aq)$$

The lead storage battery can be discharged and charged many times. But the shock of running over bumps in the road (or dead armadillos in Texas) or into the curb flakes off a little of the lead (II) sulfate and eventually causes the battery to fail.

During charging, the automobile battery acts like a second type of electrochemical cell, an *electrolytic cell,* which uses electricity to produce a desired redox reaction. This reaction may be the recharging of a battery, or it may be involved in the plating of Grandmother Grace's teapot.

Five Dollars for a Gold Chain? Electroplating

Electrolytic cells, cells that use electricity to produce a desired redox reaction, are used extensively in our society. Rechargeable batteries are a primary example of this type of cell, but there are many other applications. Ever wonder how the aluminum in that aluminum can is mined? Aluminum ore is primarily aluminum oxide (Al_2O_3). Aluminum metal is produced by reducing the aluminum oxide in a high temperature electrolytic cell using approximately 250,000 amps. That's a lot of electricity. It's far cheaper to take old aluminum cans, melt them down, and reform them into new cans than it is to extract the metal from the ore. That's why the aluminum industry is strongly behind the recycling of aluminum. It's just good business.

Water can be decomposed by the use of electricity in an electrolytic cell. This process of producing chemical changes by passing an electric current through an electrolytic cell is called *electrolysis* (yes, just like the permanent removal of hair). The overall cell reaction is

$$2\ H_2O(l) \rightarrow 2\ H_2(g) + O_2(g)$$

In a similar fashion, sodium metal and chlorine gas can be produced by the electrolysis of molten sodium chloride.

Electrolytic cells are also used in a process called *electroplating.* In electroplating, a more-expensive metal is plated (deposited in a thin layer) onto the surface of a cheaper metal by electrolysis. Back before plastic auto bumpers became popular, chromium metal was electroplated onto steel bumpers. Those five-dollar gold chains you can buy are really made of some cheap metal with an electroplated surface of gold. Figure 9-4 shows the electroplating of silver onto a teapot.

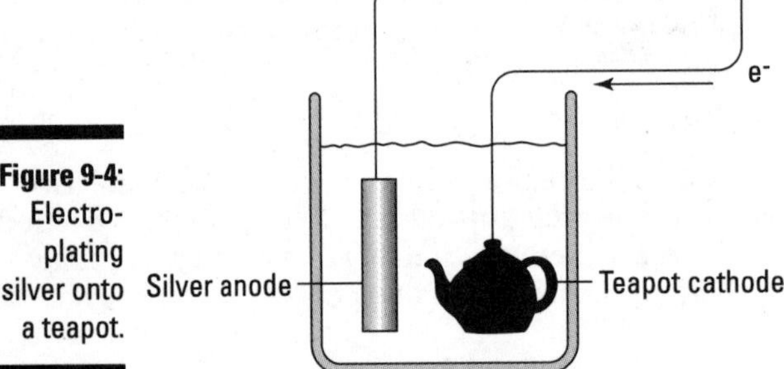

Figure 9-4: Electroplating silver onto a teapot.

A battery is commonly used to furnish the electricity for the process. The teapot acts as the cathode and a bar of silver acts as the anode. The silver bar furnishes the silver ions that are reduced onto the surface of the teapot. Many metals and even some alloys can be plated out in this fashion. Everybody loves those plated surfaces, especially without the high cost of the pure metal. (Reminds me of an Olympic athlete who was so proud of his gold metal that he had it bronzed!)

This Burns Me Up! Combustion of Fuels and Foods

Combustion reactions are types of redox reactions that are absolutely essential for life and civilization — because heat is the most important product of these reactions. The burning of coal, wood, natural gas, and petroleum heats our homes and provides the majority of our electricity. The combustion of gasoline, jet fuel, and diesel fuel powers our transportation systems. And the combustion of food powers our bodies.

Have you ever wondered how the energy content of a fuel or food is measured? An instrument called a *bomb calorimeter* is used to measure energy content. Figure 9-5 shows the major components of a bomb calorimeter.

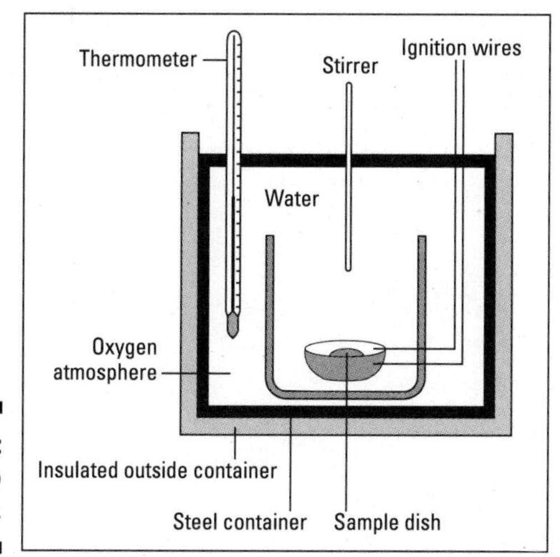

Figure 9-5: A bomb calorimeter.

To measure the energy content of fuels, a known mass of the material to be measured is placed into a sample cup and sealed. The air is removed from the sample cup and replaced with pure oxygen. The cup is then placed in the calorimeter with a known amount of water covering it. The initial temperature of the water is measured, and then the sample is ignited electrically. The rise in the temperature is measured, and the number of calories of energy that is released is calculated. A *calorie* is the amount of energy needed to raise the temperature of 1 gram of water 1 degree Celsius. The complete combustion of a large kitchen match, for example, gives you about one kilocalorie of heat. (See Chapter 2 for the basics of calories and measuring energy.)

The caloric content of foods can be determined in exactly the same fashion. Chemists report the results in calories or kilocalories, while nutritionists report the results in nutritional Calories. A nutritional Calorie is equal to a chemist's kilocalorie (1,000 calories). A 300 Calorie candy bar produces 300,000 calories of energy. Unfortunately, not all that energy is required immediately, so some is stored in compounds such as fats. I'm carrying around the result of *many* candy bars.

Part III

The Mole: The Chemist's Best Friend

"So what if you have a Ph.D. in chemistry? I used to have my own circus act."

In this part . . .

Chemists operate in the world of quantities that they can see and touch — the macroscopic world of grams, liters, and meters. They carry out chemical reactions by weighing grams of reactants, and they use grams to measure the amount of product formed. They use liters to measure the amount of gas produced. They test a solution with litmus paper to see if it's an acid or a base.

But chemists also operate in the microscopic world of atoms and molecules. Because atoms and molecules are so very small, chemists have only just recently been able to see them, thanks to advanced technology in the most powerful of microscopes. Chemists think of acids and bases in terms of the donating and accepting of protons, not just as color changes in indicators. Models help chemists understand and predict the processes that take place in the microscopic world. These models also translate into real-life applications.

These chapters show you the bridge between the macroscopic and microscopic worlds — the mole. I explain reaction stoichiometry — how much reactant it takes to produce a particular amount of product. I introduce you to solutions and colloids, acids and bases, and gases. I show you the multitude of relationships between the properties of gases and relate them back to stoichiometry. In chemistry, everything is connected.

Chapter 10

The Mole: Can You Dig It?

In This Chapter

- Figuring out how to count by weighing
- Understanding the mole concept
- Discovering how the mole is used in chemical calculations

Chemists do a lot of things. One is to make new substances, a process called *synthesis.* And a logical question they ask is "How much?"

"How much of this reactant do I need to make this much product?" "How much product can I make with this much reactant?" In order to answer these questions, chemists must be able to take a balanced chemical equation, expressed in terms of atoms and molecules, and convert it to grams or pounds or tons — some type of unit that they can actually weigh out in the lab. The mole concept enables chemists to move from the microscopic world of atoms and molecules to the real word of grams and kilograms and is one of the most important central concepts in chemistry. In this chapter, I introduce you to Mr. Mole.

Counting by Weighing

Suppose that you have a job packing 1,000 nuts and 1,000 bolts in big bags, and you get paid for each bag you fill. So what's the most efficient and quickest way of counting out the nuts and bolts? Weigh out a hundred, or even ten, of each and then figure out how much a thousand of each will weigh. Fill up the bag with nuts until it weighs the amount you figured for 1,000 nuts. After you have the correct amount of nuts, use the same process to fill the bag with bolts. In other words, count by weighing; that's one of the most efficient ways of counting large numbers of objects.

In chemistry, you count very large numbers of particles, such as atoms and molecules. To count them efficiently and quickly, you need to use the count-by-weighing method, which means you need to know how much individual atoms and molecules weigh. You can get the weights of the individual atoms on the periodic table, but what about the weights of the compounds? Well, you can simply add together the weights of the individual atoms in the compound to figure the molecular weight or formula weight. (*Molecular weights* refer to covalently bonded compounds, and *formula weights* refer to both ionic and covalent compounds. Check out Chapters 6 and 7 for details on ionic and covalent bonds.)

Here's a simple example that shows how to calculate the molecular weight of a compound: Water, H_2O, is composed of two hydrogen atoms and one oxygen atom. By looking on the periodic table, you can find that one hydrogen atom equals 1.0079 amu and one oxygen atom weighs 15.999 amu (amu stands for *atomic mass units* — see Chapter 3 for details). To calculate the molecular weight of water, you simply add together the atomic weights of two hydrogen atoms and one oxygen atom:

$2 \times 1.0079 \text{ amu} = 2.016 \text{ amu}$ (two hydrogen atoms)

$1 \times 15.999 \text{ amu} = 15.999 \text{ amu}$ (one oxygen atom)

$2.016 \text{ amu} + 15.999 \text{ amu} = 18.015 \text{ amu}$ (the weight of the water molecule)

Now try a little harder one. Calculate the formula weight of aluminum sulfate, $Al_2(SO_4)_3$. In this salt, you have 2 aluminum atoms, 3 sulfur atoms, and 12 oxygen atoms. After you find the individual weights of the atoms on the periodic table, you can calculate the formula weight like this:

$$\underbrace{[(2 \times 26.982 \text{ amu})}_{\text{for the aluminum}} + \underbrace{(3 \times 32.066 \text{ amu})}_{\text{for the sulfur}} + \underbrace{(12 \times 15.999 \text{ amu})]}_{\text{for the oxygen}} = \underbrace{315.168 \text{ amu}}_{\text{for } Al_2(SO_4)_3}$$

Pairs, Dozens, Reams, and Moles

When we humans deal with objects, we often think in terms of a convenient amount. For example, when a woman buys earrings, she normally buys a pair of them. When a man goes to the grocery store, he buys eggs by the dozen. And when I go to the office supply store, I buy copy paper by the ream.

We use words to represent numbers all the time — a pair is 2, a dozen is 12, and a ream is 500. All these words are units of measure, and they're convenient for the objects they're used to measure. Rarely would you want to buy a ream of earrings or a pair of paper.

Likewise, when chemists deal with atoms and molecules, they need a convenient unit that takes into consideration the very small size of atoms and molecules. There is such a unit. It's called a mole.

Avogadro's number: Not in the phone book

The word *mole* stands for a number — 6.022×10^{23}. It's commonly called *Avogadro's number,* named after Amedeo Avogadro, the scientist who laid the groundwork for the mole principle.

Now a mole — 6.022×10^{23} — is a really *big* number. When written in longhand notation, it's

602,200,000,000,000,000,000,000

And *that* is why I like scientific notation.

If you had a mole of marshmallows, it would cover the United States to a depth of about 600 miles. A mole of rice grains would cover the land area of the world to a depth of 75 meters. And a mole of moles . . . no, I don't even want to think about that!

Avogadro's number stands for a certain number of *things*. Normally, those things are atoms and molecules. So the mole relates to the microscopic world of atoms and molecules. But how does it relate to the macroscopic world where I work?

The answer is that a *mole* (abbreviated as mol) is also the number of particles in exactly 12 grams of a particular isotope of carbon (C-12). So if you have exactly 12 grams of ^{12}C, you have 6.022×10^{23} carbon atoms, which is also a mole of ^{12}C atoms. For any other element, a mole is the atomic weight expressed in grams. And for a compound, a mole is the formula (or molecular) weight in grams.

Using moles in the real world

The weight of a water molecule is 18.015 amu (see the section "Counting by Weighing" for how to calculate the weight of compounds). Because a mole is the formula (or molecular) weight in grams of a compound, you can now say that the weight of a mole of water is 18.015 grams. You can also say that 18.015 grams of water contains 6.022×10^{23} H_2O molecules, or a mole of water. And the mole of water is composed of two moles of hydrogen and one mole of oxygen.

The mole is the bridge between the microscopic and the macroscopic world:

$$6.022 \times 10^{23} \text{ particles} \leftrightarrow \text{mole} \leftrightarrow \text{atomic/formula weight in grams}$$

If you have any one of the three things — particles, moles, or grams — then you can calculate the other two.

For example, suppose you want to know how many water molecules there are in 5.50 moles of water. You can set up the problem like this:

$$5.50 \text{ moles} \times 6.022 \text{ x } 10^{23} \text{molecules/mol} = 3.31 \times 10^{24} \text{ molecules}$$

Or suppose that you want to know how many moles are in 25.0 grams of water. You can set up the problem like this (and see Appendix B for more on exponential arithmetic):

$$\frac{25.0\not{g}\,H_2O}{1} \times \frac{1 mol\,H_2O}{18.015\not{g}\,H_2O} = 1.39 moles H_2O$$

You can even go from grams to particles by going through the mole. For example, how many molecules are there in 100.0 grams of carbon dioxide?

The first thing you have to do is determine the molecular weight of CO_2. Look at the periodic table to find that one carbon atom equals 12.011 amu and one oxygen atom weighs 15.999 amu. Now figure the molecular weight, like this:

$$[(1 \times 12.011 \text{ g/mol}) + (2 \times 15.999 \text{ g/mol})] = 44.01 \text{ g/mol for } CO_2$$

Now you can work the problem:

$$\frac{100.0\not{g}\,CO_2}{1} \times \frac{1 m\not{o}l\,CO_2}{44.01\not{g}} \times \frac{6.022 \times 10^{23} molecules}{1 m\not{o}l} = 1.368 \times 10^{24} CO_2 molecules$$

And it's just as easy to go from particles to moles to grams.

You can also use the mole concept to calculate the empirical formula of a compound using the *percentage composition* data for that compound — the percentage by weight of each element in the compound. (The *empirical formula* indicates the different types of elements in a molecule and the lowest whole-number ratio of each kind of atom in the molecule. See Chapter 7 for details.)

When I try to determine the empirical formula of a compound, I often have percentage data available. The determination of the percentage composition is one of the first analysis that a chemist does in learning about a new compound. For example, suppose I determine that a particular compound has the following weight percentage of elements present: 26.4% Na, 36.8% S, and 36.8% O. Since I'm dealing with percentage data (amount per hundred) I will assume that I have 100 grams of the compound so that my percentages can be used as weights. I then convert each mass to moles, like this:

$$\frac{36.4g\,Na}{1} \times \frac{1mol\,Na}{22.99g} = 1.15mol\,Na$$

$$\frac{36.8g\,S}{1} \times \frac{1mol\,S}{32.07} = 1.15mol\,S$$

$$\frac{36.8g\,O}{1} \times \frac{1mol\,O}{16.00g} = 2.30mol\,O$$

Now I can write an empirical formula of $Na_{1.15}S_{1.15}O_{2.30}$. I know that my subscripts have to be whole numbers, so I divide each of these by the smallest, 1.15, to get $NaSO_2$. (If a subscript is 1, it's not shown.) I can then calculate a weight for the empirical formula, by adding together the atomic masses on the periodic table of 1 sodium (Na), 1 sulfur (S) and 2 oxygen (O). This gives me an empirical formula weight of 87.056 grams. Suppose, however, in another experiment I determined that the actual molecular weight of this compound was 174.112 grams. By dividing 174.112 grams by 87.056 grams (actual molecular weight by the empirical formula weight) I get 2. This means that the molecular formula is twice the empirical formula, so that the compound is actually $Na_2S_2O_4$.

Chemical Reactions and Moles

I think one of the reasons I enjoy being a chemist is that I like to cook. I see a lot of similarities between cooking and chemistry. A chemist takes certain things called reactants and makes something new from them. A cook does the same thing. He or she takes certain things called ingredients and makes something new from them.

For example, I like to fix Fantastic Apple Tarts (FATs). My recipe looks something like this:

apples + sugar + flour + spices = FATs

No, wait. My recipe has amounts in it. It looks more like this:

4 cups of apples + 3 cups of sugar + 2 cups of flour + 1/10 cup of spices = 12 FATs

My recipe tells me how much of each ingredient I need and how many FATs I can make. I can even use my recipe to calculate how much of each ingredient I need for a particular number of FATs. For example, suppose that I'm giving a big dinner party, and I need 250 FATs. I can use my recipe to calculate the amount of apples, sugar, flour, and spices I need. Here, for example, is how I calculate how much sugar I need:

$$\frac{250\,FATs}{1} \times \frac{3\,cups\,sugar}{12\,FATs} = 62.5\,cups\,sugar$$

And I can do the same for the apples, flour, and spices by simply changing the ratio of each *ingredient* (as a multiple of 12 FATs).

The balanced chemical equation allows you to do the same thing. For example, look at my favorite reaction, the Haber process, which is a method for preparing ammonia (NH_3) by reacting nitrogen gas with hydrogen gas:

$$N_2(g) + 3\,H_2(g) \leftrightarrow 2\,NH_3(g)$$

In Chapter 8, I use this reaction over and over again for various examples (like I said, it's my *favorite* reaction) and explain that you can read the reaction like this: 1 molecule of nitrogen gas reacts with 3 molecules of hydrogen gas to yield 2 molecules of ammonia.

$N_2(g)$	+	$3\,H_2(g)$	$\leftrightarrow$	$2\,NH_3(g)$
1 molecule		3 molecules		2 molecules

Now, you can scale everything up by a factor of 12:

$N_2(g)$	+	$3\,H_2(g)$	$\leftrightarrow$	$2\,NH_3(g)$
1 dozen molecules		3 dozen molecules		2 dozen molecules

You can even scale it up by 1,000:

$N_2(g)$	+	$3\,H_2(g)$	$\leftrightarrow$	$2\,NH_3(g)$
1,000 molecules		3,000 molecules		2,000 molecules

Or how about a factor of 6.023×10^{23}:

$N_2(g)$	+	$3\,H_2(g)$	$\leftrightarrow$	$2\,NH_3(g)$
6.023×10^{23} molecules		$3(6.023 \times 10^{23}$ molecules)		$2(6.023 \times 10^{23}$ molecules)

Wait a minute! Isn't 6.023×10^{23} a mole? So you can write the equation like this:

$N_2(g)$	+	$3\,H_2(g)$	$\leftrightarrow$	$2\,NH_3(g)$
1 mole		3 moles		2 moles

That's right — not only can those coefficients in the balanced chemical equation represent atoms and molecules, but they can also represent the number of moles.

Now take another look at my recipe for FATs:

4 cups of apples + 3 cups of sugar + 2 cups of flour + 1/10 cup of spices = 12 FATs

I have a problem. When I go to the grocery store, I don't buy fresh apples by the cup. Nor do I buy sugar or flour by the cup. I buy all these things by the pound. Now I buy a large excess, but because I'm frugal (translate as cheap), I want to figure as closely as I can the amount I really need. If I can determine the weight per cup for each ingredient, I'll be okay. So I weigh the ingredients and get

> 1 cup of apples = 0.5 lbs; 1 cup of sugar = 0.7 lbs; 1 cup of flour = 0.3 lbs; and 1 cup of spices = 0.2 lbs

Now I can substitute the measurements into my recipe:

4 cups of apples +	3 cups of sugar +	2 cups of flour +	1/10 cup of spices = 12 FATs
4(0.5lbs)	3(0.7 lbs)	2(0.3 lbs)	1/10(0.2 lbs)

Now if I want to know how many pounds of apples I need to make 250 FATs, I can set up the equation this way:

$$\frac{250\ FATs}{1} \times \frac{4\ cups\ apples}{12\ FATs} \times \frac{0.5\ lbs}{1\ cup\ apple} = 41.7\ lbs\ apples$$

I can figure out how much of each ingredient I need (based on weight), just by using the correct ingredient weight per cup.

The exact same thing is true with chemical equations. If you know the formula weight of the reactants and product, you can calculate how much you need and how much you'll get. For example, check out that Haber reaction again:

$N_2(g)$	+	$3\ H_2(g)$	$\leftrightarrow$	$2\ NH_3(g)$
1 mole		3 moles		2 moles

All you need to do is figure the molecular weights of each reactant and product, and then incorporate the weights into the equation. Use the periodic table to find the weights of the atoms and the compound (see the section "Counting by Weighing," earlier in this chapter, for the directions) and multiply those numbers by the number of moles, like this:

1(28.014 g/mol) 3(2.016 g/mol) 2(17.031 g/mol)

How much needed, how much made: Reaction stoichiometry

Once you have the weight relationships in place, you can do some stoichiometry problems. *Stoichiometry* refers to the mass relationship in chemical equations.

Look at my favorite reaction — you guessed it — the Haber process:

$$N_2(g) + 3\ H_2(g) \leftrightarrow 2\ NH_3(g)$$

Suppose that you want to know how many grams of ammonia can be produced from the reaction of 75.00 grams of nitrogen with excess hydrogen. The mole concept is the key. The coefficients in the balanced equation are not only the number of individual atoms or molecules but also the number of moles.

$N_2(g)$	+	$3\ H_2(g)$	$\leftrightarrow$	$2\ NH_3(g)$
1 mole		3 moles		2 moles
1(28.014 g/mol)		3(2.016 g/mol)		2(17.031 g/mol)

First, you can convert the 75.00 grams of nitrogen to moles of nitrogen. Then you use the ratio of the moles of ammonia to the moles of nitrogen from the balanced equation to convert to moles of ammonia. Finally you take the moles of ammonia and convert it to grams. The equation looks like this:

$$\frac{75.00 g\ N_2}{1} \times \frac{1}{28.014 g\ N_2} \times \frac{2\ mol\ NH_3}{1\ mol\ N_2}$$

$$\frac{17.031\ g\ NH_3}{1\ mol\ NH_3} \times = 91.19 g\ NH_3$$

The ratio of the mol NH_3/mol N_2 is called a *stoichiometric ratio.* This ratio enables you to convert from the moles of one substance in a balanced chemical equation to the moles of another substance.

Getting tired of the Haber process? (Me? *Never.*) Take a look at another reaction — the reduction of rust (Fe_2O_3) to iron metal by treatment with carbon (coke). The balanced chemical reaction looks like this:

$$2\ Fe_2O_3(s) + 3\ C \rightarrow 4\ Fe(s) + 3\ CO_2(g)$$

When you get ready to work stoichiometry types of problems, you *must* start with a balanced chemical equation. If you don't have it to start with, you've got to go ahead and balance the equation.

In this example, the formula weights you need are

- **Fe_2O_3:** 159.69 g/mol
- **C:** 12.01 g/mol
- **Fe:** 55.85 g/mol
- **CO_2:** 44.01 g/mol

Suppose that you want to know how many grams of carbon it takes to react with 1.000 kilogram of rust. You need to convert the kilogram of rust to grams and convert the grams to moles of rust. Then you can use a stoichiometric ratio to convert from moles of rust to moles of carbon and finally to grams. The equation looks like this:

$$\frac{1.000\cancel{Kg}Fe_2O_3}{1} \times \frac{1000\cancel{g}}{1\cancel{Kg}} \times \frac{1\cancel{mol}Fe_2O_3}{159.69\cancel{g}Fe_2O_3} \times \frac{3\cancel{mol}C}{2\cancel{mol}Fe_2O_3} \times \frac{12.01gC}{1\cancel{mol}C} = 112.8gC$$

You can even calculate the number of carbon atoms it takes to react with that 1.000 kilogram of rust. Basically, you use the same conversions, but instead of converting from moles of carbon to grams, you convert from moles of carbon to carbon atoms using Avogadro's number:

$$\frac{1.000\cancel{Kg}Fe_2O_3}{1} \times \frac{1000\cancel{g}}{1\cancel{Kg}} \times \frac{3\cancel{mol}Fe_2O_3}{2\cancel{mol}Fe_2O_3} \times \frac{1\cancel{mol}Fe_2O_3}{159.69\cancel{g}Fe_2O_3} \times \frac{6.022 \times 10^{23}\,atoms}{1\cancel{mol}C}$$

$$2.839 \times 10^{24}\,C\ atoms$$

Now I want to show you how to calculate the number of grams of iron produced from reacting 1.000 kilogram of rust with excess carbon. It's the same basic process as before — kilograms of rust to grams of rust to moles of rust to moles of iron to grams of iron:

$$\frac{1.000\cancel{Kg}Fe_2O_3}{1} \times \frac{1000\cancel{g}}{1\cancel{Kg}} \times \frac{1\cancel{mol}Fe_2O_3}{159.69\cancel{g}Fe_2O_3} \times \frac{4\cancel{mol}Fe_2O_3}{2\cancel{mol}Fe_2O_3} \times \frac{55.85g}{1\cancel{mol}Fe}\ 699.5gFe$$

So you predict that you'll get 699.5 grams of iron metal formed. What if, however, you carry out this reaction and only get 525.0 grams of iron metal formed? There may be several reasons that you produce less than you expect, such as sloppy technique or impure reactants. It may also be quite likely that the reaction is an equilibrium reaction, and you'll never get 100 percent conversion from reactants to products. (Turn to Chapter 8 for details on equilibrium reactions.) Wouldn't it be nice if there was a way to label the efficiency of a particular reaction? There is. It's called the percent yield.

Where did it go? Percent yield

In almost any reaction, you're going to produce less than expected. You may produce less because most reactions are equilibrium reactions (see Chapter 8) or because some other condition comes into play. Chemists can get an idea of the efficiency of a reaction by calculating the *percent yield* for the reaction using this equation:

$$\%yield = \frac{Actual yield}{Theoretical yield} \times 100$$

The *actual yield* is how much of the product you get when you carry out the reaction. The *theoretical yield* is how much of the product you *calculate* you'll get. The ratio of these two yields gives you an idea about how efficient the reaction is. For the reaction of rust to iron (see the preceding section), your theoretical yield is 699.5 grams of iron; your actual yield is 525.0 grams. Therefore, the percent yield is

$$\%\text{yield} = \frac{525.0\,g}{699.5\,g} \times 100 = 75.05\%$$

A percent yield of about 75 percent isn't too bad, but chemists and chemical engineers would rather see 90+ percent. One plant using the Haber reaction has a percent yield of better than 99 percent. Now that's efficiency!

Running out of something and leaving something behind: Limiting reactants

I love to cook, and I'm always hungry. So I want to talk about making some ham sandwiches. Because I'm a chemist, I can write an equation for a ham sandwich lunch:

$$2 \text{ pieces of bread} + 1 \text{ ham} + 1 \text{ cheese} \rightarrow 1 \text{ ham sandwich}$$

Suppose I check my supplies and find that I have 12 pieces of bread, 5 pieces of ham, and 10 slices of cheese. How many sandwiches can I make? I can make five, of course. I have enough bread for six sandwiches, enough ham for five, and enough cheese for ten. But I'm going to run out of ham first — I'll have bread and cheese left over. And the ingredient I run out of first really limits the amount of product (sandwiches) I'll be able to make; it can be called the *limiting ingredient*.

The same is true of chemical reactions. Normally, you run out of one of the reactants and have some others left over. (In some of the problems sprinkled throughout this chapter, I tell you which reactant is the limiting one by saying you have *an excess* of the other reactant(s).)

In this section, I show you how you can calculate which reactant is the limiting reactant.

Here is a reaction between ammonia and oxygen:

$$4\,NH_3(g) + 5\,O_2(g) \rightarrow 4\,NO(g) + 6\,H_2O(l)$$

Suppose that you start out with 100.0 grams of both ammonia and oxygen, and you want to know how many grams of NO (nitrogen monoxide, sometimes called nitric oxide) you can produce. You must determine the limiting reactant and then base your stoichiometric calculations on it.

In order to figure out which reactant is the limiting reactant, you can calculate the mole-to-coefficient ratio: You calculate the number of moles of both ammonia and oxygen, and then you divide each by their coefficient in the balanced chemical equation. The one with the smallest mole-to-coefficient ratio is the limiting reactant. For the reaction of ammonia to nitric oxide, you can calculate the mole-to-coefficient ratio for the ammonia and oxygen like this:

$$\frac{100.0gNH_3}{1} \times \frac{1molNH_3}{17.03g} = 5.87\,mol \div 4 = 1.47$$

$$\frac{100.0gNH_3}{1} \times \frac{1molO_2}{32.00g} = 3.13\,mol \div 5 = 0.625$$

Ammonia has a mole-to-coefficient ratio of 1.47, and oxygen has a ratio of 0.625. Because oxygen has the lowest ratio, oxygen is the limiting reactant, and you need to base your calculations on it.

$$\frac{100.0gO_2}{1} \times \frac{1molO_2}{32.00g} \times \frac{4molNO}{5molO_2} \times \frac{30.01gNO}{1molNO} = 75.02gNO$$

That 75.02 grams NO is your theoretical yield. But you can even calculate the amount of ammonia left over. You can figure the amount of ammonia consumed with this equation:

$$\frac{100.0gO_2}{1} \times \frac{1molO_2}{32.00g} \times \frac{4molNH_3}{5molO_2} \times \frac{17.03gNH_3}{molNH_3} = 42.58gNH_3$$

You started with 100.0 grams of ammonia, and you used 42.58 grams of it. The difference (100 grams – 42.58 grams = 57.42 grams) is the amount of ammonia left over.

Chapter 11

Mixing Matter Up: Solutions

In This Chapter

- Finding out about solutes, solvents, and solutions
- Working with the different kinds of solution concentration units
- Checking out the colligative properties of solutions
- Figuring out colloids

You encounter solutions all the time in everyday life. The air you breathe is a solution. That sports drink you use to replenish your electrolytes is a solution. That soft drink *and* that hard drink are both solutions. Your tap water is most likely a solution, too. In this chapter, I show you some of the properties of solutions. I introduce you to the different ways chemists represent a solution's concentration, and I tell you about the colligative properties of solutions and relate them to ice cream making and antifreeze. So sit back, sip on your solution of choice, and read all about solutions.

Solutes, Solvents, and Solutions

A *solution* is a homogeneous mixture, meaning that it is the same throughout. If you dissolve sugar in water and mix it really well, for example, your mixture is basically the same no matter where you sample it.

A solution is composed of a solvent and one or more solutes. The *solvent* is the substance that's present in the largest amount, and the *solute* is the substance that's present in the lesser amount. These definitions work most of the time, but there are a few cases of extremely soluble salts, such as lithium chloride, in which more than 5 grams of salt can be dissolved in 5 milliliters of water. However, water is still considered the solvent, because it's the species that has not changed state. In addition, there can be more than one solute in a solution. You can dissolve salt in water to make a brine solution, and then you can dissolve some sugar in the same solution. You then have two solutes, salt and sugar, but you still have only one solvent — water.

When I talk about solutions, most people think of liquids. But there can also be solutions of gases. Our atmosphere, for example, is a solution. Because air is almost 79 percent nitrogen, it's considered the solvent, and the oxygen, carbon dioxide, and other gases are considered the solutes. There are also solid solutions. Alloys, for example, are solutions of one metal in another metal. Brass is a solution of zinc in copper.

A discussion of dissolving

Why do some things dissolve in one solvent and not another? For example, oil and water will not mix to form a solution, but oil will dissolve in gasoline. There's a general rule of solubility that says *like-dissolves-like* in regards to polarity of both the solvent and solutes. Water, for example, is a polar material; it's composed of polar covalent bonds with a positive and negative end of the molecule. (For a rousing discussion of water and its polar covalent bonds, see Chapter 7.) Water will dissolve polar solutes, such as salts and alcohols. Oil, however, is composed of largely nonpolar bonds. So water will not act as a suitable solvent for oil.

You know from your own experiences, I'm sure, that there's a limit to how much solute can be dissolved in a given amount of solvent. Most of us have been guilty of putting far too much sugar in iced tea. No matter how much you stir, there's some undissolved sugar at the bottom of the glass. The reason is that the sugar has reached its maximum solubility in water at that temperature. *Solubility* is the maximum amount of solute that will dissolve in a given amount of a solvent at a specified temperature. Solubility normally has the units of grams solute per 100 milliliters of solvent (g/100 mL).

If you heat that iced tea, the sugar at the bottom will readily dissolve. The solubility is related to the temperature of the solvent. For solids dissolving in liquids, solubility normally increases with increasing temperature. However, for gases dissolving in liquids, such as oxygen dissolving in lake water, the solubility goes down as the temperature increases. This is the basis of *thermal pollution,* the addition of heat to water that decreases the solubility of the oxygen and affects the aquatic life.

Saturated facts

A *saturated* solution contains the maximum amount of dissolved solute possible at a given temperature. If it has less than this amount, it's called an *unsaturated* solution. Sometimes, under unusual circumstances, the solvent may actually dissolve more than its maximum amount and become *supersaturated.* This supersaturated solution is unstable, though, and sooner or later solute will precipitate (form a solid) until the saturation point has been reached.

If a solution is unsaturated, then the amount of solute that is dissolved may vary over a wide range. A couple of rather nebulous terms describe the relative amount of solute and solvent that you can use:

- You can say that the solution is *dilute,* meaning that, relatively speaking, there's very little solute per given amount of solvent. If you dissolve 0.01 grams of sodium chloride in a liter of water, for example, the solution is dilute. I once asked some students to give me an example of a dilute solution, and one replied "A $1 margarita." She was right — a lot of solvent (water) and a very little solute (tequila) are used in her example.
- A solution may be *concentrated,* containing a large amount of solute per the given amount of solvent. If you dissolve 200 grams of sodium chloride in a liter of water, for example, the solution is concentrated.

But suppose you dissolve 25 grams or 50 grams of sodium chloride in a liter of water? Is the solution dilute or concentrated? These terms don't hold up very well for most cases. And consider the case of IV solutions — they *must* have a very precise amount of solute in them, or the patient will be in danger. So you must have a quantitative method to describe the relative amount of solute and solvent in a solution. Such a method exists — solution concentration units.

Solution Concentration Units

You can use a variety of solution concentration units to quantitatively describe the relative amounts of the solute(s) and the solvent. In everyday life, percentage is commonly used. In chemistry, *molarity* (the moles of solute per liter of solution) is the solution concentration unit of choice. In certain circumstances, though, another unit, *molality* (the moles of solute per kilogram of solvent), is used. And I use parts-per-million or parts-per-billion when I discuss pollution control. The following sections cover some of these concentration units.

Percent composition

Most of us have looked at a bottle of vinegar and seen "5% acetic acid," a bottle of hydrogen peroxide and seen "3% hydrogen peroxide," or a bottle of bleach and seen "5% sodium hypochlorite." Those percentages are expressing the concentration of that particular solute in each solution. *Percentage* is the amount per one hundred. Depending on the way you choose to express the percentage, the units of amount per one hundred vary. Three different percentages are commonly used:

- Weight/weight (w/w) percentage
- Weight/volume (w/v) percentage
- Volume/volume (v/v) percentage

Unfortunately, although the percentage of solute is often listed, the method (w/w, w/v, v/v) is not. In this case, I normally assume that the method is weight/weight, but I'm sure you know about assumptions.

Most of the solutions I talk about in the following examples of these percentages are *aqueous* solutions, solutions in which water is the solvent.

Weight/weight percentage

In *weight/weight percentage,* or *weight percentage,* the weight of the solute is divided by the weight of the solution and then multiplied by 100 to get the percentage. Normally the weight unit is grams. Mathematically, it looks like this:

$$\text{w/w\%} = \frac{\text{grams solute}}{\text{grams solution}} \times 100$$

If, for example, you dissolve 5.0 grams of sodium chloride in 45 grams of water, the weight percent is

$$\text{w/w\%} = \frac{5.0\text{g NaCl}}{50\text{g solution}} \times 100 = 10\%$$

Therefore, the solution is a 10 percent (w/w) solution.

Suppose that you want to make 350.0 grams of a 5 percent (w/w) sucrose, or table sugar, solution. You know that 5 percent of the weight of the solution is sugar, so you can multiply the 350.0 grams by 0.05 to get the weight of the sugar:

$$350.0 \text{ grams} \times 0.05 = 17.5 \text{ grams of sugar}$$

The rest of the solution (350.0 grams – 17.5 grams = 332.5 grams) is water. You can simply weigh out 17.5 grams of sugar and add it to 332.5 grams of water to get your 5 percent (w/w) solution.

Weight percentage is the easiest percentage solution to make, but sometimes you may need to know the volume of the solution. In this case, you can use the weight/volume percentage.

Weight/volume percentage

Weight/volume percentage is very similar to weight/weight percentage, but instead of using grams of solution in the denominator, it uses milliliters of solution:

$$\text{w/v\%} = \frac{\text{grams solute}}{\text{mL solution}} \times 100$$

Proof reading

When it comes to ethyl alcohol solutions, another concentration unit, called *proof,* is commonly used to measure the relative amount of alcohol and water. The *proof* is simply twice the percentage. A 50 percent ethyl alcohol solution is 100 proof. Pure ethyl alcohol (100 percent) is 200 proof. This term dates back to earlier times, when the production of ethyl alcohol for human consumption was a cottage industry. (In the part of North Carolina where I grew up, it still is a cottage industry.) There was no quality control back then, so the buyer had to be sure that the alcohol he was buying was concentrated enough (or "strong" enough) for the desired purpose. Some of the alcohol solution was poured over gunpowder and then lit. If there was enough alcohol present, the gunpowder would ignite, giving "proof" that the solution was strong enough.

Suppose that you want to make 100 milliliters of a 15 percent (w/v) potassium nitrate solution. Because you're making 100 milliliters, you already know that you're going to weigh out 15 grams of potassium nitrate (commonly called saltpeter — KNO_3). Now, here comes something that's a little different: You dissolve the 15 grams of KNO_3 in a little bit of water and dilute it to exactly 100 milliliters in a volumetric flask. In other words, you dissolve and dilute 15 grams of KNO_3 to 100 milliliters. (I tend to abbreviate *dissolve and dilute* by writing *d & d,* but sometimes it gets confused with Dungeons & Dragons. Yes, chemists are really, *really* nerds.) You won't know exactly how much water you put in, but it's not important as long as the final volume is 100 milliliters.

You can also use the percentage and volume to calculate the grams of solute present. You may want to know how many grams of sodium hypochlorite are in 500 milliliters of a 5 percent (w/v) solution of household bleach. You can set up the problem like this:

$$\frac{5\text{g NaOCl}}{100\text{mL solution}} \times \frac{500\text{mL solution}}{1} = 25\text{g NaOCl}$$

You now know that you have 25 grams of sodium hypochlorite in the 500 milliliters of solution.

Sometimes both the solute and solvent are liquids. In this case, it's convenient to use a volume/volume percentage.

Volume/volume percentage

With *volume/volume percentages,* both the solute and solution are expressed in milliliters:

$$\text{v/v\%} = \frac{\text{mL solute}}{\text{mL solution}} \times 100$$

Ethyl alcohol (the drinking alcohol) solutions are commonly made using volume/volume percentages. If you want to make 100 milliliters of a 50 percent ethyl alcohol solution, you take 50 milliliters of ethyl alcohol and dilute it to 100 milliliters with water. Again, it's a case of dissolving and diluting to the required volume. You can't simply add 50 milliliters of alcohol to 50 milliliters of water — you'd get less than 100 milliliters of solution. The polar water molecules will attract the polar alcohol molecules. This tends to fill in the open framework of water molecules and prevents the volumes from simply being added together.

It's number one! Molarity

Molarity is the concentration unit most often used by chemists, because it utilizes moles. The mole concept is central to chemistry, and molarity lets chemists easily work solutions into reaction stoichiometry. (If you're cussing me out right now because you have no idea what burrowing, insect-eating mammals have to do with chemistry, let alone what stoichiometry is, just flip to Chapter 10 for the scoop. Your mother would probably recommend washing your mouth out with soap first.)

Molarity (M) is defined as the moles of solute per liter of solution. Mathematically, it looks like this:

$$M = \frac{\text{mol solute}}{\text{L solution}}$$

For example, you can take 1 mole (abbreviated as *mol*) of KCl (formula weight of 74.55 g/mol — you can get the scoop on formula and molecular weights in Chapter 10, too) and dissolve and dilute the 74.55 grams to 1 liter of solution in a volumetric flask. You then have a 1-molar solution of KCl. You can label that solution as 1 M KCl. You don't add the 74.55 grams to 1 liter of water. You want to end up with a final volume of 1 liter. When preparing molar solutions, always dissolve and dilute to the required volume. This process is shown in Figure 11-1.

Here's another example: If 25.0 grams of KCl are dissolved and diluted to 350.0 milliliters, how would you calculate the molarity of the solution? You know that molarity is moles of solute per liter of solution. So you can take the grams, convert them to moles using the formula weight of KCl (74.55 g/mol), and divide them by 0.350 liters (350.0 milliliters). You can set up the equation like this:

$$\frac{25.0\text{g}\,KCl}{1} \times \frac{1\,\text{mol}\,KCl}{74.55\text{g}} \times \frac{1}{0.350\text{L}} = 0.958\text{M}$$

Now suppose that you want to prepare 2.00 liters of a 0.550 M KCl solution. The first thing you do is calculate how much KCl you need to weigh:

$$\frac{0.550 mol\,KCl}{L} \times \frac{74.55\text{g}\,KCl}{1\,mol} \times \frac{2.00L}{1} = 82.0\text{g}\,KCl$$

You then take that 82.0 grams of KCl and dissolve and dilute it to 2.00 liters.

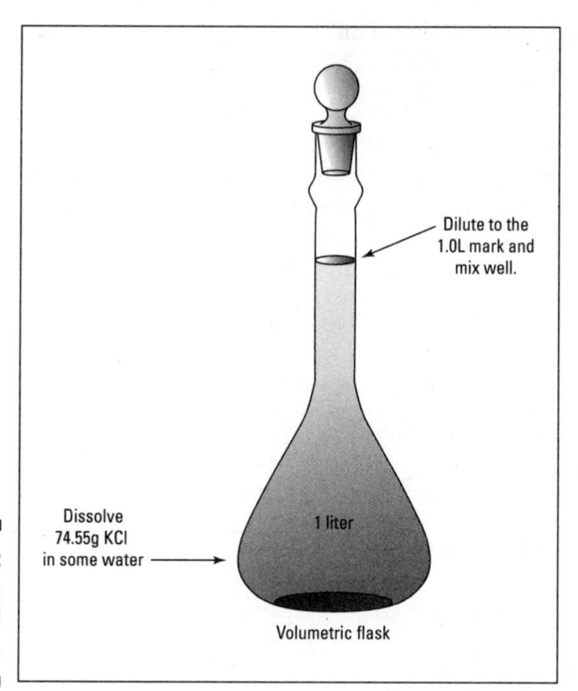

Figure 11-1: Making a 1-molar KCl solution.

There's one more way to prepare solutions — the dilution of a more concentrated solution to a less-concentrated one. For example, you can buy hydrochloric acid from the manufacturer as a concentrated solution of 12.0 M. Suppose that you want to prepare 500 milliliters of 2.0 M HCl. You can dilute some of the 12.0 M to 2.0 M, but how much of the 12.0 M HCl is needed? You can easily figure the volume (V) you need by using the following formula:

$$V_{old} \times M_{old} = V_{new} \times M_{new}$$

In the preceding equation, V_{old} is the old volume, or the volume of the original solution, M_{old} is the molarity of the original solution, V_{new} is the volume of the new solution, and M_{new} is the molarity of the new solution. After substituting the values, you have

$$V_{old} \times 12.0\ M = 500.0\ \text{milliliters} \times 2.0\ M$$

$$V_{old} = (500.0\ \text{milliliters} \times 2.0\ M)/12.0\ M = 83.3\ \text{milliliters}$$

You then take 83.3 milliliters of the 12.0 M HCl solution and dilute it to exactly 500.0 milliliters.

If you're actually doing a dilution of concentrated acids, be sure to *add the acid to the water* instead of the other way around! If the water is added to the concentrated acid, then so much heat will be generated that the solution will quite likely splatter all over you.

So to be safe, you should take about 400 milliliters of water, slowly add the 83.3 milliliters of the concentrated HCl as you stir, and then dilute to the final 500 milliliters with water.

The usefulness of the molarity concentration unit is readily apparent when dealing with reaction stoichiometry. For example, suppose that you want to know how many milliliters of 2.50 M sulfuric acid it takes to neutralize a solution containing 100.0 grams of sodium hydroxide. The first thing you must do is write the balanced chemical equation for the reaction:

$$H_2SO_4(aq) + 2\,NaOH(aq) \rightarrow 2\,H_2O(l) + Na_2SO_4(aq)$$

You know that you have to neutralize 100.0 grams of NaOH. You can convert the weight to moles (using the formula weight of NaOH, 40.00 g/mol) and then convert from moles of NaOH to moles of H_2SO_4. Then you can use the molarity of the acid solution to get the volume:

$$\frac{100.0\text{gNaOH}}{1} \times \frac{1\text{molNaOH}}{40.00\text{g}} \times \frac{1\text{mol}\,\text{H2SO4}}{2\,\text{molNaOH}} \times \frac{\text{L}}{2.50\,\text{mol}\,\text{H2SO4}} \times \frac{1000\text{ML}}{1\text{L}} = 500.0\,\text{ml}$$

It takes 500.0 milliliters of the 2.50 M H_2SO_4 solution to completely react with the solution that contains 100. grams of NaOH.

Molality: Another use for the mole

Molality is another concentration term that involves moles of solute. It isn't used very much, but I want to tell you a little about it, just in case you happen to run across it.

Molality (m) is defined as the moles of solute per kilogram of solvent. It's one of the few concentration units that doesn't use the solution's weight or volume. Mathematically, it looks like this:

$$m = \frac{\text{mol}\,\text{solute}}{\text{Kg}\,\text{solvent}}$$

Suppose, for example, you want to dissolve 15.0 grams of NaCl in 50.0 grams of water. You can calculate the molality like this (you must convert the 50.0 grams to kilograms before you use it in the equation):

$$\frac{15.0\text{g}\,\text{NaCl}}{1} \times \frac{1\,\text{mol}}{58.44\text{g}\,\text{NaCl}} \times \frac{1}{0.0500\text{Kg}} = 5.13\text{m}$$

Parts per million: The pollution unit

Percentage and molarity, and even molality, are convenient units for the solutions that chemists routinely make in the lab or the solutions that are commonly found in nature. However, if you begin to examine the concentrations of certain pollutants in the environment, you'll find that those concentrations are very, very small. Percentage and molarity will work when you're measuring solutions found in the environment, but they're not very convenient. In order to express the concentrations of very dilute solutions, scientists have developed another concentration unit — parts per million.

Percentage is parts per hundred, or grams solute per 100 grams of solution. *Parts per million (ppm)* is grams solute per one million grams of solution. It's most commonly expressed as milligrams solute per kilogram solution, which is the same ratio. The reason it's expressed this way is that chemists can easily weigh out milligrams or even tenths of milligrams, and, if you're talking about aqueous solutions, a kilogram of solution is the same as a liter of solution. (The density of water is 1 gram per milliliter, or 1 kilogram per liter. The weight of the solute in these solutions is so very small that it's negligible when converting from the mass of the solution to the volume.)

By law, the maximum contamination level of lead in drinking water is 0.05 ppm. This number corresponds to 0.05 milligrams of lead per liter of water. That's pretty dilute. But mercury is regulated at the 0.002 ppm level. Sometimes, even this unit isn't sensitive enough, so environmentalists have resorted to the parts per billion (ppb) or parts per trillion (ppt) concentration units. Some neurotoxins are deadly at the parts per billion level.

Colligative Properties of Solutions

Some properties of solutions depend on the specific nature of the solute. In other words, an effect you can record about the solution depends on the specific nature of the solute. For example, salt solutions taste salty, while sugar solutions taste sweet. Salt solutions conduct electricity (they're electrolytes — see Chapter 6), while sugar solutions don't (they're nonelectrolytes). Solutions containing the nickel cation are commonly green, while those containing the copper cation are blue.

There's also a group of solution properties that doesn't depend on the specific type of solute — just the *number* of solute particles. These properties are called *colligative properties* — properties that simply depend on the relative number of solute particles. The effect you can record about the solution depends on the number of solute particles present. These colligative properties — these effects — include

- Vapor pressure lowering
- Boiling point elevation
- Freezing point depression
- Osmotic pressure

Vapor pressure lowering

If a liquid is contained in a closed container, the liquid eventually evaporates, and the gaseous molecules contribute to the pressure above the liquid. The pressure due to the gaseous molecules of the evaporated liquid is called the liquid's *vapor pressure.*

If you take that same liquid and make it the solvent in a solution, the vapor pressure due to the solvent evaporation will be lower. This is because the solute particles in the liquid take up space at the surface and so the solvent can't evaporate as easily. And many times there may be an attraction between the solute and solvent that also makes it more difficult for the solvent to evaporate. And that lowering is independent of what kind of solute you use. Instead, it depends on the number of solute particles.

In other words, if you add one mole of sucrose to a liter of water and add one mole of dextrose to another liter of water, the amount that the pressure lowers will be the same, because you're adding the same *number* of solute particles. If, however, you add a mole of sodium chloride, the vapor pressure will be lowered by twice the amount of sucrose or glucose. The reasons is that the sodium chloride breaks apart into two ions, so adding a mole of sodium chloride yields two moles of particles (ions).

This lowering of vapor pressure partially explains why the Great Salt Lake has a lower evaporation rate than you may expect. The salt concentration is so high that the vapor pressure (and evaporation) has been significantly lowered.

Why use antifreeze in the summer? Boiling point elevation

Each individual liquid has a specific temperature at which it boils (at a given atmospheric pressure). This temperature is the liquid's *boiling point.* If you use a particular liquid as a solvent in a solution, you find that the boiling point of the solution is always higher than the pure liquid. This is called the *boiling point elevation.*

It explains why you don't replace your antifreeze with pure water in the summer. You want the coolant to boil at a higher temperature so that it will absorb as much engine heat as possible *without* boiling. You also use a pressure cap on your radiator, because the higher the pressure, the higher the boiling point. It also explains why a pinch of salt in the cooking water will cause foods to cook a little faster. The salt raises the boiling point so that more energy can be transferred to cooking the food during a given amount of time.

As an FYI, you can actually calculate the amount of boiling point elevation by using this formula:

$$\Delta T_b = K_b m$$

ΔT_b is the *increase* in the boiling point, K_b is the boiling point elevation constant (0.512°C kg/mol for water), and *m* is the molality of particles. (For molecular substances, the molality of particles is the same as the molality of the substance; for ionic compounds, you have to take into consideration the formation of ions and calculate the molality of the ion particles.) Solvents other than water have a different boiling point elevation constant (K_b).

Making ice cream: Freezing point depression

Each individual liquid has a specific temperature at which it freezes. If you use a particular liquid as a solvent in a solution, though, you find that the freezing point of the solution is always lower than the pure liquid. This is called the *freezing point depression,* and it's a colligative property of a solution.

The depression of the freezing point of a solution relative to the pure solvent explains why you put rock salt in the ice/water mix when making homemade ice cream. The rock salt forms a solution with a lower freezing point than water (or the ice cream mix that's to be frozen). The freezing point depression effect also explains why a salt (normally calcium chloride, $CaCl_2$) is spread on ice to melt it. The dissolving of calcium chloride is highly exothermic (it gives off a lot of heat). When the calcium chloride dissolves, it melts the ice. The salt solution that's formed when the ice melts has a lowered freezing point that keeps the solution from refreezing. Freezing point depression also explains the use of antifreeze in your cooling system during the winter. The more you use (up to a concentration of 50/50), the lower the freezing point.

In case you're interested, you can actually calculate the amount the freezing point will be depressed:

$$\Delta T_f = K_f m$$

ΔT_f is the amount the freezing point will be lowered, K_f is the freezing point depression constant (1.86°C kg/mol for water), and *m* is the molality of the particles.

Figure 11-2 shows the effect of a solute on both the freezing point and boiling point of a solvent.

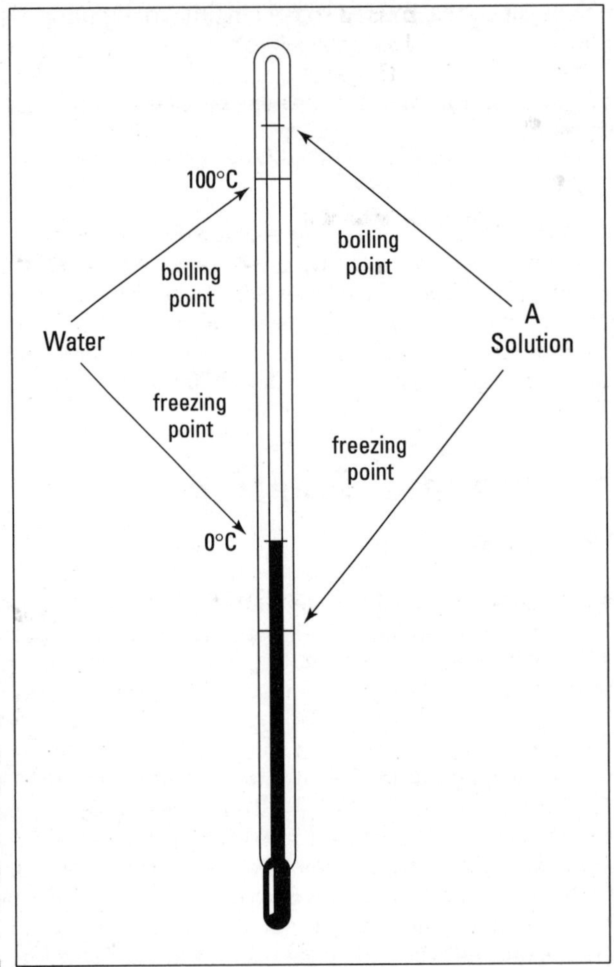

Figure 11-2: Boiling point elevation and freezing point depression of a solution.

Keeping blood cells alive and well: Osmotic pressure

Suppose that you take a container and divide it into two compartments with a thin membrane containing microscopic pores large enough to allow water molecules but not solute particles to pass through. This membrane type is called a *semipermeable membrane;* it lets some small particles pass through but not other, larger particles.

You then add a concentrated salt solution to one compartment and a more dilute salt solution to the other. Initially, the two solution levels start out the same. But after a while, you notice that the level on the more concentrated side has risen, and the level on the more dilute side has dropped. This change in levels is due to the passage of water molecules from the more dilute side to the more concentrated side through the semipermeable membrane. This process is called *osmosis*, the passage of a solvent through a semipermeable membrane into a solution of higher solute concentration. The pressure that you have to exert on the more concentrated side in order to stop this process is called *osmotic pressure*. This process is shown in Figure 11-3.

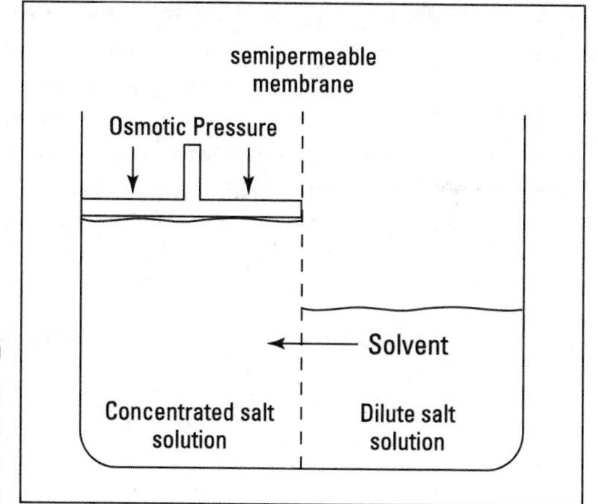

Figure 11-3: Osmotic pressure.

The solvent always flows through the semipermeable membrane from the more dilute side to the more concentrated side. In fact, you can have pure water on one side and any salt solution on the other, and water always goes from the pure-water side to the salt-solution side. The more concentrated the salt solution, the more pressure it takes to stop the osmosis (the higher the osmotic pressure).

But what if you apply more pressure than is necessary to stop the osmotic process, exceeding the osmotic pressure? Water is forced through the semipermeable membrane from the more concentrated side to the more dilute side, a process called *reverse osmosis*. Reverse osmosis is a good, relatively inexpensive way of purifying water. My local "water store" uses this process to purify drinking water (so-called "RO water"). There are many reverse osmosis plants in the world, extracting drinking water from seawater. Navy pilots even carry small reverse osmosis units with them in case they have to eject at sea.

The process of osmosis is important in biological systems. Cell walls often act as semipermeable membranes. Do you ever eat pickles? Cucumbers are soaked in a brine solution in order to make pickles. The concentration of the solution inside the cucumber is less than the concentration of the brine solution, so water migrates through the cell walls into the brine, causing the cucumber to shrink.

One of the most biologically important consequences of osmotic pressure involves the cells within our own body. You can look at red blood cells as an example. There's an aqueous solution inside the blood cell and another aqueous solution outside the cell (intercellular fluid). When the solution outside the cell has the same osmotic pressure as the solution inside the cell, it's said to be *isotonic.* Water can be exchanged in both directions, helping to keep the cell healthy. However, if the intercellular fluid becomes more concentrated and has a higher osmotic pressure *(hypertonic),* water flows primarily out of the blood cell, causing it to shrink and become irregular in shape. This is a process called *crenation.* The process may occur if the person becomes seriously dehydrated, and the crenated cells are not as efficient in carrying oxygen. If, on the other hand, the intercellular fluid is more dilute than the solution inside the cells and has a lower osmotic pressure *(hypotonic),* the water flows mostly into the cell. This process, called *hemolysis,* causes the cell to swell and eventually rupture. Figure 11-4 shows crenation and hemolysis.

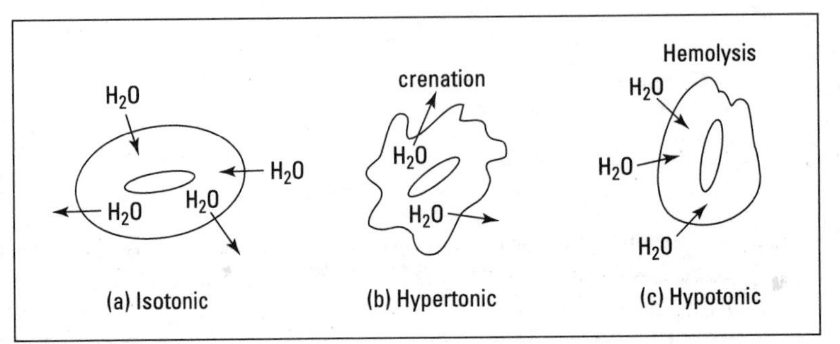

Figure 11-4: Crenation and hemolysis of red blood cells.

The processes of crenation and hemolysis explain why the concentration of IV solutions is so very critical. If they're too dilute, then hemolysis can take place, and if they're too concentrated, crenation is a possibility.

Smoke, Clouds, Whipped Cream, and Marshmallows: Colloids All

If you dissolve table salt in water, you form an aqueous solution. The solute particle size is very small — around 1 nanometer (nm), which is 1×10^{-9} meters. This solute doesn't settle to the bottom of a glass, and it can't be filtered out of the solution.

If, however, you go down to your local stream and dip out a glass of water, you'll notice that there's a lot of material in it. Many of the solute particles are larger than 1,000 nm. They quickly settle to the bottom of the glass and can be filtered out. In this case, you have a *suspension* and not a solution. Whether you have one or the other depends on the size of the solute particles.

But there's also something in the intermediate range between solutions and suspensions. When the solute particle size is 1 to 1,000 nanometers, you have a *colloid.* Solutes in colloids don't settle out like they do in suspensions. In fact, it's sometimes difficult to distinguish colloids from true solutions. One of the few ways to distinguish between them is to shine a light through the suspected liquid. If it's a true solution, with very small solute particles, the light beam will be invisible. If you have a colloid, however, you'll be able to see the light beam as it reflects off the relatively large solute particles. This is called the *Tyndall effect,* and it's shown in Figure 11-5.

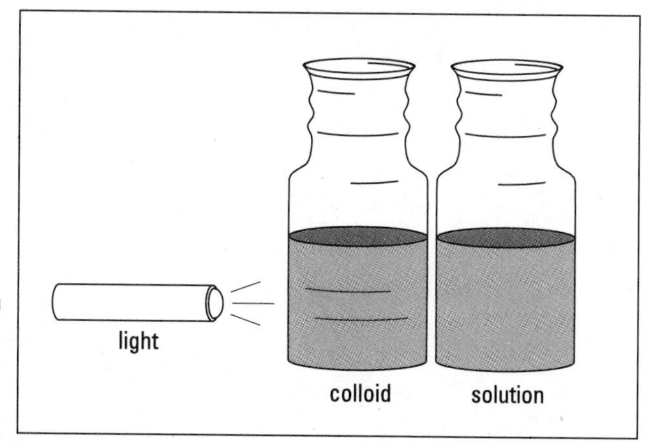

Figure 11-5: The Tyndall effect.

There are many types of colloids. Have you ever eaten a marshmallow? It's a colloid of a gas in a solid. Whipped cream is a colloid of a gas in a liquid. Have you ever driven through the fog and seen your headlight beams? You were experiencing the Tyndall effect of a liquid-in-a-gas colloid. Smoke is a colloid of a solid (ash or soot) in a gas (air). Air pollution problems are often caused by the stability of this type of colloid.

Chapter 12

Sour and Bitter: Acids and Bases

In This Chapter

- Discovering the properties of acids and bases
- Finding out about the two acid-base theories
- Differentiating between strong and weak acids and bases
- Understanding indicators
- Taking a look at the pH scale
- Figuring out buffers and antacids

Walk into any kitchen or bathroom, and you'll find a multitude of acids and bases. Open the refrigerator, and you'll find soft drinks full of carbonic acid. In the pantry, there's vinegar and baking soda, an acid and a base. Peek under the sink, and you'll notice the ammonia and other cleaners, most of which are bases. Check out that can of lye-based drain opener — it's highly basic. In the medicine cabinet, you'll find aspirin, an acid, and antacids of all types. Our everyday world is full of acids and bases. And so is the everyday world of the industrial chemist. In this chapter, I cover acids and bases, indicators and pH, and some good *basic* chemistry.

Properties of Acids and Bases: Macroscopic View

Look at the properties of acids and bases that can be observed in the world around us.

Acids:

- Taste sour (but remember, in the lab, you test, not taste)
- Produce a painful sensation on the skin
- React with certain metals (magnesium, zinc, and iron) to produce hydrogen gas

- React with limestone and baking soda to produce carbon dioxide
- React with litmus paper and turn it red

Bases:

- Taste bitter (again, in the lab, you test, not taste)
- Feel slippery on the skin
- React with oils and greases
- React with litmus paper and turn it blue
- React with acids to produce a salt and water

Quite a number of acids and bases are found in our everyday life. Tables 12-1 and 12-2 show some common acids and bases found around the home.

Table 12-1 Common Acids Found in the Home

Chemical Name	*Formula*	*Common Name or Use*
hydrochloric acid	HCl	muratic acid
acetic acid	CH_3COOH	vinegar
sulfuric acid	H_2SO_4	auto battery acid
carbonic acid	H_2CO_3	carbonated water
boric acid	H_3BO_3	antiseptic; eye drops
acetylsalicylic acid	$C_{16}H_{12}O_6$	aspirin

Table 12-2 Common Bases Found in the Home

Chemical Name	*Formula*	*Common Name or Use*
ammonia	NH_3	cleaner
sodium hydroxide	$NaOH$	lye
sodium bicarbonate	$NaHCO_3$	baking soda
magnesium hydroxide	$Mg(OH)_2$	milk of magnesia
calcium carbonate	$CaCO_3$	antacid
aluminum hydroxide	$Al(OH)_3$	antacid

What Do Acids and Bases Look Like? — Microscopic View

If you look at Tables 12-1 and 12-2 closely, you may recognize the fact that all the acids contain hydrogen, while most of the bases contain the hydroxide ion (OH^-). Two main theories use these facts in their descriptions of acids and bases and their reactions:

- Arrhenius theory
- Bronsted-Lowery theory

The Arrhenius theory: Must have water

The Arrhenius theory was the first modern acid-base theory developed. In this theory, an acid is a substance that, when dissolved in water, yields H^+ (hydrogen) ions, and a base is a substance that, when dissolved in water, yields OH^- (hydroxide) ions. HCl(g) can be considered as a typical Arrhenius acid, because when this gas dissolves in water, it *ionizes* (forms ions) to give the H^+ ion. (Chapter 6 is where you need to go for the riveting details about ions.)

$$HCl(aq) \rightarrow H^+ + Cl^-$$

According to the Arrhenius theory, sodium hydroxide is classified as a base, because when it dissolves, it yields the hydroxide ion:

$$NaOH(aq) \rightarrow Na^+ + OH^-$$

Arrhenius also classified the reaction between an acid and base as a *neutralization* reaction, because if you mix an acidic solution with a basic solution, you end up with a neutral solution composed of water and a salt.

$$HCl(aq) + NaOH(aq) \rightarrow H_2O(l) + NaCl(aq)$$

Look at the ionic form of this equation (the form showing the reaction and production of ions) to see where the water comes from:

$$\mathit{H^+} + Cl^- + Na^+ + \mathit{OH^-} \rightarrow \mathit{H_2O}(l) + Na^+ + Cl^-$$

As you can see, the water is formed from combining the hydrogen and hydroxide ions. In fact, the net-ionic equation (the equation showing only those chemical substances that are changed during the reaction) is the same for all Arrhenius acid-base reactions:

$$H^+ + OH^- \rightarrow H_2O(l)$$

The Arrhenius theory is still used quite a bit. But, like all theories, it has some limitations. For example, look at the gas phase reaction between ammonia and hydrogen chloride gases:

$$NH_3(g) + HCl(g) \rightarrow NH_4^+ + Cl^- \rightarrow NH_4Cl(s)$$

The two clear, colorless gases mix, and a white solid of ammonium chloride forms. I show the intermediate formation of the ions in the equation so that you can better see what's actually happening. The HCl transfers an H^+ to the ammonia. That's basically the same thing that happens in the HCl/NaOH reaction, but the reaction involving the ammonia can't be classified as an acid-base reaction, because it doesn't occur in water, and it doesn't involve the hydroxide ion. But again, the same basic process is taking place in both cases. In order to account for these similarities, a new acid-base theory was developed, the Bronsted-Lowery theory.

The Bronsted-Lowery acid-base theory: Giving and accepting

The Bronsted-Lowery theory attempts to overcome the limitations of the Arrhenius theory by defining an acid as a proton (H^+) donor and a base as a proton (H^+) acceptor. The base accepts the H^+ by furnishing a lone pair of electrons for a *coordinate-covalent bond,* which is a covalent bond (shared pair of electrons) in which one atom furnishes both of the electrons for the bond. Normally, one atom furnishes one electron for the bond and the other atom furnishes the second electron (see Chapter 7). In the coordinate-covalent bond, one atom furnishes both bonding electrons.

Figure 12-1 shows the NH_3/HCl reaction using the electron-dot structures of the reactants and products. (Electron-dot structures are covered in Chapter 7, too.)

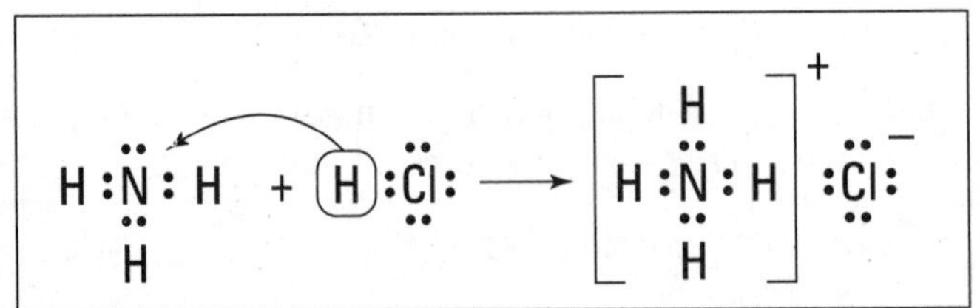

Figure 12-1: Reaction of NH_3 with HCl.

HCl is the proton donor, and the acid and ammonia are the proton acceptor, or the base. Ammonia has a lone pair of nonbonding electrons that it can furnish for the coordinate-covalent bond.

I discuss acid-base reactions under the Bronsted-Lowery theory in the section "Give me that proton: Bronsted-Lowery acid-base reactions," later in this chapter.

Acids to Corrode, Acids to Drink: Strong and Weak Acids and Bases

I want to introduce you to a couple of different categories of acids and bases — strong and weak. However, it's important to remember that acid-base strength is not the same as concentration. *Strength* refers to the amount of ionization or breaking apart that a particular acid or base undergoes. *Concentration* refers to the amount of acid or base that you initially have. You can have a concentrated solution of a weak acid, or a dilute solution of a strong acid, or a concentrated solution of a strong acid or . . . well, I'm sure you get the idea.

Strong acids

If you dissolve hydrogen chloride gas in water, the HCl reacts with the water molecules and donates a proton to them:

$$HCl(g) + H_2O(l) \rightarrow Cl^- + H_3O^+$$

The H_3O^+ ion is called the hydronium ion. This reaction goes essentially to completion, meaning the reactants keep creating the product until they're all used up. In this case, all the HCl ionizes to H_3O^+ and Cl^-; there's no more HCl present. Acids such as HCl, which ionizes essentially 100 percent in water, are called *strong acids*. Note that water, in this case, acts as a base, accepting the proton from the hydrogen chloride.

Because strong acids ionize completely, it's easy to calculate the concentration of the hydronium ion and chloride ion in solution if you know the initial concentration of the strong acid. For example, suppose that you bubble 0.1 moles (see Chapter 10 to get a firm grip on moles) of HCl gas into a liter of water. You can say that the initial concentration of HCl is 0.1 M (0.1 mol/L). *M* stands for molarity, and *mol/L* stands for moles of solute per liter. (For a detailed discussion of molarity and other concentration units, see Chapter 11.)

You can represent this 0.1 M concentration for the HCl in this fashion: [HCl] = 0.1. Here, the brackets around the compound indicate molar concentration, or mol/L. Because the HCl completely ionizes, you see from the balanced equation that for every HCl that ionizes, you get one hydronium ion and one chloride ion. So the concentration of ions in that 0.1 M HCl solution is

$[H_3O^+] = 0.1$ and $[Cl^-] = 0.1$

This idea is valuable when you calculate the pH of a solution. (And you can do just that in the section "How Acidic Is That Coffee: The pH Scale," later in this chapter.)

Table 12-3 lists the most common strong acids you're likely to encounter.

Table 12-3 Common Strong Acids

Name	*Formula*
Hydrochloric acid	HCl
Hydrobromic acid	HBr
Hydroiodic acid	HI
Nitric acid	HNO_3
Perchloric acid	$HClO_4$
Sulfuric acid (first ionization only)	H_2SO_4

Sulfuric acid is called a *diprotic* acid. It can donate 2 protons, but only the first ionization goes 100 percent. The other acids listed in Table 12-3 are *monoprotic* acids, because they donate only one proton.

Strong bases

You'll normally see only one strong base, and that's the hydroxide ion, OH^-. Calculating the hydroxide ion concentration is really straightforward. Suppose that you have a 1.5 M (1.5 mol/L) NaOH solution. The sodium hydroxide, a salt, completely *dissociates* (breaks apart) into ions:

$$NaOH \rightarrow Na^+ + OH^-$$

If you start with 1.5 mol/L NaOH, then you have the same concentration of ions:

$[Na^+] = 1.5$ and $[OH^-] = 1.5$

Weak acids

Suppose that you dissolve acetic acid (CH_3COOH) in water. It reacts with the water molecules, donating a proton and forming hydronium ions. It also establishes equilibrium, where you have a significant amount of unionized acetic acid. (In reactions that go to completion, the reactants are completely used up creating the products. But in equilibrium systems, two exactly opposite chemical reactions — one on each side of the reaction arrow — are occurring at the same place, at the same time, with the same speed of reaction. For a discussion of equilibrium systems, see Chapter 8.)

If you want to see whether a person is a chemist, ask him to pronounce *union-ized.* A chemist pronounces it *un-ionized,* meaning "not ionized." Everyone else pronounces it *union-ized,* meaning "being part of a union."

The acetic acid reaction with water looks like this:

$$CH_3COOH(l) + H_2O(l) \leftrightarrow CH_3COO^- + H_3O^+$$

The acetic acid that you add to the water is only partially ionized. In the case of acetic acid, about 5 percent ionizes, while 95 percent remains in the molecular form. The amount of hydronium ion that you get in solutions of acids that don't ionize completely is much less than it is with a strong acid. Acids that only partially ionize are called *weak acids.*

Calculating the hydronium ion concentration in weak acid solutions isn't as straightforward as it is in strong solutions, because not all of the weak acid that dissolves initially has ionized. In order to calculate the hydronium ion concentration, you must use the equilibrium constant expression for the weak acid. Chapter 8 covers the K_{eq} expression that represents the equilibrium system. For weak acid solutions, you use a modified equilibrium constant expression called the K_a — the *acid ionization constant.* Take a look at the generalized ionization of some weak acid HA:

$$HA + H_2O \leftrightarrow A^- + H_3O^+$$

The K_a expression for this weak acid is

$$K_a = \frac{[H_3O^+][A^-]}{[HA]}$$

Note that the *[HA]* represents the molar concentration of HA *at equilibrium,* not initially. Also, note that the concentration of water doesn't appear in the K_a expression, because there's so much that it actually becomes a constant incorporated into the K_a expression.

Now go back to that acetic acid equilibrium. The K_a for acetic acid is 1.8×10^{-5}. The K_a expression for the acetic acid ionization is

$$K_a = 1.8 \times 10^{-5} = \frac{[H_3O^+][CH_3COO^-]}{[CH_3COOH]}$$

You can use this K_a when calculating the hydronium ion concentration in, say, a 2.0 M solution of acetic acid. You know that the initial concentration of acetic acid is 2.0 M. You know that a little bit has ionized, forming a little hydronium ion and acetate ion. You also can see from the balanced reaction that for every hydronium ion that's formed, an acetate ion is also formed — so their concentrations are the same. You can represent the amount of $[H_3O^+]$ and $[CH_3COO^-]$ as *x*, so

$$[H_3O^+] = [CH_3COO^-] = x$$

In order to produce the *x* amount of hydronium and acetate ion, the same amount of ionizing acetic acid is required. So you can represent the amount of acetic acid remaining at equilibrium as the amount you started with, 2.0 M, minus the amount that ionizes, *x*:

$$[CH_3COOH] = 2.0 - x$$

For the vast majority of situations, you can say that *x* is very small in comparison to the initial concentration of the weak acid. So you can say that $2.0 - x$ is approximately equal to 2.0. This means that you can often approximate the equilibrium concentration of the weak acid with its initial concentration. The equilibrium constant expression now looks like this:

$$K_a = 1.8 \times 10^{-5} = \frac{[X][X]}{[2.0]} = \frac{[X]^2}{[2.0]}$$

At this point, you can solve for *x*, which is the $[H_3O^+]$:

$$(1.8 \times 10^{-5})[2.0] = [X]^2$$
$$\sqrt{3.6 \times 10^{-5}} = [X] = [H_3O^+]$$
$$6.0 \times 10^{-3} = [H_3O^+]$$

Table 12-3 shows some common strong acids. Most of the other acids you encounter are weak.

One way to distinguish between strong and weak acids is to look for an acid ionization constant (K_a) value. If the acid has a K_a value, then it's weak.

Weak bases

Weak bases also react with water to establish an equilibrium system. Ammonia is a typical weak base. It reacts with water to form the ammonium ion and the hydroxide ion:

$$NH_3(g) + H_2O(l) \leftrightarrow NH_4^+ + OH^-$$

Like a weak acid, a weak base is only partially ionized. There's a modified equilibrium constant expression for weak bases — the K_b. You use it exactly the same way you use the K_a (see "Weak acids" for the details) except you solve for the $[OH^-]$.

Give me that proton: Bronsted-Lowery acid-base reactions

With the Arrhenius theory, acid-base reactions are neutralization reactions. With the Bronsted-Lowery theory, acid-base reactions are a competition for a proton. For example, take a look at the reaction of ammonia with water:

$$NH_3(g) + H_2O(l) \leftrightarrow NH_4^+ + OH^-$$

Ammonia is a base (it accepts the proton), and water is an acid (it donates the proton) in the forward (left to right) reaction. But in the reverse reaction (right to left), the ammonium ion is an acid and the hydroxide ion is a base. If water is a stronger acid than the ammonium ion, then there is a relatively large concentration of ammonium and hydroxide ions at equilibrium. If, however, the ammonium ion is a stronger acid, much more ammonia than ammonium ion is present at equilibrium.

Bronsted and Lowery said that an acid reacts with a base to form conjugate acid-base pairs. Conjugate acid base pairs differ by a single H+. NH_3 is a base, for example, and NH_4^+ is its conjugate acid. H_2O is an acid in the reaction between ammonia and water, and OH^- is its conjugate base. In this reaction, the hydroxide ion is a strong base and ammonia is a weak base, so the equilibrium is shifted to the left — there's not much hydroxide at equilibrium.

Make up your mind: Amphoteric water

When acetic acid reacts with water, water acts as a base, or a proton acceptor. But in the reaction with ammonia (see the preceding section), water acts as an acid, or a proton donor. Water can act as either an acid or a base, depending on what it's combined with. Substances that can act as either an acid or a base are called *amphoteric.* If you put water with an acid, it acts as a base, and vice versa.

But can it react with itself? Yes, it can. Two water molecules can react with each other, with one donating a proton and the other accepting it:

$$H_2O(l) + H_2O(l) \leftrightarrow H_3O^+ + OH^-$$

This reaction is an equilibrium reaction. A modified equilibrium constant, called the K_w (which stands for *water dissociation constant*) is associated with this reaction. The K_w has a value of 1.0×10^{-14} and has the following form:

$$1.0 \times 10^{-14} = K_w = [H_3O^+] [OH^-]$$

In pure water, the $[H_3O^+]$ equals the $[OH^-]$ from the balanced equation, so $[H_3O^+] = [OH^-] = 1.0 \times 10^{-7}$. The K_w value is a constant. This value allows you to convert from $[H^+]$ to $[OH^-]$, and vice versa, in *any* aqueous solution, not just pure water. In aqueous solutions, the hydronium ion and hydroxide ion concentrations are rarely going to be equal. But if you know one of them, the K_w allows you to figure out the other one.

Take a look at the 2.0 M acetic acid solution problem in the section "Weak acids," earlier in this chapter. You find that the $[H_3O^+]$ is 6.0×10^{-3}. Now you have a way to calculate the $[OH^-]$ in the solution by using the K_w relationship:

$$K_w = 1.0 \times 10^{-14} = [H_3O^+] [OH^-]$$

$$1.0 \times 10^{-14} = [6.0 \times 10^{-3}] [OH^-]$$

$$1.0 \times 10^{-14}/6.0 \times 10^{-3} = [OH^-]$$

$$1.7 \times 10^{-12} = [OH^-]$$

An Old Laxative and Red Cabbage: Acid-Base Indicators

Indicators are substances (organic dyes) that change color in the presence of an acid or base. You may be familiar with an acid-base indicator plant — the hydrangea. If it's grown in acidic soil, it turns pink; if it's grown in alkaline soil, it turns blue. Another common substance that acts as a good acid-base indicator is red cabbage. I have my students chop some up and boil it (most of them really *love* this part). They then take the liquid that is left over and use it to test substances. When mixed with an acid, the liquid turns pink; when mixed with a base, it turns green. In fact, if you take some of this liquid, make it slightly basic, and then exhale your breath into it through a straw, the solution eventually turns blue, indicating that the solution has turned slightly acidic. The carbon dioxide in your breath reacts with the water, forming carbonic acid:

$$CO_2(g) + H_2O(l) \leftrightarrow H_2CO_3(aq)$$

Carbonated beverages are slightly acidic due to this reaction. Carbon dioxide is injected into the liquid to give it fizz. A little of this carbon dioxide reacts with the water to form carbonic acid. This reaction also explains why rainwater is slightly acidic. It absorbs carbon dioxide from the atmosphere as it falls to earth.

In chemistry, indicators are used to indicate the presence of an acid or base. Chemists have many indicators that change at slightly different pHs. (You've probably heard the term *pH* used in various contexts. Me, I even remember it being used to sell deodorant on TV. If you want to know what it actually stands for, check out the section "How Acidic Is That Coffee: The pH Scale.") Two indicators are used most often:

- Litmus paper
- Phenolphthalein

Good old litmus paper

Litmus is a substance that is extracted from a type of lichen and absorbed into porous paper. (In case you're scheduled for a hot game of *Trivial Pursuit* this weekend, a *lichen* is a plant — found in the Netherlands — that's made up of an alga and a fungus that are intimately living together and mutually benefiting from the relationship. Sounds kind of sordid to me.) There are three different types of litmus — red, blue, and neutral. Red litmus is used to test for bases, and blue litmus is used to test for acids, while neutral litmus can be used to test for both. If a solution is acidic, both blue and neutral litmus will turn red. If a solution is basic, both red and neutral litmus will turn blue. Litmus paper is a good, quick test for acids and bases. And you don't have to put up with the smell of boiling cabbage.

Phenolphthalein: Helps keep you regular

Phenolphthalein (pronounced *fe-nul-tha-Leen*) is another commonly used indicator. Until a few years ago, phenolphthalein was used as the active ingredient in a popular laxative. In fact, I used to extract the phenolphthalein from the laxative by soaking it in either rubbing alcohol or gin (being careful not to drink it). I'd then use this solution as an indicator.

Phenolphthalein is clear and colorless in an acid solution and pink in a basic solution. It's commonly used in a procedure called a *titration,* where the concentration of an acid or base is determined by its reaction with a base or acid of known concentration.

Suppose, for example, that you want to determine the molar concentration of an HCl solution. First, you place a known volume (say, 25.00 milliliters measured accurately with a pipette) in an Erlenmeyer flask (that's just a flat-bottomed, conical-shaped flask) and add a couple drops of phenolphthalein solution. Because you're adding the indicator to an acidic solution, the solution in the flask remains clear and colorless. You then add small amounts of a standardized sodium hydroxide solution of known molarity (for example, 0.100 M) with a buret. (A *buret* is a graduated glass tube with a small opening and a stopcock, which helps you measure precise volumes of solution.) You keep adding base until the solution turns the faintest shade of pink detectable. I call this the *endpoint* of the titration, the point in which the acid has been exactly neutralized by the base. Figure 12-2 shows the titration setup.

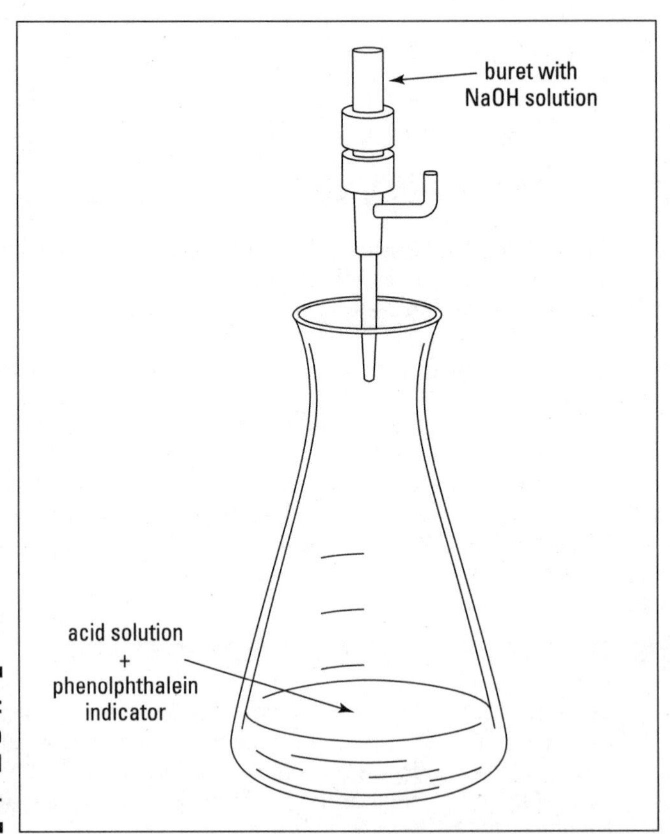

Figure 12-2: Titration of an acid with a base.

Suppose that it takes 35.50 milliliters of the 0.100 M NaOH to reach the endpoint of the titration of the 25.00 milliliters of the HCl solution. Here's the reaction:

$$HCl(aq) + NaOH(aq) \rightarrow H_2O(l) + NaCl(aq)$$

From the balanced equation, you can see that the acid and base react in a 1:1 mole ratio. So if you can calculate the moles of bases added, you'll also know the number of moles of HCl present. Knowing the volume of the acid solution then allows you to calculate the molarity (note that you convert the milliliters to liters so that your units cancel nicely):

$$\frac{0.100\,\cancel{mol\,NaOH}}{\cancel{L}} \times \frac{0.03550\cancel{L}}{1} \times \frac{1\,mol\,HCl}{1\,\cancel{mol\,NaOH}} \times \frac{1}{0.02500L} = 0.142\,M\ \ HCl$$

The titration of a base with a standard acid solution (one of known concentration) can be calculated in exactly the same way, except the endpoint is the first disappearance of the pink color.

How Acidic Is That Coffee: The pH Scale

The amount of acidity in a solution is related to the concentration of the hydronium ion in the solution. The more acidic the solution is, the larger the concentration of the hydronium ion. In other words, a solution in which the $[H_3O^+]$ equals 1.0×10^{-2} is more acidic than a solution in which the $[H_3O^+]$ equals 1.0×10^{-7}. The *pH scale,* a scale based on the $[H_3O^+]$, was developed to more easily tell, at a glance, the relative acidity of a solution. *pH* is defined as the negative logarithm (abbreviated as *log*) of the $[H_3O^+]$. Mathematically, it looks like this:

$$pH = -\log [H_3O^+]$$

Based on the water dissociation constant, K_w (see "Make up your mind: Amphoteric water," earlier in this chapter), in pure water the $[H_3O^+]$ equals 1.0×10^{-7}. Using this mathematical relationship, you can calculate the pH of pure water:

$$pH = -\log [H_3O^+]$$

$$pH = -\log [1.0 \times 10^{-7}]$$

$$pH = -[-7]$$

$$pH = 7$$

The pH of pure water is 7. Chemists call this point on the pH scale *neutral.* A solution is called *acidic* if it has a larger $[H_3O^+]$ than water and a smaller pH value than 7. A *basic* solution has a smaller $[H_3O^+]$ than water and a larger pH value than 7.

The pH scale really has no end. You can have a solution of pH that registers less than 0. (A 10 M HCl solution, for example, has a pH of –1.) However, the 0 to 14 range is a convenient range to use for weak acids and bases and for dilute solutions of strong acids and bases. Figure 12-3 shows the pH scale.

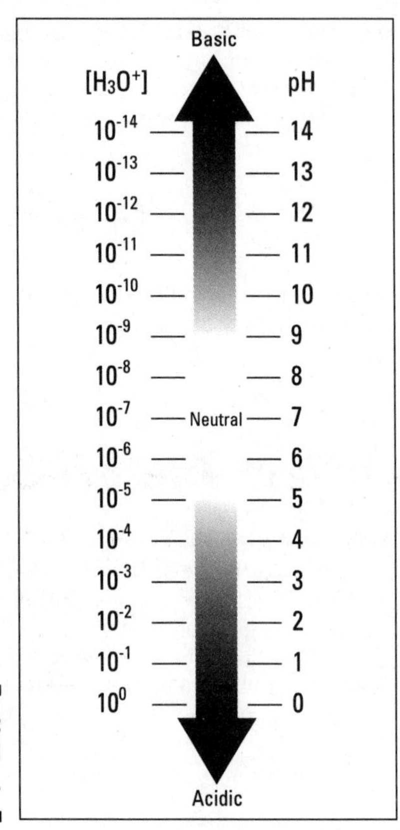

Figure 12-3: The pH scale.

The $[H_3O^+]$ of a 2.0 M acetic acid solution is 6.0×10^{-3}. Looking at the pH scale, you see that this solution is acidic. Now calculate the pH of this solution:

$$pH = -\log [H_3O^+]$$

$$pH = -\log [6.0 \times 10^{-3}]$$

$$pH = -[-2.22]$$

$$pH = 2.22$$

In the section "Make up your mind: Amphoteric water," I explain that the K_w expression enables you to calculate the $[H_3O^+]$ if you have the $[OH^-]$. Another equation, called the *pOH,* can be useful in calculating the pH of a solution. The pOH is the negative logarithm of the $[OH^-]$. You can calculate the pOH of a solution just like the pH by taking the negative log of the hydroxide ion concentration. If you use the K_w expression and take the negative log of both sides, you get 14 = pH + pOH. This equation makes it easy to go from pOH to pH.

Just as you can you convert from $[H_3O^+]$ to pH, you can also go from pH to $[H_3O^+]$. To do this, you use what's called the *antilog relationship,* which is

$$[H_3O^+] = 10^{-pH}$$

Human blood, for example, has a pH of 7.3. Here's how you calculate the $[H_3O^+]$ from the pH of blood:

$$[H_3O^+] = 10^{-pH}$$

$$[H_3O^+] = 10^{-7.3}$$

$$[H_3O^+] = 5.01 \text{ x } 10^{-8}$$

The same procedure can be used to calculate the $[OH^-]$ from the pOH.

Substances commonly found in our surroundings cover a wide range of pH values. Table 12-4 lists some common substances and their pH values.

Table 12-4 Average pH Values of Some Common Substances

Substance	*PH*
Oven cleaner	13.8
Hair remover	12.8
Household ammonia	11.0
Milk of magnesia	10.5
Chlorine bleach	9.5
Seawater	8.0
Human blood	7.3
Pure water	7.0
Milk	6.5
Black coffee	5.5
Soft drinks	3.5
Aspirin	2.9
Vinegar	2.8

(continued)

Table 12-4 *(continued)*

Substance	*PH*
Lemon juice	2.3
Auto battery acid	0.8

Human blood has a pH of around 7.3. There's a narrow range in which blood pH can change and still sustain life, about +/– 0.2 pH units. Many things in our environment, such as foods and hyperventilation, can act to change the pH of our blood. Buffers help to regulate blood pH and keep it in the 7.1 to 7.5 range.

Buffers: Controlling pH

Buffers, or *buffer solutions* as they're sometimes called, resist a change in pH caused by the addition of acids or bases. Obviously, the buffer solution must contain something that reacts with an acid — a base. Something else in the buffer solution reacts with a base — an acid. There are, in general, two types of buffers:

- Mixtures of weak acids and bases
- Amphoteric species

The mixtures of weak acids and bases may be conjugate acid-base pairs (such as H_2CO_3/HCO_3^-) or nonconjugate acid-base pairs (such as NH_4+/CH_3COO^-). (For more info about conjugate acid-base pairs, see "Give me that proton: Bronsted-Lowery acid-base reactions," earlier in this chapter.)

In the body, conjugate acid-base pairs are more common. In the blood, for example, the carbonic acid/bicarbonate pair helps to control the pH. This buffer can be overcome, though, and some potentially dangerous situations can arise. If a person exercises strenuously, lactic acid from the muscles is released into the bloodstream. If there's not enough bicarbonate ion to neutralize the lactic acid, the blood pH drops, and the person is said to be in *acidosis*. Diabetes may also cause acidosis. On the other hand, if a person hyperventilates (breathes too fast), she breathes out too much carbon dioxide. The carbonic acid level in the blood is reduced, causing the blood to become too basic. This condition, called *alkalosis,* can be very serious.

Amphoteric species may also act as buffers by reacting with an acid or a base. (For an example of an amphoteric species, see "Make up your mind: Amphoteric water," earlier in this chapter) The bicarbonate ion (HCO_3^-) and the monohydrogen phosphate ion (HPO_4^{-2}) are amphoteric species that neutralize both acids and bases. Both of these ions are also important in controlling the blood's pH.

Antacids: Good, Basic Chemistry

Go to any drugstore or grocery store and look at the shelves upon shelves of antacids. They represent acid-base chemistry in action!

The stomach secretes hydrochloric acid in order to activate certain enzymes (biological catalysts) in the digestion process. But sometimes the stomach produces too much acid, or the acid makes its way up into the esophagus (leading to heartburn), making it necessary to neutralize the excess acid with — you guessed it — a base. The basic formulations that are sold to neutralize this acid are called *antacids.* Antacids include the following compounds as active ingredients:

- **Bicarbonates** — $NaHCO_3$ and $KHCO_3$
- **Carbonates** — $CaCO_3$ and $MgCO_3$
- **Hydroxides** — $Al(OH)_3$ and $Mg(OH)_2$

Acids with bad press: An introduction to acid rain

Over the past few years, acid rain has emerged as a great environmental problem. Natural rainwater is somewhat acidic (around pH 5.6) due to the absorption of carbon dioxide from the atmosphere and the creation of carbonic acid. However, when acid rain is mentioned in the press, it usually refers to rain in the pH 3 to 3.5 range.

The two major causes of acid rain are automotive and industrial pollution. In the automobile's internal combustion engine, nitrogen in the air is oxidized to various oxides of nitrogen. These nitrogen oxides, when released into the atmosphere, react with water vapor to form nitric acid (HNO_3).

In fossil fuel power plants, oxides of sulfur are formed from the burning of the sulfur impurities commonly found in coal and petroleum. These oxides of sulfur, if released into the atmosphere, combine with water vapor to form both sulfuric and sulfurous acids (H_2SO_4 and H_2SO_3). Oxides of nitrogen are also produced in these power plants.

These acids fall to earth in the rain and cause a multitude of problems. They dissolve the calcium carbonate of marble statues and monuments. They decrease the pH of lake water to such a degree that fish can no longer live in the lakes. They cause whole forests to die or become stunted. They react with the metals in cars and buildings.

Industrial controls have been somewhat effective in reducing the problem, but it's still a major environmental issue. (See Chapter 18 for more info about acid rain.)

Trying to select the "best" antacid for occasional use can be complicated. Certainly price is a factor, but the chemical nature of the bases can also be a factor. For example, individuals with high blood pressure may want to avoid antacids containing sodium bicarbonate because the sodium ion tends to increase blood pressure. Individuals concerned about loss of calcium from the bones, or *osteoporosis,* may want to use an antacid containing calcium carbonate. However, both calcium carbonate and aluminum hydroxide can cause constipation if used in large doses. On the other hand, large doses of both magnesium carbonate and magnesium hydroxide can act as laxatives. Selecting an antacid can really be a balancing act!

Chapter 13

Balloons, Tires, and Scuba Tanks: The Wonderful World of Gases

In This Chapter

- Accepting the Kinetic Molecular Theory of Gases
- Figuring out pressure
- Coming to understand the gas laws

Gases are all around you. Because gases are generally invisible, you may not think of them directly, but you're certainly aware of their properties. You breathe a mixture of gases that you call air. You check the pressure of your automobile tires, and you check the atmospheric pressure to see if a storm is coming. You burn gases in your gas grill and lighters. You fill birthday balloons for your loved ones.

The properties of gases and their interrelationships are important to you. Is there enough pressure in my tires? How big is that balloon going to be? Is there enough air in my scuba tanks? The list goes on and on.

In this chapter, I introduce you to gases at both the microscopic and macroscopic levels. I show you one of science's most successful theories — the Kinetic Molecular Theory of Gases. And I explain the macroscopic properties of gases and show you the important interrelationships among them. I also show you how these relationships come into play in reaction stoichiometry. This chapter is a real gas!

Microscopic View of Gases: The Kinetic Molecular Theory

A theory is useful to scientists if it describes the physical system they're examining and allows them to predict what will happen if they change some variable. The Kinetic Molecular Theory of Gases does just that. It has limitations — all theories do — but it's one of the most useful theories in chemistry. This section describes the theory's basic *postulates* — assumptions, hypotheses, axioms (pick your favorite word) you can accept as being true.

- **Postulate 1: Gases are composed of tiny particles, either atoms or molecules.**

  Unless you're discussing matter at greatly elevated temperatures, the particles referred to as gases tend to be relatively small. The more massive particles clump together to form liquids or even solids. So gases are normally small with relatively low atomic and molecular weights.

- **Postulate 2: The gas particles are so small when compared to the distances between them that the volume the gas particles themselves take up is negligible and is assumed to be zero.**

  These gas particles do take up some volume — that's one of the properties of matter. But if the gas particles are small (which they are), and there aren't many of them in a container, you say that their volume is negligible when compared to the volume of the container or the space between the gas particles. This explains why gases are compressible. There's a lot of space between the gas particles, so they can be squeezed together. This isn't true in solids and liquids, where the particles are *much* closer together. (Chapter 2 covers the various states of matter, if you want to have a look-see at the differences between solids, liquids, and gases.)

  The concept of a quantity being *negligible* is used a lot in chemistry. An example is in Chapter 12 where I use the acid ionization constant (K_a) of a weak acid, ignoring the amount of weak acid that has ionized compared to the initial concentration of the acid.

  In the real world, I like to compare this negligible concept to finding a dollar in the street. If you have no money at all, then that dollar represents a sizable quantity of cash (perhaps your next meal). But if you're a multimillionaire, then that dollar doesn't represent much at all. It may as well be a piece of scrap paper. You may not even pick it up. (I really can't imagine being that rich.) Its value is negligible when compared to the rest of your wealth. This is what I'm saying with respect to the volume of the gas particles — sure, they have a volume, but it's so small that it's insignificant when compared with the distance between the gas particles and the volume of the container.

- **Postulate 3: The gas particles are in constant random motion, moving in straight lines and colliding with the inside walls of the container.**

  The gas particles are always moving in a straight-line motion. (Gases have a higher kinetic energy — energy of motion — associated with them when compared to solids or liquids; see Chapter 2.) They continue to move in these straight lines until they collide with something — either with each other or with the inside walls of the container. The particles also all move in different directions, so the collisions with the inside walls of the container tend to be uniform over the entire inside surface. You can observe this uniformity by simply blowing up a balloon. The balloon is spherical because the gas particles are hitting all points of the inside walls the same. The collision of the gas particles with the inside walls of the container is called *pressure*. The idea that the gas particles are in constant, random, straight-line motion explains why gases uniformly mix if put in the same container. It also explains why, when you drop a bottle of cheap perfume at one end of the room, the people at the other end of the room are able to smell it right away.

- **Postulate 4: The gas particles are assumed to have negligible attractive or repulsive forces between each other.**

  In other words, the gas particles are assumed to be totally independent, neither attracting nor repelling each other. That said, it's hair-splitting time. This statement is actually false; if it were true, chemists would never be able to liquefy a gas, which they can. But the reason you can accept this assumption as true is that the attractive and repulsive forces are generally so small that they can safely be ignored. The assumption is most valid for nonpolar gases, such as hydrogen and nitrogen, because the attractive forces involved are London forces. However, if the gas molecules are polar, as in water and HCl, this assumption can become a problem. (Turn to Chapter 7 for the scoop about London forces and polar things — all related to the attraction between molecules.)

- **Postulate 5: The gas particles may collide with each other. These collisions are assumed to be elastic, with the total amount of kinetic energy of the two gas particles remaining the same.**

  Not only do the gas particles collide with the inside walls of the container, they also collide with each other. If they hit each other, no kinetic energy is lost, but kinetic energy may be transferred from one gas particle to the other. For example, imagine two gas particles — one moving fast and the other moving slow — colliding. Kinetic energy is transferred from the faster particle to the slower particle. The one that's moving slow bounces off the faster particle and moves away at a greater speed than before, while the one that's moving fast bounces off the slower particle and moves away at a slower speed. The total amount of kinetic energy remains the same, but one gas particle loses energy and the other gains energy. This is the principle behind pool — you transfer kinetic energy from your moving pool stick to the cue ball to the ball you're aiming at.

- **Postulate 6: The Kelvin temperature is directly proportional to the *average* kinetic energy of the gas particles.**

  The gas particles aren't all moving with the same amount of kinetic energy. A few are moving relatively slow and a few are moving very fast, but most are somewhere in between these two extremes. Temperature, particularly as measured using the Kelvin temperature scale, is directly related to the *average* kinetic energy of the gas. If you heat the gas so that the Kelvin temperature (K) increases, the average kinetic energy of the gas also increases. (To calculate the Kelvin temperature, add 273 to the Celsius temperature: K = °C + 273. Temperature scales and average kinetic energy are all tucked neatly into Chapter 2.)

A gas that obeys all the postulates of the Kinetic Molecular Theory is called an *ideal gas.* Obviously, no real gas obeys the assumptions made in the second and fourth postulates *exactly.* But a nonpolar gas at high temperatures and low pressure (concentration) approaches ideal gas behavior.

I'm Under Pressure — Atmospheric Pressure, That Is

Although you're not in a container, the gas molecules of the atmosphere are constantly hitting you, your books, your computer, and everything, and exerting a force called *atmospheric pressure.* Atmospheric pressure is measured using an instrument called a *barometer.*

Measuring atmospheric pressure: The barometer

If you get a complete weather report, the atmospheric pressure is normally included. You can get an idea about changes in the weather by observing whether the atmospheric pressure is rising or falling. The atmospheric pressure is measured using a barometer, and Figure 13-1 shows the components of one.

A barometer is composed of a long glass tube that's closed at one end and totally filled with a liquid. You can use water, but the tube would have to be *very* long (about 35 feet long), making for a rather inconvenient barometer. So it makes more sense to use mercury because it's a very dense liquid. The tube filled with mercury is inverted into an open container of mercury so that the open end of the tube is under the surface of the mercury in the container. A couple of things now take place. The force of gravity pulls the mercury in the tube *down,* causing it to drain out into the container. The weight of the gases in the atmosphere exert a force downward on the mercury in the open container and force it *up* into the tube. Sooner or later, these forces balance, and the mercury in the tube comes to rest at a certain height from the top of the pool of mercury in the container. The greater the pressure of the atmosphere, the higher the mercury column that can be measured in the tube; the lower the pressure of the atmosphere (for example, at the top of a tall mountain), the shorter the column. At sea level, the column is 760 millimeters high, the so-called normal *atmospheric pressure.*

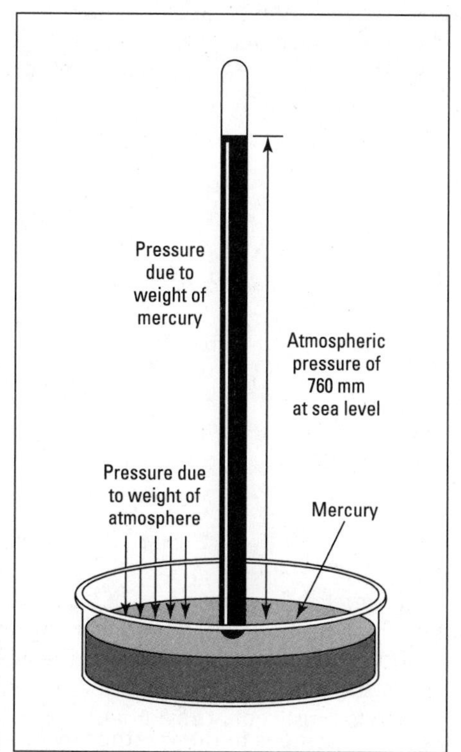

Figure 13-1: A barometer.

Atmospheric pressure can be expressed a number of different ways. It can be expressed in millimeters of mercury (mm Hg); atmospheres (atm), a unit of pressure where 1 atmosphere is the pressure at sea level; torr, a unit of pressure where 1 torr equals 1 millimeter of mercury; pounds per square inch (psi); pascals (Pa), a unit of pressure where 1 pascal equals 1 newton per square meter (don't worry about what a newton is; just trust me that this stuff is a way to express pressure); or kilopascals (kPa), where 1 kilopascal equals 1,000 pascals.

So you can express the atmospheric pressure at sea level as

$$760 \text{ mm Hg} = 1 \text{ atm} = 760 \text{ torr} = 14.69 \text{ psi} = 101{,}325 \text{ Pa} = 101.325 \text{ kPa}$$

Note that sometimes you also hear atmospheric pressure reported as inches of mercury (1 atm = 29.921 in Hg). In this book, I primarily use atmospheres and torr, with an occasional millimeter of Hg. Variety is the spice of life.

Measuring confined gas pressure: The manometer

You can measure the pressure of a gas confined in a container by using an apparatus called a *manometer.* (It's pronounced *man-AH-muh-ter,* not *man-o-meter.*) Figure 13-2 shows the components of a manometer.

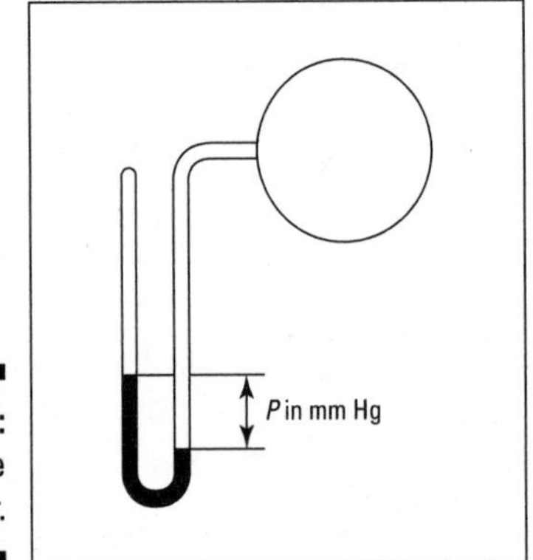

Figure 13-2: The manometer.

A manometer is kind of like a barometer. The container of gas is attached to a U-shaped piece of glass tubing that's partially filled with mercury and sealed at the other end. Gravity pulls down the mercury column at the closed end. The mercury is then balanced by the pressure of the gas in the container. The difference in the two mercury levels represents the amount of gas pressure.

Gases Obey Laws, Too — Gas Laws

Various scientific laws describe the relationships between four of the important physical properties of gases:

- Volume
- Pressure
- Temperature
- Amount

This section covers those various laws. Boyle's, Charles's, and Gay-Lussac's Laws each describe the relationship between two properties while keeping the other two properties constant. (In other words, you take two properties, change one, and then see its effect on the second — while keeping the remaining properties constant.) Another law — a combo of Boyle's, Charles's, and Gay-Lussac's individual laws — enables you to vary more than one property at a time.

But that combo law doesn't let you vary the physical property of amount. Avogadro's Law, however, does. And there's even an ideal gas law, which lets you take into account variations in all four physical properties.

Yes, this section is so chock-full of laws, it'll probably *give* you gas just trying to digest it.

Boyle's Law: Nothing to do with boiling

Boyle's Law, named after Robert Boyle, a seventeenth-century English scientist, describes the pressure-volume relationship of gases if the temperature and amount are kept constant. Figure 13-3 illustrates the pressure-volume relationship using the Kinetic Molecular Theory.

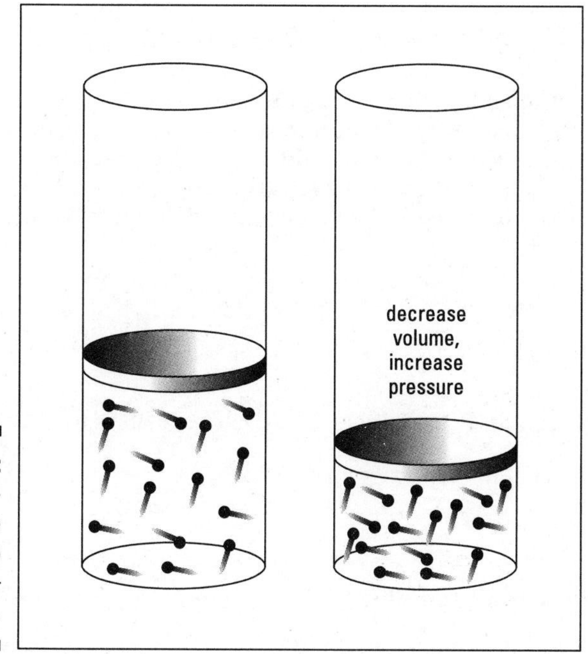

Figure 13-3: Pressure-volume relationship of gases — Boyle's Law.

The left-hand cylinder in the figure contains a certain volume of gas at a certain pressure. (*Pressure* is the collision of the gas particles with the inside walls of the container.) When the volume is decreased, the same number of gas particles are now contained in a much smaller volume and the number of collisions increases significantly. Therefore, the pressure is greater.

Boyle's Law states that there's an inverse relationship between the volume and pressure. As the volume decreases, the pressure increases, and vice versa.

Boyle determined that the product of the pressure and the volume is a constant (k):

$$PV = k$$

Now consider a case where you have a gas at a certain pressure (P_1) and volume (V_1). If you change the volume to some new value (V_2), the pressure also changes to a new value (P_2). You can use Boyle's Law to describe both sets of conditions:

$$P_1V_1 = k$$

$$P_2V_2 = k$$

The constant, *k,* is going to be the same in both cases. So you can say

$$P_1V_1 = P_2V_2 \qquad \text{(with temperature and amount constant)}$$

This equation is another statement of Boyle's Law — and it's really a more useful one, because you'll normally deal with changes in pressure and volume. If you know three of the preceding quantities, you can calculate the fourth one. For example, suppose that you have 5.00 liters of a gas at 1.00 atm pressure, and then you decrease the volume to 2.00 liters. What's the new pressure?

To find the answer, use the following setup:

$$P_1V_1 = P_2V_2$$

Substituting 1.00 atm for P_1, 5.00 liters for V_1, and 2.00 liters for V_2, you get

$$(1.00 \text{ atm})(5.00 \text{ liters}) = P_2(2.00 \text{ liters})$$

Now solve for P_2:

$$(1.00 \text{ atm})(5.00 \text{ liters})/2.00 \text{ liters} = P_2 = 2.50 \text{ atm}$$

The answer makes sense, because you decreased the volume and the pressure increased, which is exactly what Boyle's law says.

Charles's Law: Don't call me Chuck

Charles's Law, named after Jacques Charles, a nineteenth-century French chemist, has to do with the relationship between volume and temperature, keeping the pressure and amount constant. You run across situations dealing with this relationship in everyday life, especially in terms of the heating and cooling of balloons.

Figure 13-4 shows the temperature-volume relationship.

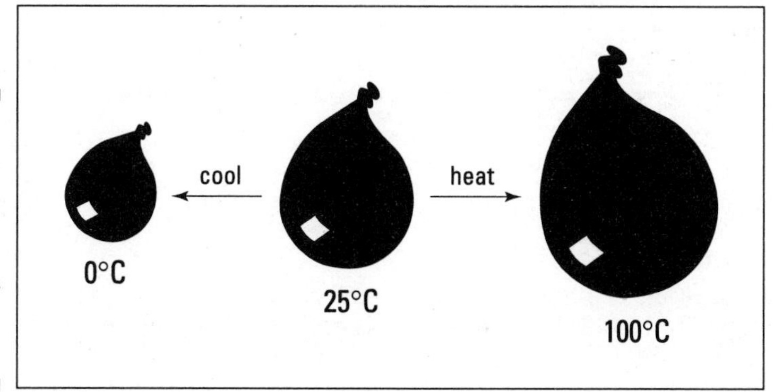

Figure 13-4: The temperature-volume relationship of gases — Charles's Law.

Look at the balloon in the middle of Figure 13-4. What do you think would happen to the balloon if you placed it in the freezer? It'd get smaller. Inside the freezer, the external pressure, or atmospheric pressure, is the same, but the gas particles inside the balloon aren't moving as fast, so the volume shrinks to keep the pressure constant. If you heat the balloon, the balloon expands and the volume increases. This is a *direct relationship* — as the temperature increases, the volume increases, and vice versa.

Jacques Charles developed the mathematical relationship between temperature and volume. He also discovered that you must use the Kelvin (K) temperature when working with gas law expressions.

In gas law calculations, the Kelvin temperature *must* be used.

Charles's Law says that the volume is directly proportional to the Kelvin temperature. Mathematically, the law looks like this:

$V = bT$ or $V/T = b$ (where *b* is a constant)

If the temperature of a gas with a certain volume (V_1) and Kelvin temperature (T_1) is changed to a new Kelvin temperature (T_2), the volume also changes (V_2).

$V_1/T_1 = b \qquad V_2/T_2 = b$

The constant, *b,* is the same, so

$V_1/T_1 = V_2/T_2$ (with the pressure and amount of gas held constant and temperature expressed in K)

If you have three of the quantities, you can calculate the fourth. For example, suppose you live in Alaska and are outside in the middle of winter, where the temperature is –23 degrees Celsius. You blow up a balloon so that it has a volume of 1.00 liter. You then take it inside your home, where the temperature is a toasty 27 degrees Celsius. What's the new volume of the balloon?

First, convert your temperatures to Kelvin by adding 273 to the Celsius temperature:

$-23\ °C + 273 = 250\ K$ (outside)

$27\ °C + 273 = 300\ K$ (inside)

Now you can solve for V_2, using the following setup:

$V_1/T_1 = V_2/T_2$

Multiply both sides by T_2 so that V_2 is on one side of the equation by itself:

$[V_1T_2]/T_1 = V_2$

Then substitute the values to calculate the following answer:

$[(1.00\ \text{liter})(300\ K)]/250\ K = V_2 = 1.20\ \text{liters}$

It's a reasonable answer, because Charles's Law says that if you increase the Kelvin temperature, the volume increases.

Gay-Lussac's Law

Gay-Lussac's Law, named after the 19th century French scientist Joseph-Louis Gay-Lussac deals with the relationship between the pressure and temperature of a gas if its volume and amount are held constant. Imagine, for example, that you have a metal tank of gas. The tank has a certain volume, and the gas inside has a certain pressure. If you heat the tank, you increase the kinetic energy of the gas particles. So they're now moving much faster, and they're not only hitting the inside walls of the tank more often but also with more force. The pressure has increased.

Gay-Lussac's Law says that the pressure is directly proportional to the Kelvin temperature. Figure 13-5 shows this relationship.

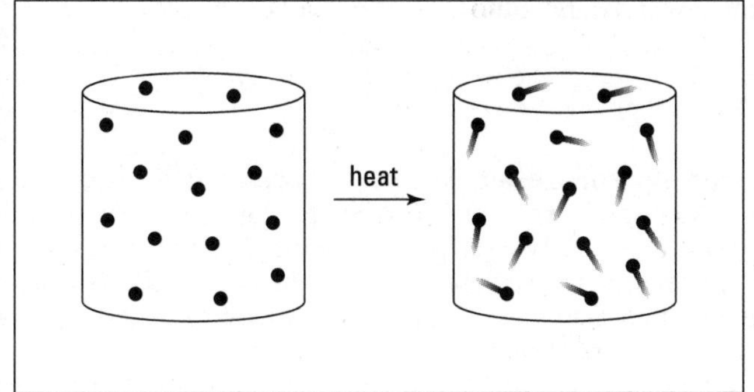

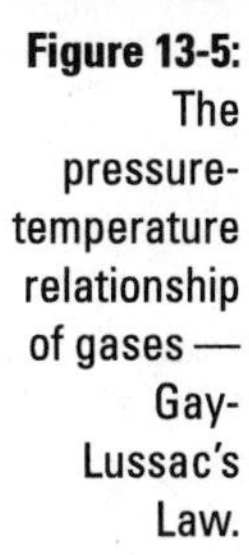

Figure 13-5: The pressure-temperature relationship of gases — Gay-Lussac's Law.

Mathematically, Gay-Lussac's Law looks like this:

$P = kT$ (or $P/T = k$ at constant volume and amount)

Consider a gas at a certain Kelvin temperature and pressure (T_1 and P_1), with the conditions being changed to a new temperature and pressure (T_2 and P_2):

$$P_1/T_1 = P_2/T_2$$

If you have a tank of gas at 800 torr pressure and a temperature of 250 Kelvin, and it's heated to 400 Kelvin, what's the new pressure?

Starting with $P_1/T_1 = P_2/T_2$, multiply both sides by T_2 so you can solve for P_2:

$$[P_1T_2]/T_1 = P_2$$

Now substitute the values to calculate the following answer:

$$[(800 \text{ torr})(400 \text{ K})]/250 \text{ K} = P_2 = 1{,}280 \text{ torr}$$

This is a reasonable answer because if you heat the tank, the pressure should increase.

The combined gas law

All the preceding examples assume that two properties are held constant and one property is changed to see its effect on a fourth property. But life is rarely that simple. How do you handle situations in which two or even three properties change? You can treat each one separately, but it sure would be nice if you had a way to combine things so that wouldn't be necessary.

Actually, there is a way. You can combine Boyle's Law, Charles's Law, and Gay-Lussac's Law into one equation. Trust me, you don't want me to show you exactly how it's done, but the end result is called the *combined gas law,* and it looks like this:

$$P_1V_1/T_1 = P_2V_2/T_2$$

Just like in the preceding examples, *P* is the pressure of the gas (in atm, mm Hg, torr, and so on), *V* is the volume of the gas (in appropriate units), and *T* is the temperature (in Kelvin). The *1* and *2* stand for the initial and final conditions, respectively. The amount is still held constant: No gas is added, and no gas escapes. There are six quantities involved in this combined gas law; knowing five allows you to calculate the sixth.

For example, suppose that a weather balloon with a volume of 25.0 liters at 1.00 atm pressure and a temperature of 27 degrees Celsius is allowed to rise to an altitude where the pressure is 0.500 atm and the temperature is –33 degrees Celsius. What's the new volume of the balloon?

Before I show you how to work this problem, do a little reasoning. The temperature is decreasing, so that should cause the volume to decrease (Charles's Law). However, the pressure is also decreasing, which should cause the balloon to expand (Boyle's Law). These two factors are competing, so at this point, you don't know which will win out.

You're looking for the new volume (V_2), so rearrange the combined gas law to obtain the following equation (by multiplying each side by T_2 and dividing each side by P_2, which puts V_2 by itself on one side):

$$[P_1V_1T_2]/[P_2T_1] = V_2$$

Now identify your quantities:

P_1 = 1.00 atm; V_1 = 25.0 liters; T_1 = 27°C + 273 = 300. K

P_2 = 0.500 atm; T_2 = –33°C + 273 = 240. K

Now substitute the values to calculate the following answer:

[(1.00 atm)(25.0 liters)(240. K)]/[(0.500 atm)(300. K)] = V_2 = 40.0 liters

Because the volume increased overall in this case, Boyle's Law had a greater effect than Charles's Law.

Avogadro's Law

The combined gas equation gives you a way to calculate changes involving pressure, volume, and temperature. But you still have the problem of amount to deal with. In order to account for amount, you need to know another law.

Amedeo Avogadro (the same Avogadro that gave us his famous number of particles per mole — see Chapter 10) determined, from his study of gases, that equal volumes of gases at the same temperature and pressure contain equal numbers of gas particles. So Avogadro's Law says that the volume of a gas is directly proportional to the number of moles of gas (number of gas particles) at a constant temperature and pressure. Mathematically, Avogadro's Law looks like this:

$$V = kn \text{ (at constant temperature and pressure)}$$

In this equation, *k* is a constant and *n* is the number of moles of gas. If you have a number of moles of gas (n_1) at one volume (V_1), and the moles change due to a reaction (n_2), the volume also changes (V_2), giving you the equation

$$V_1/n_1 = V_2/n_2$$

I'm not going to work any problems with this law, because it's basically the same idea as the other gas laws covered in this chapter.

A very useful consequence of Avogadro's Law is that the volume of a mole of gas can be calculated at any temperature and pressure. An extremely useful form to know when calculating the volume of a mole of gas is *1 mole of any gas at STP occupies 22.4 liters. STP* in this case is not an oil or gas additive. It stands for Standard Temperature and Pressure.

- **Standard Pressure:** 1.00 atm (760 torr or mm Hg)
- **Standard Temperature:** 273 K

This relationship between moles of gas and liters gives you a way to convert the gas from a mass to a volume. For example, suppose that you have 50.0 grams of oxygen gas (O_2), and you want to know its volume at STP. You can set up the problem like this (see Chapters 10 and 11 for the nuts and bolts of using moles in chemical equations):

$$\frac{50.0\,\cancel{g}\,O_2}{1} \times \frac{1\,mol\,\cancel{O_2}}{32.0\,\cancel{g}} \times \frac{22.4\,L}{1\,mol\,\cancel{O_2}} = 35.0\,L$$

You now know that the 50.0 grams of oxygen gas occupies a volume of 35.0 liters at STP. But what if the gas is not at STP? What's the volume of 50.0 grams of oxygen at 2.00 atm and 27.0 degrees Celsius? In the next section, I show you a really easy way of doing this problem. But right now, you can use the combined gas law, because you know the volume at STP:

$$P_1V_1/T_1 = P_2V_2/T_2$$

$P_1 = 1.00$ atm; $V_1 = 35.0$ liters; $T_1 = 273$ K

$P_2 = 2.00$ atm; $T_2 = 300.$ K (27 °C + 273)

Solving for V_2, you calculate the following answer:

$[P_1V_1T_2]/[P_2T_1] = V_2$

$[(1.00\ atm)(35.0\ liters)(300\ K)]/[(2.00\ atm)(273\ K)] = V_2 = 19.2\ liters$

The ideal gas equation

If you take Boyle's Law, Charles's Law, Gay-Lussac's Law, and Avogadro's Law and throw them into a blender, turn the blender on high for a minute, and then pull them out, you get the *ideal gas equation* — a way of working in volume, temperature, pressure, *and* amount. The ideal gas equation has the following form:

$PV = nRT$

The *P* represents pressure in atmospheres (atm), the *V* represents volume in liters (L), the *n* represents moles of gas, the *T* represents the temperature in Kelvin (K), and the *R* represents the ideal gas constant, which is 0.0821 liters atm/K mol.

Using the value of the ideal gas constant, the pressure must be expressed in atm, and the volume must be expressed in liters. You can calculate other ideal gas constants if you really want to use torr and milliliters, for example, but why bother? It's easier to memorize one value for *R* and then remember to express the pressure and volume in the appropriate units. Naturally, you'll *always* express the temperature in Kelvin when working any kind of gas law problem.

That said, now I want to show you an easy way to convert a gas from a mass to a volume if the gas is not at STP. What's the volume of 50.0 grams of oxygen at 2.00 atm and 27.0 degrees Celsius?

The first thing you have to do is convert the 50.0 grams of oxygen to moles using the molecular weight of O_2:

$(50.0\ grams)(1\ mol/32.0\ grams) = 1.562\ mol$

Now take the ideal gas equation and rearrange it so you can solve for V:

$PV = nRT$

$V = nRT/P$

Add your known quantities to calculate the following answer:

$V = [(1.562\ mol)(0.0821\ liters\ atm/K\ mol)(300\ K)]/2.00\ atm = 19.2\ liters$

This is the exact same answer you get in the preceding section, but it's calculated in a much more straightforward way.

Stoichiometry and the Gas Laws

The ideal gas equation (and even the combined gas equation) allows chemists to work stoichiometry problems involving gases. (Chapter 10 is your key to the world of stoichiometry.) In this section, you're going to use the ideal gas equation to do such a problem, using a classic chemistry experiment — the decomposition of potassium chlorate to potassium chloride and oxygen by heating:

$$2\ KClO_3(s) \rightarrow 2\ KCl(s) + 3\ O_2(g)$$

Here's your mission: Figure out the volume of oxygen gas produced at 700 torr and 27 degrees Celsius from the decomposition of 25.0 grams of $KClO_3$.

First, you need to calculate the number of moles of oxygen gas produced:

$$\frac{50.0g\ O_2}{1} \times \frac{50.0g\ O_2}{1} \times \frac{3\ mol\ O_2}{1} = 0.3059\ mol\ O_2$$

Next, convert the temperature to Kelvin and the pressure to atm:

27 °C + 273 = 300 K

700 torr/760 torr/atm = 0.9211 atm

Now put everything in the ideal gas equation:

PV = nRT

V = nRT/P

V = [(0.3059 mol)(0.0821 L·atm/K·mol)(300 K)]/0.9211 atm = 8.18 liters

Mission accomplished.

Dalton's and Graham's Laws

This section covers a couple of miscellaneous but fine gas laws you should have a nodding acquaintance with. One relates to partial pressures and the other to gaseous effusion/diffusion. Party on.

Dalton's Law

Dalton's Law of partial pressures says that in a mixture of gases, the total pressure is the sum of the partial pressures of each individual gas.

If you have a mixture of gases — gas A, gas B, gas C, and so on — then the total pressure of the system is simply the sum of the pressures of the individual gases. Mathematically, the relationship can be expressed like this:

$$P_{Total} = P_A + P_B + P_C + \ldots$$

When working stoichiometry problems like the one in the preceding section involving the decomposition of potassium chlorate, the oxygen is normally collected over water by displacement and the volume is then measured. However, in order to get the pressure of just the oxygen, you have to subtract the pressure due to the water vapor. You have to mathematically "dry out" the gas.

Suppose, for example, that a sample of oxygen is collected over water at a total pressure of 755 torr at 20 degrees Celsius. And suppose that your job, you lucky dawg, is to calculate the pressure of the oxygen.

You know that the total pressure is 755 torr. Your first task is to reference a table of vapor pressures of water versus temperature. (You can find such a table in a variety of places, such as the *Chemical Rubber Company (CRC) Handbook.*) After looking at the table, you determine that the partial pressure of water at 20 degrees Celsius is 17.5 torr. Now you're ready to calculate the pressure of the oxygen:

$$P_{Total} = P_{Oxygen} + P_{water\ vapor}$$

$$755\text{ torr} = P_{oxygen} + 17.5\text{ torr}$$

$$P_{oxygen} = 755\text{ torr} - 17.5\text{ torr} = 737.5\text{ torr}$$

Knowing the partial pressure of gases like oxygen is important in deep sea diving and the operation of respirators in hospitals.

Graham's Law

Place a few drops of a strong perfume on a table at one end of a room, and soon people at the other end of the room can smell it. This process is called *gaseous diffusion,* the mixing of gases due to their molecular motion.

Place a few drops of that same perfume inside an ordinary rubber balloon and blow it up. Very soon you'll be able to smell the perfume outside of the balloon as it makes its way through the microscopic pores of the rubber. This process is called *gaseous effusion,* the movement of a gas through a tiny opening. The same process of effusion is responsible for the helium being quickly lost from rubber balloons.

Thomas Graham determined that the rates of diffusion and effusion of gases are inversely proportional to the square roots of their molecular or atomic weights. This is Graham's Law. In general, it says that the lighter the gas, the faster it will effuse (or diffuse). Mathematically, Graham's Law looks like this:

$$\frac{V_1}{V_2} = \sqrt{\frac{M_2}{M_1}}$$

Suppose that you fill two rubber balloons to the same size, one with hydrogen (H_2) and the other with oxygen (O_2). The hydrogen, being lighter, should effuse through the balloon pores faster. But how much faster? Using Graham's Law, you can determine the answer:

$$\frac{V_{H_2}}{V_{O_2}} = \sqrt{\frac{M_{O_2}}{M_{H_2}}}$$

$$\frac{V_{H_2}}{V_{O_2}} = \sqrt{\frac{32.0\,glmol}{2.0\,glmol}}$$

$$\frac{V_{H_2}}{V_{O_2}} = \sqrt{16}$$

$$\frac{V_{H_2}}{V_{O_2}} = 4$$

The hydrogen should effuse out four times as fast as the oxygen.

Part IV

Chemistry in Everyday Life: Benefits and Problems

In this part . . .

Chemistry isn't just something that's done in an academic or industrial lab. Professional chemists aren't the only individuals who do chemistry. *You* do chemistry, too. Chemistry touches your life each and every day.

Chemistry gives us great benefits, but it can also give us great problems. Our modern society is complex. Chemistry holds the promise of solving many of the problems facing society, making our lives easier and more meaningful.

In the chapters of this part, I show you some applications of chemistry. I cover the chemistry of carbon and show you how it applies to petroleum and the process of making gasoline. I show you how that very same petroleum can be used to make plastics and synthetic fibers. I zip you home to look at the chemistry behind cleaners and detergents, medicines, and cosmetics of all kinds. And I show you some problems that society, technology, and science have created — air and water pollution.

Chapter 14

The Chemistry of Carbon: Organic Chemistry

In This Chapter

- Taking a look at hydrocarbons
- Seeing how to name some simple hydrocarbons
- Checking out the different functional groups
- Discovering organic chemistry's place in society

The largest and most systematic area of chemistry is *organic chemistry,* the chemistry of carbon. Of the 11 to 12 million chemical compounds known, about 90 percent are organic compounds. We burn organic compounds as fuel. We eat organic compounds. We wear organic compounds. We're made of organic compounds. Our whole world is built of organic compounds.

In this chapter, I give you a brief introduction to organic chemistry. I spend some time showing you the *hydrocarbons,* compounds of carbon and hydrogen, as well as some other classes of organic compounds and their uses in everyday life. As you read this chapter, you'll find that a lot of chemistry can be found in carbon.

Organic synthesis: Where it all began

In the early years of chemistry, it was thought that organic compounds could only be produced from living organisms. People thought that there had to be a "vital force" involved. But in 1828, the German scientist Friedrich Wohler changed the field of chemistry forever by developing an organic compound, urea, by accident while trying to make an inorganic compound. This was the beginning of our modern field of organic synthesis.

Hydrocarbons: From Simple to Complex

A natural question chemistry students ask is "Why are there so many compounds of carbon?" The answer: Carbon contains four valence electrons and so can form four covalent bonds to other carbons or elements. (A common mistake organic chemistry students make when drawing structures is not ensuring that every carbon has four bonds attached to it.) The bonds that carbon forms are strong covalent bonds (Chapter 7 covers covalent bonds), and carbon has the ability to bond to itself in long chains and rings. It can form double and triple bonds to another carbon or to another element. No other element, with the possible exception of silicon, has this ability. (And the bonds silicon makes aren't nearly as strong as carbon's.) These properties allow carbon to form the vast multitude of compounds needed to make an amoeba or a butterfly or a baby.

The simplest organic compounds are called the *hydrocarbons,* compounds composed of carbon and hydrogen. Economically, the hydrocarbons are extremely important to us — primarily as fuels. Gasoline is a mixture of hydrocarbons. We use methane (natural gas) and propane and butane, all hydrocarbons, for their ability to burn and release a large amount of energy. Hydrocarbons may contain only single bonds (the alkanes) or double bonds (the alkenes) or triple bonds (the alkynes). And they may form rings containing single or double bonds (cycloalkanes, cycloalkenes, and aromatics).

Even compounds containing only carbon and hydrogen have a great deal of diversity; imagine what can happen when a few more elements are mixed in!

From gas grills to gasoline: Alkanes

The simplest of the hydrocarbons are the *alkanes*. Alkanes are called *saturated* hydrocarbons — that is, each carbon is bonded to four other atoms. Carbon can form a maximum of four covalent bonds. If those four covalent bonds are to different atoms, then chemists say that the carbon is saturated. There are no double or triple bonds in the alkanes.

Alkanes have the general formula of C_nH_{2n+2}, where *n* is a whole number. If n = 1, then there are four hydrogen atoms, and the result is CH_4, methane.

Table 14-1 lists the names of the first ten *normal,* or *straight-chained,* alkanes. They really aren't straight; that's just what they're called. When I draw the structures, though, I often show them in a straight line. (Technically, they're carbon bonds in a tetrahedral fashion with bond angles of 109.5 degrees. See Chapter 7 for a discussion of this stuff, called *molecular geometry.*) Every carbon, except the end ones, is bonded to two other carbons. Figure 14-1 shows models of the first four listed in the table.

Table 14-1 The First Ten Normal Alkanes (C_nH_{2n+2})

n	*Formula*	*Name*
1	CH_4	Methane
2	C_2H_6	Ethane
3	C_3H_8	Propane
4	C_4H_{10}	Butane
5	C_5H_{12}	Pentane
6	C_6H_{14}	Hexane
7	C_7H_{16}	Heptane
8	C_8H_{18}	Octane
9	C_9H_{20}	Nonane
10	$C_{10}H_{22}$	Decane

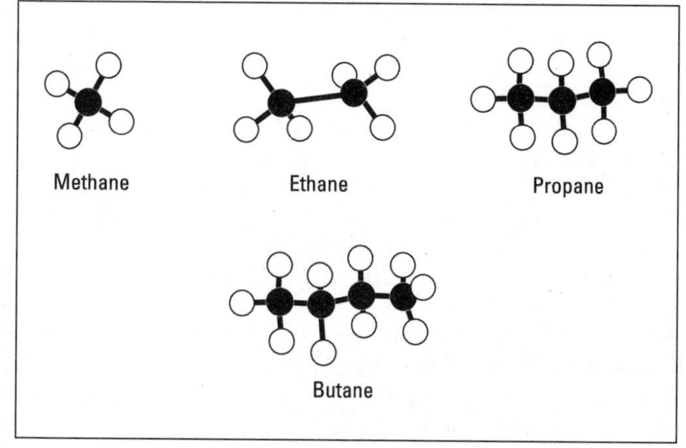

Figure 14-1: The first four alkanes.

Molecular and structural formulas

Table 14-1 shows the molecular formula of some of the alkanes. The *molecular formula* shows what atoms are present in the compound and the actual number of each. They're all normal, or straight-chained, hydrocarbons, but the bonding pattern can be illustrated better by using a structural formula. The *structural formula* shows the atoms present, the actual number of each, and the bonding pattern, or what is bonded to what.

The structural formula can be shown in a number of different ways. One is called the *expanded structural formula,* which basically shows each covalent bond as a line. With organic compounds such as the hydrocarbons, if you're really just interested in showing the way the carbons are bonded, you can omit the hydrogen atoms on the expanded form and just indicate them by the covalent-bond line. You can also use a condensed form, which groups parts of the molecule and still indicates the bonding pattern. The condensed form can be done several ways. Figure 14-2 shows a couple of expanded and three condensed forms of the structural formula of butane, C_4H_{10}.

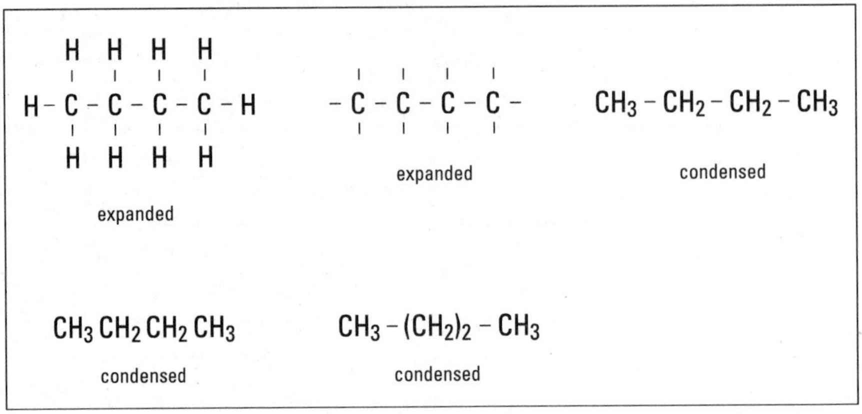

```
    H   H   H   H
    |   |   |   |
H - C - C - C - C - H        - C - C - C - C -
    |   |   |   |              |   |   |   |
    H   H   H   H
                                  expanded
      expanded
```

$CH_3 - CH_2 - CH_2 - CH_3$

condensed

$CH_3 CH_2 CH_2 CH_3$

condensed

$CH_3 - (CH_2)_2 - CH_3$

condensed

Figure 14-2: Structural formulas of butane.

Naming problems

Sometimes two entirely different compounds — with two entirely different sets of properties — have the same molecular structure. The difference is in the way the atoms are bonded — what's bonded to what. These types of compounds are called *isomers,* compounds that have the same molecular formula but different structural formulas. Simply knowing the molecular formula isn't enough to distinguish between them.

An isomer of butane, for example, has the same molecular formula as the straight-chained compound shown in Figure 14-2, C_4H_{10}, but a different bonding pattern. This isomer is mostly referred to by the common name *isobutane* and is what I call a branched hydrocarbon. Check out Figure 14-3 to see it shown in a variety of ways.

So how do you differentiate which butane you're talking about when faced with the formula C_4H_{10}? Use a unique name that stands only for that one compound. For the straight-chained compound, you can say butane or normal-butane or, better yet, n-butane. The *n-* makes it perfectly clear to a chemist that you're talking about the straight-chained isomer.

But what about the other isomer, isobutane? You can use the common name, but it isn't really accepted everywhere. Chemists all over the world need to agree on the name to help communication among scientists of all nations.

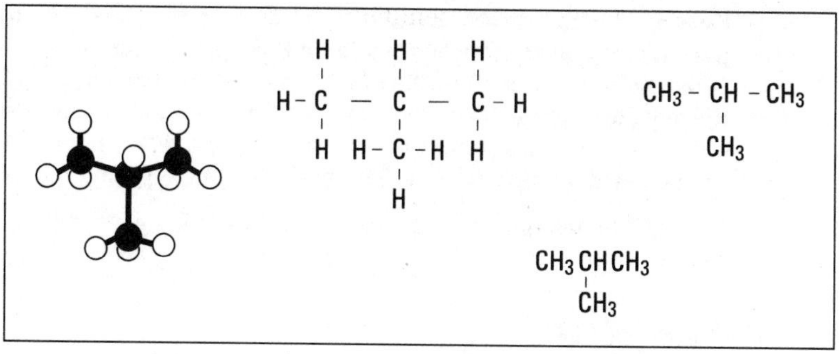

$CH_3-CH-CH_3$ (with CH_3 branch)

CH_3CHCH_3 (with CH_3 branch)

Figure 14-3: Isobutane.

An international group of chemists sets rules for things such as the naming of organic compounds. This group is called IUPAC, the International Union of Pure and Applied Chemistry. These chemists have developed an extremely systematic set of rules for the naming of compounds, and they meet regularly to decide how to name new types of compounds discovered in nature or made in the laboratory.

Systematically naming all the various types of organic compounds would probably take another book, *IUPAC Nomenclature For Dummies.* Here, I just show you the rules for naming simple alkanes (even the naming of alkanes can get complicated, so it's important to use the KISS Rule — Keep It Simple, Silly):

- **Rule 1:** Locate the longest continuous carbon chain in the alkane (*longest* means the greatest number of carbon atoms, and *continuous* means starting at one end of the chain and connecting the carbons with your pencil without picking it up or backtracking). The straight-chained hydrocarbon that has the greatest number of carbons is the parent, or base, name of the alkane. The parent name ends in the suffix *-ane.*
- **Rule 2:** The parent name is modified by adding the names of substituent groups that are attached as branches to the parent compound. *Substituent groups* are those groups that have been substituted for a hydrogen atom in the alkane parent. For alkane hydrocarbons, these substituent groups are alkane branches that attach to the parent. They are named by taking the alkane name, dropping the *–ane*, and substituting *-yl.* So, for example, methane becomes methyl, ethane becomes ethyl, and so on.
- **Rule 3:** The position of a particular substituent group on the parent carbon chain is indicated by location numbers. They're assigned by consecutively numbering the carbons of the parent carbon chain from one end to the other so that the sum of the location numbers of all substituent groups will be as small as possible. (If this doesn't make any sense — and I feel your pain here — see "Naming examples" for, well, some examples. They make this stuff pretty clear.) The location number of the carbon to which the group is attached is placed in front of the substituent group name and separated from the name by a hyphen.

- **Rule 4:** The names of the substituent groups are placed in front of the parent name in alphabetical order. If there are a number of identical substituent groups, then the numbers of all the carbons to which these groups are attached — separated by commas — are used, and the common Greek prefixes — such as *di-*, *tri-*, *tetra-*, *penta-* — are used. These prefix names are not used to determine alphabetical order.
- **Rule 5:** The last substituent alkyl group is used as a prefix to the parent alkane name.

Naming examples

Okay. You're ready to throw down this book in disgust, aren't you? I know the alkane naming system sounds ridiculous, but it's really much easier than it looks. In fact, most of my students find that naming organic compounds is one of the most fun things to do in organic chemistry.

To show you how easy it can be, I'm going to walk you through the process of naming the compound shown in Figure 14-4 (cross-referencing the rule numbers as I go along).

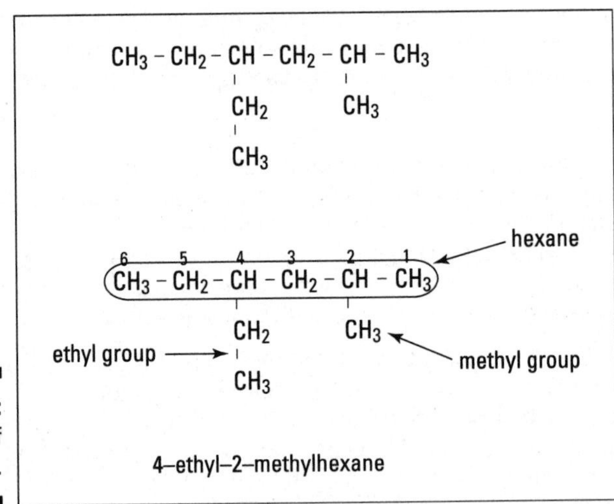

Figure 14-4: Naming of an alkane.

Using the condensed structural formula, the longest continuous carbon chain is composed of six carbons. Three different 6-carbon chains can actually be used (with the same name resulting eventually), but start with the horizontal one. The chain has six carbons, so the parent name is hexane (Rule 1). You have two substituent groups, one composed of two carbons (ethyl) and another with one carbon (methyl) (Rule 2). Number the parent chain from right to left, giving you alkyl groups at carbons 2 and 4 (sum of 6). Now do the same thing numbering from left to right on the parent chain, giving you groups at carbons 3 and 5 (sum of 8). Compare the right-to-left sum with the left-to-right sum and go with the smallest sum's numbers. So you have a 4-ethyl and a 2-methyl (Rule 3). Placing them alphabetically and with the last substituent alkyl group used as a prefix to the parent alkane name, the result is 4-ethyl-2-methylhexane (Rules 4 and 5).

Got it?

Now you try one: Name the alkane shown in Figure 14-5.

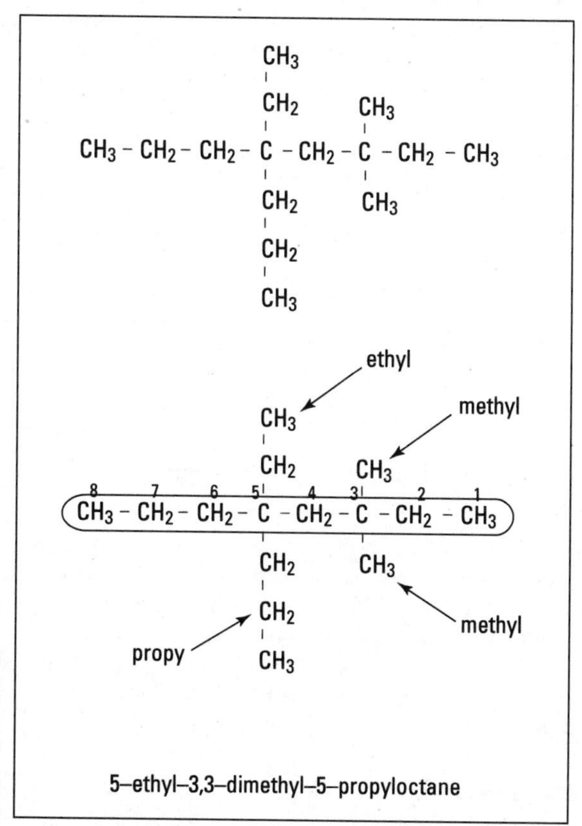

Figure 14-5: Naming another alkane.

The longest carbon chain has eight carbon atoms, so you have an octane parent name. You have two methyl groups (dimethyl), an ethyl group, and a propyl group. Again, number right to left (3+3+5+5=16) instead of left to right (4+4+6+6=20) so that you have 3,3-dimethyl (remember that if you have identical substituent groups, then you include the numbers of all the carbons to which these groups are attached, separated by commas), 5-ethyl, and 5-propyl groups. Then arrange them alphabetically, remembering that the *di-* of dimethyl doesn't count: 5-ethyl-3,3-dimethyl-5-propyloctane.

Now, that wasn't as hard as you thought it'd be, was it?

As you may figure, the more carbons you have, the more isomers are possible. For an alkane with the formula $C_{20}H_{42}$, there are over 300,000 possible isomers, and for $C_{40}H_{82}$, there are about 62 *trillion* possible isomers!

Ring in the cycloalkanes

Alkanes may also form ring systems to make compounds called *cycloalkanes*. The naming of these compounds is very similar to the branched alkanes, except the *cyclo-* prefix is used on the parent name. In the condensed structural formula, the ring is often drawn as lines where the intersection of two straight lines indicates a carbon atom, and the hydrogen atoms aren't shown at all. Figure 14-6 shows both the expanded and condensed form of 1,3-dimethyl cyclohexane.

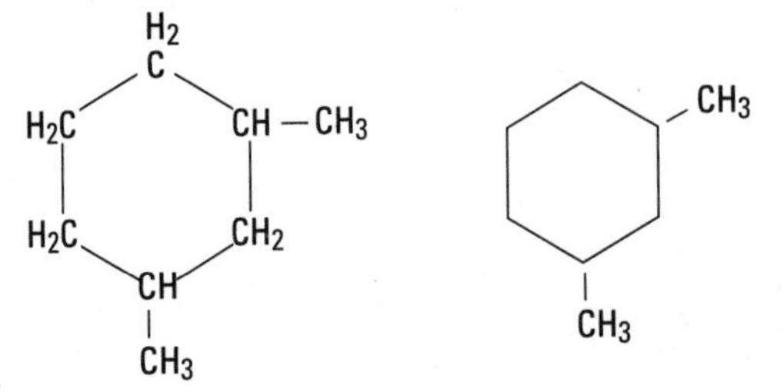

Figure 14-6: 1,3-dimethyl cyclohexane.

Straight-chained alkanes and some cycloalkanes are primarily used as fuels. Methane is the primary component in natural gas and, like most hydrocarbons, is odorless. The gas companies add a stinky organic compound containing sulfur, called a *mercaptan*, to the natural gas to help alert you to gas leaks. Butane is used in lighters, and propane is used in gas grills. Some of the heavier hydrocarbons are found in petroleum. Combustion is the primary reaction of alkanes.

Say hello to halogenated hydrocarbons

The *halogenated hydrocarbons* are a related class of compounds. These are hydrocarbons, including alkanes, in which one or more of the hydrogen atoms have been replaced by some halogen — normally chlorine or bromine. Halogen substituents are named as *chloro-*, *bromo-*, and so on. Members of this class of compound include chloroform, once used as an anesthetic; carbon tetrachloride, used at one time in dry cleaning solvent; and freons (chlorofluorocarbons, CFCs), elements that have played a major role in the depletion of the ozone layer. See Chapter 18 for a discussion of CFCs and ozone.

Unsaturated hydrocarbons: Alkenes

Alkenes are hydrocarbons that have at least one carbon-to-carbon double bond (C=C). Alkenes that have only one double bond have the general formula C_nH_{2n}. For every additional double bond, subtract two hydrogen atoms.

These compounds are called *unsaturated hydrocarbons* because they don't have the maximum possible number of hydrogen atoms attached to the carbons. (I'm sure that you've heard the terms *saturated* and *unsaturated* used in regard to fats and oils in nutrition discussions. They mean exactly the same thing there — saturated fats and oils contain no carbon-to-carbon double bonds, unsaturated fats and oils do, and polyunsaturated fats and oils have more than one C=C per molecule.)

Naming alkenes

Alkenes have a parent name ending with the *-ene* suffix. You find the longest carbon chain containing the double bond and number it so that the carbon atoms involved in the double bond have the lowest location numbers.

Ethene, written as $H_2C{=}CH_2$ or $CH_2{=}CH_2$, and *propene,* $CH_3CH{=}CH_2$, are the first two members of the alkene family. These two alkenes are often called by their common names, ethylene and propylene, respectively. They are two of the most important chemicals produced by the chemical industry in the United States. Ethylene is used in the production of *polyethylene,* one of the most useful plastics produced, and in the production of *ethylene glycol,* the principal ingredient in most antifreeze. Propylene is used in the production of *isopropyl alcohol* (rubbing alcohol) and some plastics. Figure 14-7 shows a couple of ways to represent the structural formula of ethene (ethylene).

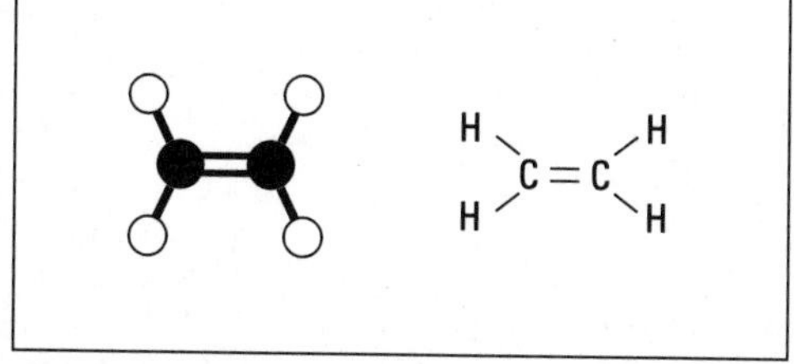

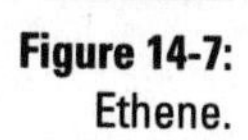

Figure 14-7: Ethene.

Alkene reactions

Although the alkenes will readily undergo combustion, their primary reaction is *addition reactions.* A double bond is very reactive. One of the bonds can easily be broken, and the two carbons can then form new single bonds to other atoms. One of the most economically important addition reactions is the process of *hydrogenation,* in which hydrogen is added across the double bond. Here's the hydrogenation of propene: $CH_3CH{=}CH_2 + H_2 \rightarrow CH_3CH_2CH_3$

This hydrogenation reaction is used in the food industry to convert unsaturated vegetable oils to solid fats (vegetable oil to margarine, for example) and requires the use of a nickel metal catalyst.

Another important addition reaction of alkenes is *hydration,* the addition of a water molecule across the double bonding, yielding an alcohol. Here's the hydration of ethylene that gives ethyl alcohol (notice that I show the water molecule in a slightly different way so you can tell where the -OH ends up):

$$H_2C{=}CH_2 + H\text{-}OH \rightarrow H_3C\text{-}CH_2OH$$

The ethyl alcohol produced in this way is identical to the ethyl alcohol produced by the fermentation process, but, by federal law, it can't be sold for human consumption in alcoholic beverages.

Undoubtedly, the most important reaction of the alkenes is *polymerization,* in which the double bond reacts to produce long chains of the once-alkenes bonded together. This is the process used to produce plastics (see Chapter 16).

It takes alkynes to make the world

Alkynes are hydrocarbons that have at least one carbon-to-carbon triple bond. These compounds have the IUPAC suffix *-yne.* Hydrocarbons with only a single triple bond have the general formula of C_nH_{2n-2}. The simplest alkyne is ethyne, commonly called *acetylene.* Figure 14-8 shows the structure of acetylene.

Figure 14-8: Ethyne (Acetylene).

$$H-C\equiv C-H$$

Acetylene is produced in a variety of ways. One way is to react coal with calcium oxide to produce calcium carbide, CaC_2. Calcium carbide is then reacted with water to produce acetylene. Miners' lamps used to be powered by this reaction. Water was dripped on calcium carbide, and the acetylene burned to produce light. Today, most of the acetylene produced is either used in oxyacetylene torches in cutting and welding or to make a variety of polymers (plastics).

Aromatic compounds: Benzene and other smelly compounds

Aromatic hydrocarbons are hydrocarbons that contain a cyclohexene type of ring system that has alternating single and double bonds. The simplest aromatic compound is *benzene,* C_6H_6. Benzene is far less reactive than you'd imagine, having those three sets of double bonds. In the current model for benzene, six electrons, two from each of the three double bonds, are donated to an electron cloud associated with the entire benzene molecule. These electrons are *delocalized* over the entire ring instead of simply located between two carbon atoms. This electron cloud is above and below the planar ring system. Figure 14-9 shows a couple of traditional ways to represent the benzene molecule and a couple of ways to represent the delocalized structure.

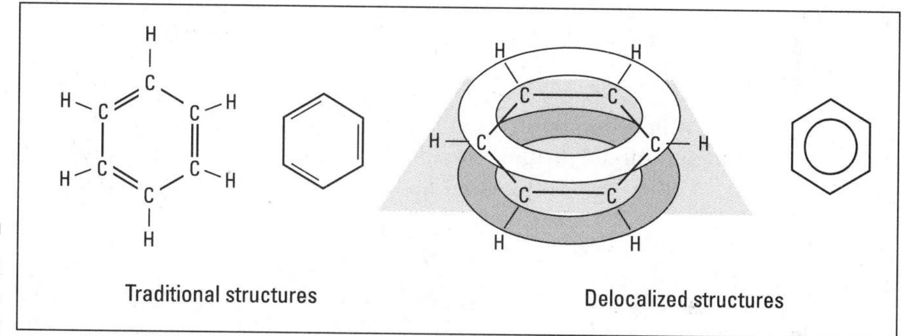

Figure 14-9: Benzene.

Many groups may be attached to this benzene ring, making many new aromatic compounds. For example, an -OH may replace a hydrogen atom. The resulting compound is called *phenol.* Phenol is used as a disinfectant and in the manufacture of plastics, drugs, and dyes. Two benzene rings fused together make *naphthalene,* which is commonly called mothballs.

Benzene and its related compounds burn, but they burn with a sooty flame. It's also been shown that benzene and some of its related compounds are either known or suspected carcinogens.

Functional Groups: That Special Spot

The preceding section covers hydrocarbons, or compounds of just carbon and hydrogen. Can you imagine how many new organic compounds can be generated if a nitrogen atom, halogen atom, sulfur atom, or some other element is thrown in?

Consider some alcohols. Ethyl alcohol (drinking alcohol), methyl alcohol (wood alcohol), and isopropyl alcohol (rubbing alcohol) are quite different and yet remarkably the same in terms of the kinds of chemical reactions they undergo. The reactions all involve the -OH group on the molecule, the part of the molecule that really defines the identity of an alcohol, just as the double bond really defines the identity of an alkene. In many cases, it doesn't really matter what the rest of the molecule turns out to be. In reactions, one alcohol is pretty much the same as another.

The atom or group of atoms that defines the reactivity of the molecule is called the *functional group.* For alcohols, it's the –OH; for alkenes, it's the C=C; and so on. This makes it much easier to study and classify the properties of compounds. You can learn the general properties of all alcohols instead of the properties of every individual one, for example. The use of functional groups makes the study of organic chemistry *much* easier.

This section places a spotlight on a few functional groups. What can really make things complex in the lab is that a molecule may have two, three, or more functional groups present, which leads to a wide array of reactions. But this is one of the things that makes organic chemistry challenging — and fun.

Alcohols (rubbing to drinking): R-OH

Alcohols are a group of organic compounds that contain the -OH functional group. In fact, alcohols are often generalized as *R-OH,* where the *R* stands for the *R*est of the molecule (like that radio newscaster's "rest of the story"). Alcohols are named using the *-ol* suffix replacing the *-ane* of the corresponding alkane.

Methanol, methyl alcohol, is sometimes called wood alcohol because years ago, its primary synthesis involved heating wood in the absence of air. The more current method of synthesis of methanol involves reacting carbon monoxide and hydrogen with a special catalyst at elevated temperatures:

$$CO(g) + 2\,H_2(g) \rightarrow CH_3OH(l)$$

About half of the methanol produced in the United States is used in the production of *formaldehyde,* which is used as embalming fluid and in the plastics industry. It's also sometimes added to ethanol to make it unfit for human consumption, a process called *denaturing.* Methanol is also being considered as a replacement to gasoline, but some major problems still need to be overcome. A process that uses methanol in the production of gasoline does exist. New Zealand currently has such a plant that produces about a third of its gasoline.

Ethanol, ethyl alcohol or grain alcohol, is produced primarily in one of two ways. If the ethanol is to be used in alcoholic beverages, it's produced by the fermentation of carbohydrates and sugars by the enzymes in yeast:

$$C_6H_{12}O_6(aq) \rightarrow 2\ CH_3CH_2OH(l) + CO_2(g)$$

As a brewer of beer and mead, I can attest that the yeast beasties certainly know how to make good alcohol!

If the ethanol is to be used for industrial purposes, such as a solvent in perfumes and medicines or as an additive to gasoline (making it *gasohol*), it's produced by the hydration of ethylene using an acid catalyst:

$$H_2C{=}CH_2 + H_2O \rightarrow CH_3\text{-}CH_2\text{-}OH$$

Carboxylic acids (smelly things): R-COOH

Figure 14-10 shows the structure of the carboxylic acid functional group.

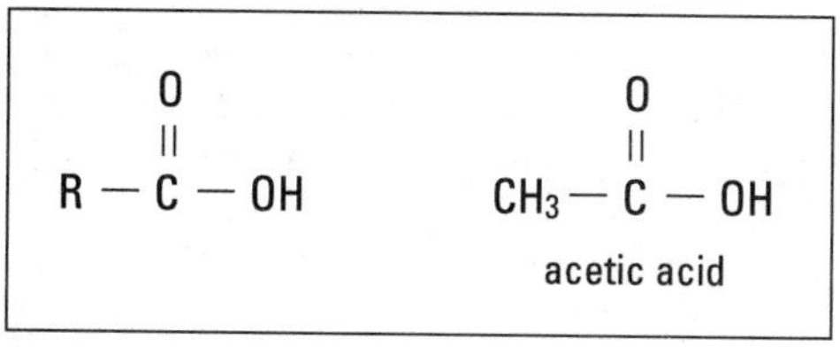

Figure 14-10: The carboxylic acid functional group and acetic acid.

```
      O                      O
      ||                     ||
R  —  C  —  OH       CH3  —  C  —  OH
                          acetic acid
```

Chemists often use -COOH or $-CO_2H$ to indicate this functional group. These compounds are named with an *-oic acid* suffix. Acetic acid, shown in Figure 14-10, is also called *ethanoic acid.*

Carboxylic acids can be prepared by the oxidation of an alcohol. For example, leave a bottle of wine in contact with the air or some other oxidizing agent, and the ethanol oxidizes to acetic acid:

$$CH_3CH_2OH(l) + O_2(g) \rightarrow CH_3COOH(l) + H_2O(l)$$

This is something that really breaks my heart, especially if I paid a lot for that particular bottle.

Formic acid, or methanoic acid, can be isolated by the distillation of ants. Yes, as in the critters that make mountainous sand piles in the cracks of the sidewalk in front of your house. The sting resulting from the bite of an ant is due to formic acid. That's why applying some base, such as baking soda, helps to neutralize the acid and relieve the pain. (Chapter 12 is a lively read about acids and bases, if you're interested.)

Many of these organic acids have a distinct odor associated with them. I'm sure you're familiar with the odor of vinegar, or acetic acid, but other acids have distinct odors, such as those mentioned in Table 14-2.

Table 14-2 Nasty Smells and What They Are

Formula	Name	Odor
$CH_3(CH_2)_2COOH$	Butyric acid	Odor of rancid butter
$CH_3(CH_2)_3COOH$	Pentanoic acid	Odor of manure
$CH_3(CH_2)_4COOH$	Hexanoic acid	Odor of goats

Esters (more smelly things, but mostly good odors): R-COOR'

The ester functional group is very similar to the carboxylic acid functional group except that another -R group has replaced the hydrogen atom. Esters are made by reacting a carboxylic acid with an alcohol, producing an ester and water. Figure 14-11 shows the synthesis of an ester.

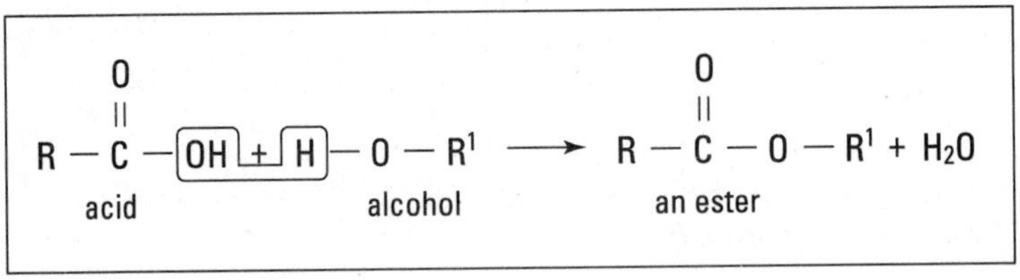

$$\underset{\text{acid}}{R-\overset{O}{\overset{\|}{C}}-\boxed{OH}} + \underset{\text{alcohol}}{\boxed{H}-O-R^1} \longrightarrow \underset{\text{an ester}}{R-\overset{O}{\overset{\|}{C}}-O-R^1} + H_2O$$

Figure 14-11: Synthesis of an ester.

Although many of the carboxylic acids that esters are made from have foul odors, many esters have pleasant odors. Oil of wintergreen is an ester. Other esters have the odor of bananas, apples, rum, roses, and pineapples. Esters are often used in the flavoring and perfume industry.

Aldehydes and ketones: Related to alcohols

Both aldehydes and ketones are produced by the oxidation of alcohols. These functional groups are shown in Figure 14-12.

Eewww — what's that smell?

I liked organic chemistry when I was in college and enjoyed the lab experiments a bunch — especially the synthesis labs where I got to build complex molecules from simpler ones. I wasn't really wild about the odors, though. Organic chemistry is one of the main reasons chemistry has the reputation for being stinky.

Figure 14-12: Aldehyde and ketone functional groups.

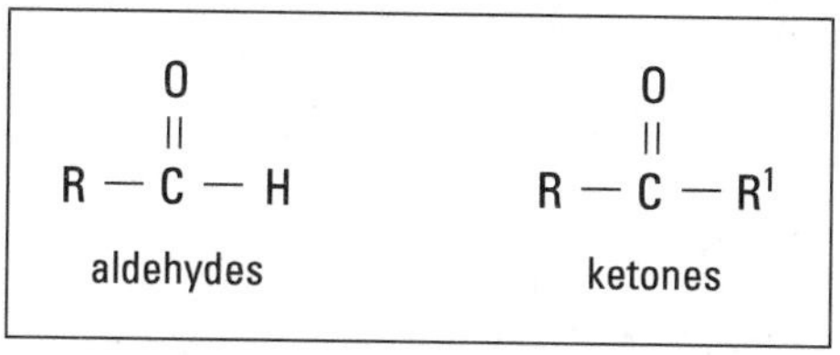

$$\underset{\text{aldehydes}}{R-\overset{\overset{\displaystyle O}{\|}}{C}-H} \qquad \underset{\text{ketones}}{R-\overset{\overset{\displaystyle O}{\|}}{C}-R^1}$$

Formaldehyde, HCHO, is an economically important aldehyde. It's used as a solvent and for preservation of biological specimens. Formaldehyde is also used in the synthesis of certain polymers, such as Bakelite and *Melmac* (used in melamine dishes). Other aldehydes, especially those with a benzene ring in their structure, have pleasing odors and, like esters, are used in the perfume and flavoring industry.

Acetone, CH_3-CO-CH_3, is the simplest ketone and has many uses as a solvent, especially for paint. Many of us are familiar with acetone-based fingernail polish remover. And methyl ethyl ketone is the solvent in model airplane glue.

Ethers (sleepy time): R-O-R

Ethers contain an oxygen atom bonded to two hydrocarbon groups, R-O-R. This reminds me of a corny joke that chemists are wont to tell: Did you hear about the ether that was found in the Playboy Mansion? She was an ether bunny!

Diethyl ether was once used as an anesthetic, but its high flammability has caused it to be largely replaced in operating rooms. Because ethers are fairly unreactive (except for combustion), they're commonly used as solvents in organic reactions. They will, however, slowly react with the oxygen in the atmosphere to form explosive compounds called *peroxides.*

You can synthesize ethers by the reaction of alcohols with the loss of water (a *dehydration* reaction). Diethyl ether can be made by reacting ethyl alcohol in the presence of sulfuric acid:

$$2\ CH_3CH_2OH(l) \rightarrow CH_3CH_2\text{-}O\text{-}CH_2CH_3(l) + H_2O(l)$$

If you use two different alcohols, you get what's called a mixed ether, where the two R groups are not the same.

Amines and amides: Organic bases

Amines and amides are derived from ammonia and contain nitrogen in their functional groups. Figure 14-13 shows the amine and amide functional groups.

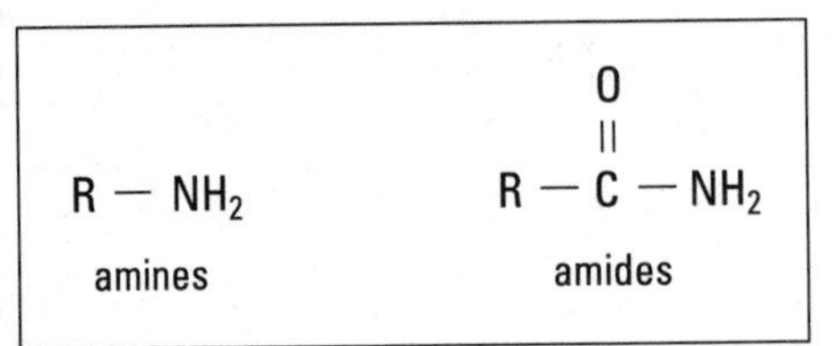

Figure 14-13: Amine and amide functional groups.

Take another look at the figure. Any of the hydrogen atoms attached to the nitrogen on both the amine and amide can be replaced by some other R group.

Amines and amides, like ammonia, tend to be weak bases (see Chapter 12). Amines are used in the synthesis of disinfectants, insecticides, and dyes. They're in many drugs, both naturally occurring and synthetic. *Alkaloids* are naturally occurring amines found in plants. Most amphetamines are amines.

Chapter 15

Petroleum: Chemicals for Burning or Building

In This Chapter

- Discovering how petroleum is refined
- Taking a look at gasoline

Petroleum is the basis of our modern society. Our automobiles run on gasoline, which is produced in part from petroleum, and many of our homes are heated with petroleum. It provides the feedstock for the expansive petrochemical industry. It's used to make plastics, paints, medicines, textiles, herbicides, and pesticides. The list is almost endless. Every year, the United States consumes over six billion barrels of petroleum. States and whole nations have risen to prosperity thanks to petroleum.

In this chapter, I show you how petroleum is refined and converted into useful products. I concentrate on the production of gasoline, because it's one of the most economically important uses of petroleum. I show you some of the problems that have been caused by our reliance on the internal combustion engine. This is one chapter in which I can talk crude — oil, that is.

Don't Be Crude, Get Refined

Petroleum, or crude oil (sometimes referred to as "Black Gold" or "Texas Tea"), as it comes out of the ground, is a complex mixture of hydrocarbons (see Chapter 14) of varying molecular weights. The lighter hydrocarbons are gases dissolved in the liquid mixture, while the heavier hydrocarbons are higher molecular weight solids that are also dissolved in the liquid mixture. The mixture was formed from decaying animal and plant material that was in the earth's crust for a very long time. Because it takes an extremely long time for petroleum to form (many millions of years), it's called a *nonrenewable resource.*

Before the hydrocarbon mixture can really be of much economic value, it must be *refined,* freed from impurities or unwanted material. The mixture is separated into groups of hydrocarbons, and in some cases, the molecular structure of the hydrocarbons is changed. The refining process occurs at a plant called a *refinery,* which produces the refined mixtures and individual compounds that are used for gasoline and feedstock for the vast petrochemical industry. A number of processes occur at the refinery, starting with the fractional distillation of the crude petroleum.

Fractional distillation: Separating chemicals

You've probably simmered a liquid in a covered pot on the stove. And you've probably noticed that when you remove the lid, water is on the inside of the lid. The heat has caused the water to evaporate from the liquid, and the vapors have condensed back into a liquid on the inside of the cooler lid. This is the most basic example of a process called *distillation.*

In the laboratory, you can take a mixture of liquids and carefully heat them. The liquid with the lowest boiling point boils first. You can then condense this vapor back to a liquid and collect it. The substance with the next highest boiling point then begins to boil, and so on. You can use this process of distillation as a means for separating the components of a mixture and purifying them. Distillation is an important procedure in organic chemistry, and it's the first step in the refining process. The distillation process that's commonly used in the refining industry is called *fractional distillation.* In this process, the petroleum mixture is heated and different *fractions* (groups of hydrocarbons with similar boiling points) are collected. Figure 15-1 shows the fractional distillation of crude oil.

The crude oil is brought into the refinery by pipeline and is initially heated and vaporized in a furnace. The hot vapors are then allowed to enter a huge distillation column, called a *fractional distillation tower.* The vapors containing the lightest molecular weight hydrocarbons rise to the top of the tower. The higher the molecular weight of the hydrocarbons, the lower the level to which they rise. The various fractions are then collected as each hydrocarbon reaches its distinct boiling point. The hydrocarbons within a fraction are all somewhat similar in size and complexity and can be used for the same purposes in the chemical industry. Six fractions are commonly collected:

- The first fraction is composed of the lightest hydrocarbons, which are gases with a boiling point of less than 40 degrees Celsius. A major component of this fraction is methane (CH_4), a gas that's sometimes called "marsh gas" because it was first found in marshes. Its primary use is as a fuel, *natural gas,* because it's a very clean-burning gas. Propane (C_3H_8) and butane (C_4H_{10}) are also found in this fraction. These two gases are normally collected and put under pressure, a process that causes them to liquefy. They can then be transported by truck as liquefied petroleum (LP) gas and used as fuel. This fraction is also used as starting materials in the synthesis of plastics.

- The second fraction is composed of hydrocarbons of C_5H_{12} (pentane) to $C_{12}H_{26}$ (dodecane), with boiling points below the 200 degrees Celsius range. This fraction is commonly called *natural gasoline* or *straight-run gasoline*, because it can be used in automobile engines with little additional refining. With each barrel (42 gallons) of crude oil that starts out in the tower, less than a quarter of a barrel of straight-run gasoline is produced.
- The third fraction is composed of hydrocarbons of 12 to 16 carbon atoms in the boiling range of 150 to 275 degrees Celsius. This fraction is used as *kerosene* and *jet fuel*. In the next section, I tell you how this fraction is also used to make additional gasoline.
- The fourth fraction is composed of hydrocarbons in the 12 to 20 carbon-atom chains, with a boiling range of 250 to 400 degrees Celsius. This fraction is used for *heating oil* and *diesel fuel*. Again, it can be used in the production of additional gasoline.
- The fifth fraction is composed of hydrocarbons in the 20 to 36 carbon-atom range, with boiling points of 350 to 550 degrees Celsius. They're used as *greases, lubricating oils,* and *paraffin-based waxes.*
- The sixth fraction is composed of the residue of semisolid and solid materials that has a boiling point well above 550 degrees Celcius. It's used as *asphalt* and *tar.*

This cracks me up: Catalytic cracking

A barrel of crude oil yields a wide variety of products, but they don't all have the same value to us. Gasoline is the product of petroleum that's in the highest demand. The straight-run gasoline fraction that comes directly from the crude oil can't keep pace with the demand for gasoline.

With the high demand for gasoline, somebody got the bright idea that if you take a fraction of higher molecular weight hydrocarbons and break it down into smaller chains, the lower molecular weight hydrocarbons can be used for gasoline. The idea of *catalytic cracking* was born.

In a catalytic cracking plant ("cat crackers," as they're called in Texas), fractions in the C_{12} to C_{20} range are heated in the absence of air with a catalyst. This process causes the long *alkanes* (compounds of carbon and hydrogen with only carbon-to-carbon single bonds, which are covered in glorious detail in Chapter 14) to break apart into smaller alkanes and *alkenes* (hydrocarbons with at least one carbon-to-carbon double bond, covered in equally glorious detail in Chapter 14).

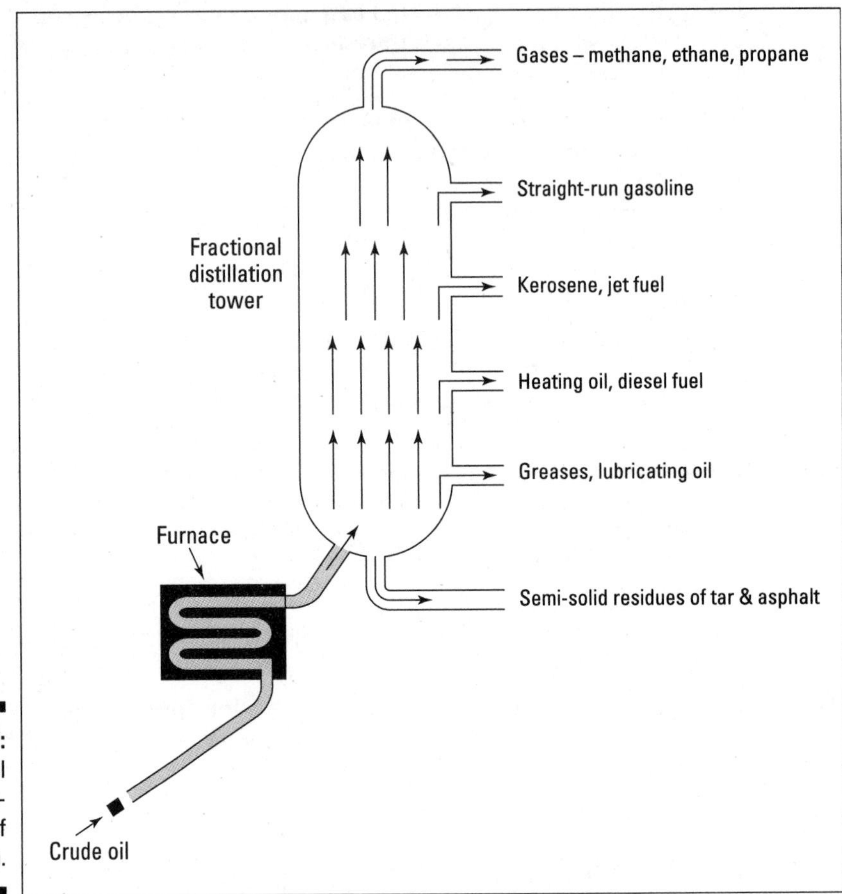

Figure 15-1: Fractional distillation of petroleum.

For example, suppose that you take $C_{20}H_{42}$ and "crack it":

$$CH_3\text{-}(CH_2)_{18}\text{-}CH_3 \rightarrow CH_3\text{-}(CH_2)_8\text{-}CH_3 + CH_2 = CH\text{-}(CH_2)_7\text{-}CH_3$$

This process yields hydrocarbons that are useful in the production of gasoline. In fact, the double bonds actually give it a higher octane rating, as I explain in "The Gasoline Story," later in this chapter.

Catalytic cracking is done on the fraction that's used for kerosene and jet fuel. But in order to produce even more gasoline, catalytic cracking is also done on the fraction used for heating oil. Using this fraction can present a problem, though, if a severe winter hits and the demand for heating oil skyrockets. Oil companies watch the long-range weather forecasts closely. In the summer, when the demand for gasoline is high, fractions that can be used for heating oil are converted to gasoline to meet the demand. Then, as fall arrives, refineries shift their production schedule somewhat. They reduce the amount of gasoline they produce and increase the amount of heating oil so that the winter demand for heating oil can be met. But the refineries don't want to overproduce heating oil and have to store large amounts, so they try to second-guess the weather to develop a supply that will meet the demand. It's a real balancing act.

Moving molecular parts around: Catalytic reforming

As the internal combustion engine gained popularity as a mode of transportation, chemists noted that if the gasoline contained only straight-chained hydrocarbons, it didn't burn properly; it had a tendency to knock or ping. They found that hydrocarbons with branched structures burned much better. In order to increase the amount of branching in the petroleum hydrocarbon fraction being used for gasoline, a process called *catalytic reforming* was developed. In this process, the hydrocarbon vapors are passed over a metal catalyst such as platinum, and the molecule is rearranged into one with a branched structure or even a cyclic structure. Figure 15-2 shows the catalytic reforming of n-hexane to 2-methylpentane and to cyclohexane.

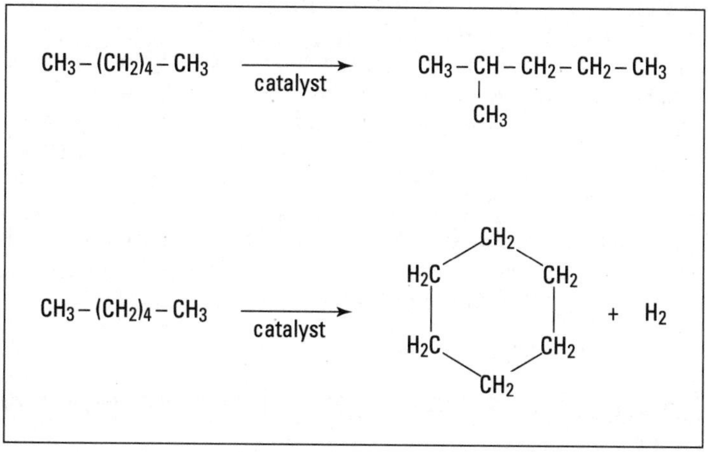

$$CH_3-(CH_2)_4-CH_3 \xrightarrow[\text{catalyst}]{} CH_3-\underset{\displaystyle CH_3}{\underset{|}{CH}}-CH_2-CH_2-CH_3$$

$$CH_3-(CH_2)_4-CH_3 \xrightarrow[\text{catalyst}]{} \text{cyclo-}(CH_2)_6\ (H_2C,\ CH_2,\ CH_2,\ CH_2,\ CH_2,\ H_2C \text{ ring}) + H_2$$

Figure 15-2: Catalytic reforming of n-hexane.

This same process is used extensively to produce benzene and other aromatic compounds for use in the manufacture of plastics, medicines, and synthetic materials. (For a discussion of aromatic compounds, as well as branched and cyclic structures, see Chapter 14. What a treasure trove it is.)

The Gasoline Story

In order for you to better understand the properties of gasoline, I want to tell you a little about how gasoline is reacted in an internal combustion engine. The gasoline is mixed with air (a mixture of nitrogen, oxygen, and so on) and injected into the cylinder as the piston moves to the bottom of the cylinder. The piston then begins to move upward, compressing the gasoline-air mixture. At just the right moment, the spark plug fires, igniting the mixture. The hydrocarbons react with the oxygen in the cylinder, producing water vapor, carbon dioxide, and, unfortunately, large amounts of carbon monoxide.

This reaction is an example of converting the potential energy contained in the hydrocarbon bonds to the kinetic energy of the hot gas molecules. The increase in the number of gas molecules boosts the pressure tremendously, shoving the piston down. The linear motion is then converted to a rotary motion, which powers the wheels. And off you go!

The gasoline-air mixture must ignite at exactly the right moment in order for the engine to operate properly. This process is largely a property of the gasoline and not the engine itself (assuming the timing is set correctly, the spark plugs are good, the compression ratio is okay, and so on). The *volatility* of the hydrocarbon fuel (that is, how easily it's converted into a vapor) is important. Volatility is related to the boiling point of the hydrocarbon. In fact, manufacturers *blend* (adjust the hydrocarbon mixture) their gasoline to match the climate. (They don't do that in the part of Texas I live in — it's summer almost year-round.) Winter gas is more volatile than summer gas. Some fuels are prone to produce knocking or pinging in an engine. This propensity to cause knocking or pinging may be a result of *preignition,* where the igniting of the gasoline occurs before the compression of the fuel-air mixture is complete, or *spotty ignition,* where combustion starts taking place at a number of sites in the cylinder instead of right around the spark plug electrode. Again, this is a property of the gasoline and not the engine. The energy content of the fuel is important, but how efficiently it burns in the cylinder is just as important. The octane rating scale was developed to rate the burning characteristics of a gasoline.

How good is your gas: Octane ratings

In the early stages of the development of the internal combustion engine, scientists and engineers found that certain hydrocarbons burned well in an internal combustion engine. They also found that certain hydrocarbons did not burn well in these engines. A hydrocarbon that did not burn well was n-heptane (straight-chained heptane). However, 2,2,4-trimethylpentane (commonly called *isooctane*) had excellent burning characteristics. These two compounds were chosen to define the *octane rating scale.* The hydrocarbon n-heptane was assigned an octane rating of zero, while isooctane was given a value of 100. Blends of gasoline are then burned in a standard engine and are rated according to this scale. For example, if a particular gasoline blend burns 90 percent as well as isooctane, then it's assigned an octane value of 90. Figure 15-3 shows the octane scale and the octane values of certain pure compounds.

Look carefully at Figure 15-3. A couple of things are useful to note in terms of octane rating and chemical structure. The n-pentene has an octane value of 62. Its octane value can be increased to 91 by introducing a double bond (making it 1-pentene) and making it an unsaturated compound. The octane value increases by almost 30 points with the introduction of the double bond.

The process of catalytic reforming introduces chains, and catalytic cracking introduces double bonds. Not only do these two processes increase the amount of gasoline that's produced, but they also improve the quality of the gasoline's burning characteristics. Also notice that benzene, an aromatic compound, has an octane value of 106. Its burning characteristics are better than isooctane. Other substituted aromatic compounds have octane ratings of almost 120. However, benzene and some related compounds are health hazards, so they're not used.

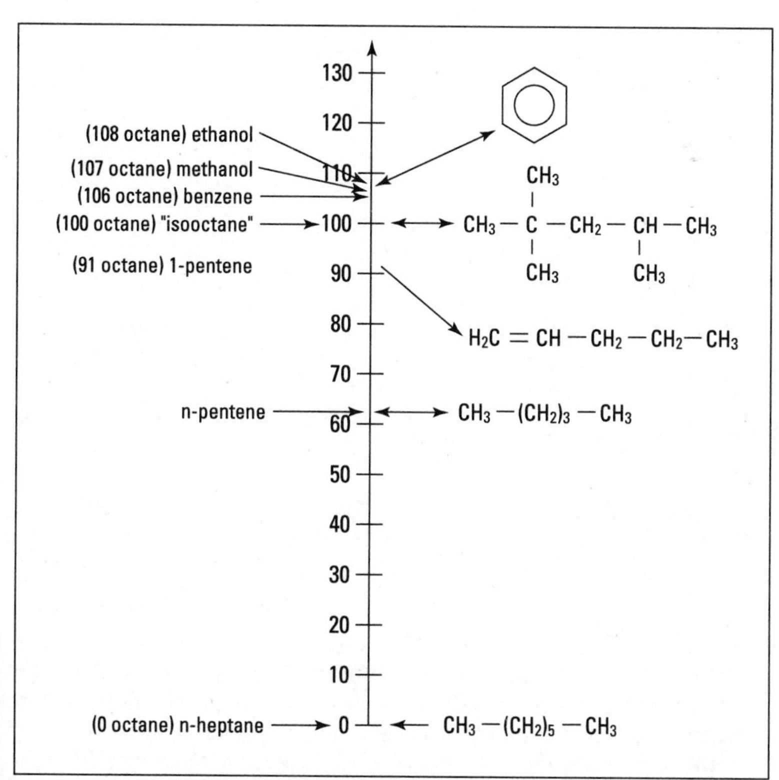

Figure 15-3: The octane rating scale.

The octane rating that is posted on gas pumps is really an average of two kinds of ratings. The *Research octane rating (R)* relates to the burning characteristics of the fuel in a cold engine. The *Motoring octane value (M)* refers to how the fuel behaves while you're cruising down the interstate. If you average R and M — (R+M)/2 — you get the posted octane rating.

Additives: Put the lead in, get the lead out

The first gasoline engines had a compression ratio that was much lower than today's automobile engines, and they required lower octane gas. However, as engines became more powerful, gasoline with a higher octane rating was required. Catalytic cracking and reforming added significant cost to the gasoline. The search was on for something cheap that could be added to gasoline to effectively increase the octane rating. The substance tetraethyllead, or TEL, was found.

In the early 1920s, scientists discovered that adding a little bit of TEL to gasoline (1 milliliter per liter of gasoline) increased the octane rating by 10 to 15 points.

Tetraethyllead is basically a lead atom with four ethyl groups attached to it. Figure 15-4 shows the structure of TEL.

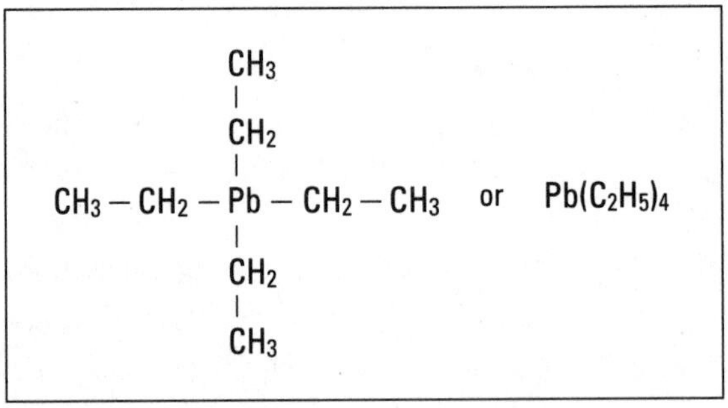

$$
\begin{array}{c}
CH_3 \\
| \\
CH_2 \\
| \\
CH_3 - CH_2 - Pb - CH_2 - CH_3 \quad \text{or} \quad Pb(C_2H_5)_4 \\
| \\
CH_2 \\
| \\
CH_3
\end{array}
$$

Figure 15-4: The composition of tetraethyllead.

TEL was quite effective as an additive to increase the octane rating and prevent engine knocking. It was used for many years. However, the Clean Air Act of 1970 indirectly did it in.

Oops! We're polluting the air

Hydrocarbon fuel burns in the cylinders of internal combustion engines. During this process, not all of the hydrocarbon molecules are converted to water and CO/CO_2. Before the Clean Air Act of 1970 came into play, unburned hydrocarbons and oxides of both sulfur and nitrogen were being released into the environment from automobiles (along with lead from the TEL, which was later discovered to be very toxic). These gaseous pollutants dramatically increased the amount and severity of air pollution and gave rise to health hazards such as photochemical smog. (For a more complete discussion of air pollution, see Chapter 18.)

Bring on the catalytic converter

In the United States, The Clean Air Act of 1970 mandated the reduction of automotive pollutant emissions. The most effective way to accomplish the reduction of emissions was through the use of a *catalytic converter.* It's shaped like a muffler and connected to the exhaust system of an automobile. It has a solid catalyst, either palladium or platinum, inside. When the exhaust gases pass over the catalyst, the catalytic converter helps to complete the oxidation of the hydrocarbons and carbon monoxide to carbon dioxide and water. In other words, it helps to change the harmful gases from gasoline to mostly harmless products.

Lose the lead

The catalytic converter worked well at reducing the automotive emissions as long as there was no lead in the fuel. But if leaded gasoline was used, the lead vapor in the exhaust gases would coat the catalyst, rendering it useless. So there was a big push by the government and environmental groups to "get the lead out." Now it's very difficult to find leaded gasoline in the United States, although it's still available in some foreign countries.

With TEL no longer available as an octane booster, chemists tried to find other compounds to replace it. Aromatic compounds were effective in enhancing the octane value, but they were discovered to be serious health hazards. Recently, methyl alcohol, tert-butyl alcohol, and methyl tert-butyl ether (MTBE) have been used as octane boosters.

MTBE (see Figure 15-5) showed great promise because it not only boosted the octane rating but also acted as an *oxygenate,* a compound containing oxygen that increases the efficiency of the complete hydrocarbon combustion. But it has already been removed from gasoline due to increasing evidence that it's related to respiratory illnesses and possible cancers in humans. As for the other compounds, although none of them are as effective as TEL, the partial redesign of the internal combustion engine has allowed the use of slightly lower octane fuels.

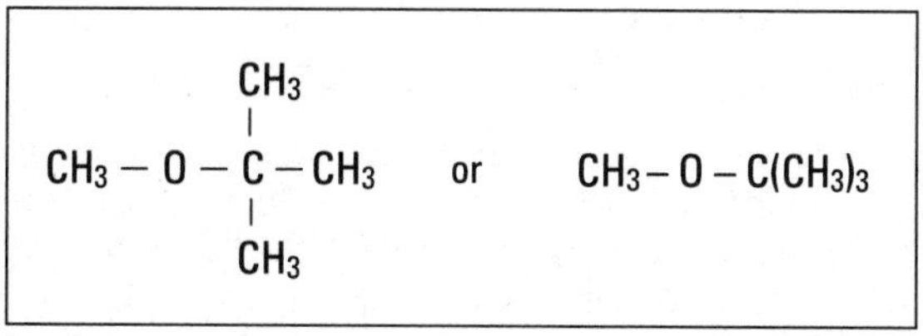

$$CH_3-O-\underset{\displaystyle CH_3}{\overset{\displaystyle CH_3}{\overset{|}{\underset{|}{C}}}}-CH_3 \quad \text{or} \quad CH_3-O-C(CH_3)_3$$

Figure 15-5: Methyl tert-butyl ether (MTBE).

Chapter 16

Polymers: Making Big Ones from Little Ones

In This Chapter

- Understanding polymerization
- Differentiating the different types of plastics
- Finding out about recycling plastics

I once heard someone say that man never really invents anything new; he just copies nature. I'm not sure I believe that, with all the new inventions that have been developed recently. But I certainly think it's true in the case of polymers. Nature has been building polymers forever. Proteins, cotton, wool, and cellulose are all polymers. They all fall into a class of compounds called *macromolecules* — very large molecules. Man has learned to produce macromolecules in the lab, changing the face of our society forever.

When I was a child, my father, very much a traditionalist, said that he wanted things made of metal, not that cheap imported plastic stuff. Wow, would he be shocked today. I'm surrounded by synthetic textiles (clothing and carpet, for example), I ride around in autos that are fast becoming cocoons of plastic, my home is filled with plastic bottles of all shapes, sizes, and hardness, and I have friends with knees or other parts that have been either replaced or enhanced with polymers. I cook with a skillet that has a nonstick surface, I use a nylon spatula, I watch a TV with a plastic case, and I go to sleep on a foam pillow. Our world is truly part of the Age of Plastics.

In this chapter, I show you how the process of polymerization takes place and how chemists go about designing polymers with certain desired characteristics. I also show you some different kinds of polymers and how they're created. And I discuss some ways for getting rid of plastics before we bury ourselves in a mountain of milk jugs and disposable diapers. Welcome to the wonderful world of polymers!

Natural Monomers and Polymers

Nature has been building polymers for a long time. Cellulose (wood) and starch are prime examples of naturally occurring polymers. Take a look at Figure 16-1, which shows, the structures of cellulose and starch.

Notice anything similar about the two structures in the figure? They're both made up of repeating units. In fact, the repeating unit in both cases is a glucose unit. Both starch and cellulose are natural *macromolecules* (large molecules), but they're also examples of naturally occurring *polymers,* macromolecules in which there's a repeating unit called a *monomer.* (Polymer should stand for "many mer." The *mer* in this case is the mono*mer.*) In the case of starch and cellulose, the monomer is a glucose unit. The structure of polymers is similar to taking a bunch of paper clips (monomers) and hooking them together to make a big long chain (polymer).

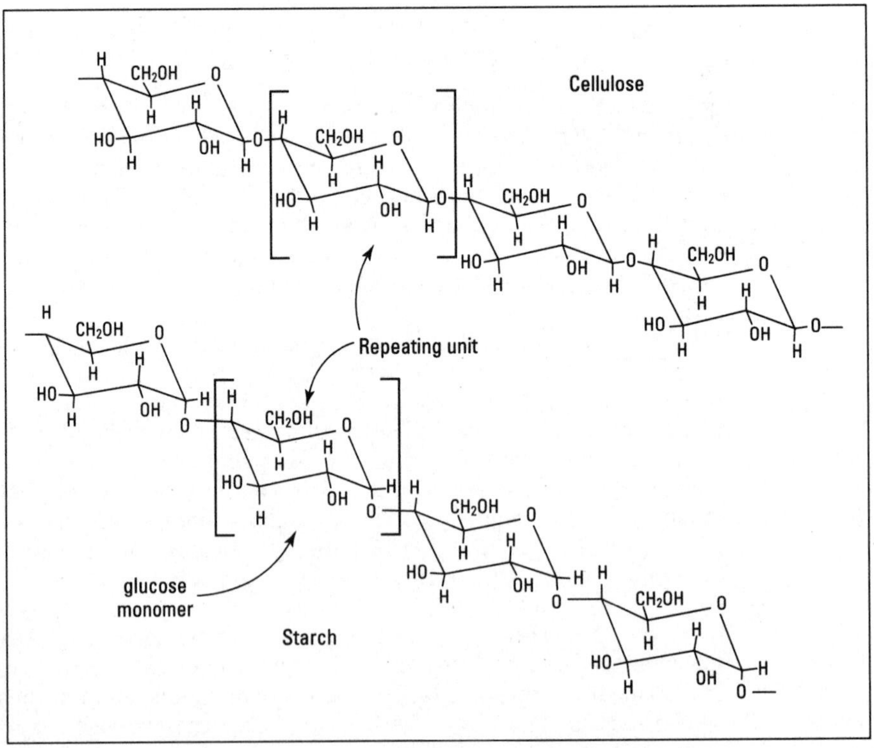

Figure 16-1: Cellulose and starch.

Notice another thing about cellulose and starch. The only way they differ is in how the glucose units are attached to each other. This minor change makes the difference between a potato and a tree. (Okay, it's not *quite* that simple.) Human beings can digest (metabolize) starches but not cellulose. A termite can digest cellulose just fine. In natural polymers, just like in synthetic ones, a minor change sometimes makes a big difference in the properties of the polymer.

Classifying Unnatural (Synthetic) Monomers and Polymers

Chemists took this idea of hooking together small units into very large ones from nature and developed a number of different ways of doing it in the lab. Now there are many different types of synthetic polymers. In this section, I introduce you to some of them and talk about their structures, properties, and uses.

Because chemists are big on grouping things together, they've put polymers into different classes. That works out just fine. Grouping gives chemists something to do and makes it easier for normal folks to get familiar with the various kinds of polymers out there.

We all need a little structure

One way of classifying polymers is by the structure of their polymer chain. Some polymers are *linear.* They're composed of many long strands thrown together like pieces of rope. *Branched* polymers have short branches coming off the main polymer strand. Imagine taking those long pieces of rope and tying short pieces of rope to them along the entire length. *Crosslinked* polymers have the individual polymer chains linked together by side chains. Imagine taking those pieces of rope and making them into a hammock.

Feel the heat

Another way of classifying polymers is by their behavior under heat. *Thermoplastic* polymers become soft when they're heated. Polymers of this type are composed of long linear or branched strands of monomer units hooked together. Have you ever left a pair of plastic sunglasses or a child's plastic toy on the dashboard of your car in the middle of the summer? These plastics become really soft. Because they soften and melt, they can be remolded time and time again. This makes thermoplastics much easier to recycle. A vast majority of the plastics produced in the United States are of the thermoplastic type.

Thermosetting polymers don't soften when heated, and they can't be remolded. During production of this type of polymer, crosslinking (bridges between the polymer strands) is created in the plastic by heating it. Bakelite is a good example of a thermosetting plastic. It's a hard, strong nonconductor. These properties make it ideal as an insulator and as a handle for frying pans and toasters.

Used and abused

A third way of classifying polymers is by their use by the consumer.

A *plastic* refers to the polymer's ability to be molded. Whether they're of the thermoplastic or thermosetting type, these polymers are molded during the manufacture of the end product. And they're used to make our dishes, toys, and so on.

Fibers are linear strands held together by intermolecular forces such as hydrogen bonding between the polymer strands. These polymers are generally called textiles. They're used to make our clothes and our carpets.

Elastomers, sometimes called rubber, are thermoplastic materials that become slightly crosslinked during their formation. Because of this, they stretch and bounce. Natural rubber (latex) is classified as an elastomer along with its synthetic counterparts. These types of polymers are used for things like latex gloves and rubber bands and balls.

Chemical process

One of the best ways of classifying synthetic polymers is by the chemical processes used to create them. These processes normally fall into one of two categories:

- Addition polymerization
- Condensation polymerization

Let's hook up: Addition polymerization

Many of the common polymers you come into contact with every day are called *addition polymers* — polymers that are formed in a reaction called *addition polymerization.* In this type of reaction, all the atoms that start out in the monomer are incorporated into the polymer chain. The monomers involved in this type of polymerization normally have a carbon-to-carbon double bond that's partially broken during polymerization. This broken bond forms a *radical reactive site,* or *radical,* which is a highly reactive atom that has an unpaired electron. The radical then gains an electron by joining up with another radical, and a chain is started, which eventually becomes the polymer. Scratching your head a bit? Looking at examples often helps folks understand chemical processes, so following are some examples of addition polymerization.

Polyethylene: Sandwich wrap and milk jugs

Polyethylene is the simplest of the addition polymers. It's also one of the most economically important. Ethane is heated at a high temperature in the presence of a metal catalyst, like palladium. Ethane loses two atoms of hydrogen (which make hydrogen gas) and forms a double bond:

$$CH_3\text{-}CH_3(g) + \text{heat and catalyst} \rightarrow CH_2\text{=}CH_2(g) + H_2(g)$$

The ethylene (ethene) that's produced here is the monomer used in the production of polyethylene. The ethylene is then subjected to high heat with a catalyst in the absence of air. The high heat and catalytic action causes one of the carbon-to-carbon double bonds (C=C) to break, with one electron going to each carbon. Both carbons now have an unpaired electron, so they become radicals. Radicals are extremely reactive and attempt to gain an electron. In terms of this polymerization reaction, the radicals can gain an electron by joining up with another radical to form a covalent bond. This happens at both ends of the molecule, and the chain begins to grow. Polyethylene molecules up to a molecular weight of 1 million grams/mol may be produced in this way (see Figure 16-2).

Different catalysts and pressures are used to control the structure of the final product. The polymerization of ethylene can yield three different types of polyethylene:

- Low-density polyethylene (LDPE)
- High-density polyethylene (HDPE)
- Crosslinked polyethylene (CLPE)

Low-density polyethylene (LDPE) has some branches on the carbon chain, so it doesn't pack together as closely and tightly as the linear polymer. It forms a tangled network of branched polymer strands. This type of polyethylene is soft and flexible. It can be used for food wrap, sandwich bags, grocery bags, and trash bags. And it, like all the forms of polyethylene, is resistant to chemicals.

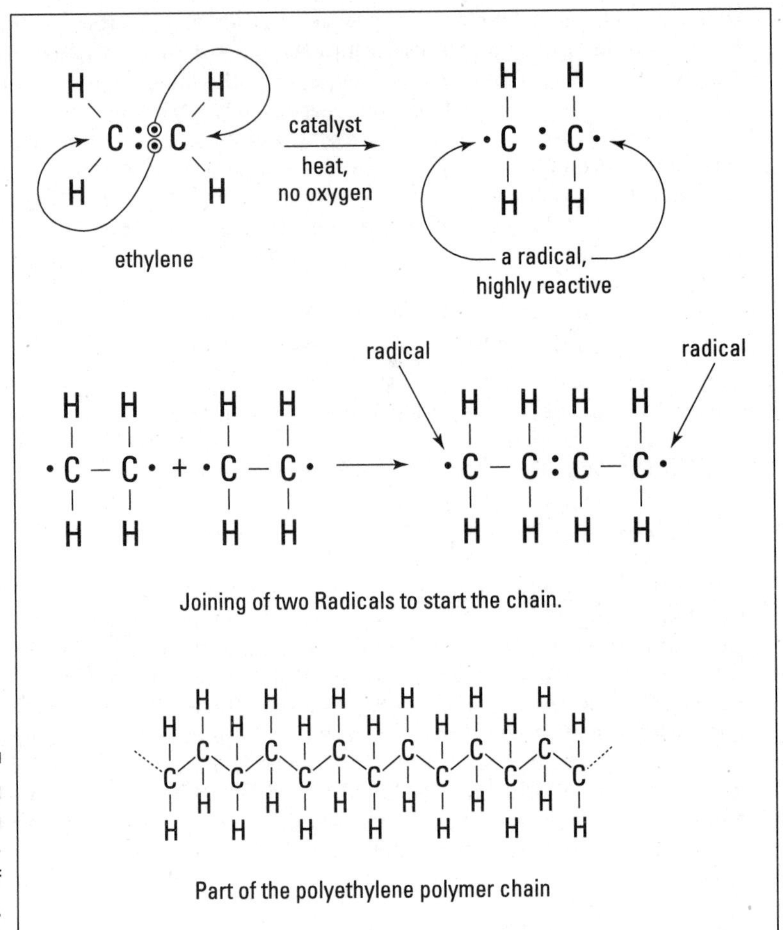

Figure 16-2: The addition polymerization of ethylene.

High-density polyethylene (HDPE) is composed of linear chains that are closely packed. This type of polymer is rigid, hard, and tough. Milk jugs, toys, and TV cabinets are made from HDPE. The Hula-Hoop was one of the first products ever made from this form of polyethylene.

Crosslinked polyethylene (CLPE) has crosslinking between the linear strands of monomers that are bonded together, producing a polymer that's extremely tough. The lid on that HDPE milk jug is probably CLPE. Soft drink bottle caps are also CLPE. The soft drink bottles are made of another type of polymer that I discuss a little later in the chapter.

Polypropylene: Plastic ropes

If you substitute another atom for a hydrogen atom on ethylene, you can produce a different polymer with different properties. If you substitute a methyl group for a hydrogen atom, you get propylene. Propylene, just like ethylene, has a double bond, so it can undergo addition polymerization in the same way as ethylene. The result is polypropylene (see Figure 16-3).

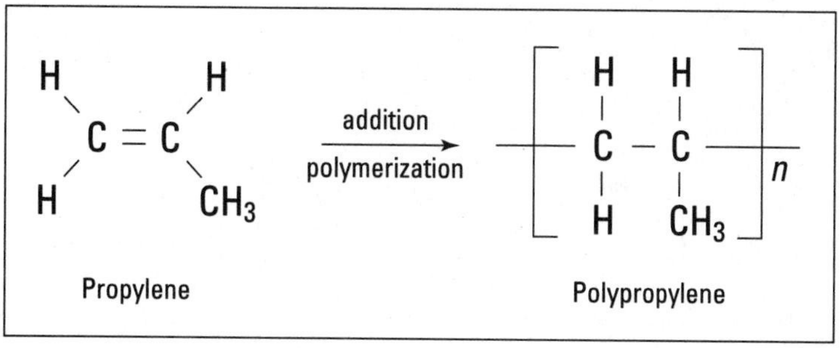

addition polymerization

Propylene

Polypropylene

Figure 16-3: Propylene and polypropylene.

The small *n* in Figure 16-3 indicates that there are a number of the repeating units. Notice that this polymer has a methyl group side chain. Any time that the structure of a molecule is changed, the properties of the molecule change. By carefully adjusting the reaction conditions, chemists can construct polymers that have the side chains on the same side of the molecule, on alternating sides of the molecule, or distributed randomly. The position of these side chains changes the properties of the polymer somewhat so that polypropylene can be used for a wide variety of purposes, such as indoor-outdoor carpeting, battery cases, ropes, bottles, and automotive trim.

Polystyrene: Styrofoam cups

If you substitute a benzene ring for one of the hydrogen atoms on ethylene, you make styrene. Addition polymerization gives you polystyrene, as shown in Figure 16-4.

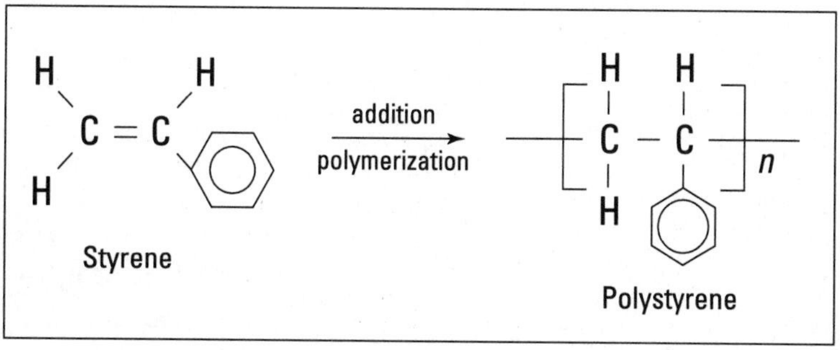

addition polymerization

Styrene

Polystyrene

Figure 16-4: Styrene and polystyrene.

Polystyrene (Styrofoam) is a rigid polymer used for making foam drink cups, egg cartons, clear rigid drinking glasses, insulating materials, and packing materials. Environmentalists have criticized its use because it's more difficult to recycle than some other plastics and is so widely used.

Polyvinyl chloride: Pipes and simulated leather

Substituting a chloride for one of the hydrogen atoms on ethylene gives you the vinyl chloride monomer that can polymerize to polyvinyl chloride (PVC), as shown in Figure 16-5.

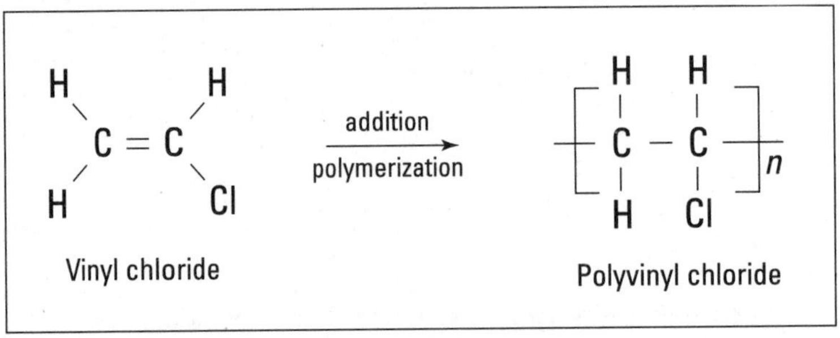

$$H_2C=CHCl \xrightarrow[\text{polymerization}]{\text{addition}} \left[-CH_2-CHCl- \right]_n$$

Vinyl chloride | Polyvinyl chloride

Figure 16-5: Vinyl chloride and polyvinyl chloride.

PVC is a tough polymer. It's used extensively in rigid pipes of all types, flooring, garden hoses, and toys. Thin sheets of PVC used as simulated leather crack easily, so a *plasticizer* is added (a liquid that's mixed with plastics to soften them and allow them to more closely resemble leather). However, after many years, plasticizers can evaporate from plastic, making it brittle and allowing it to crack.

Polytetrafluoroethylene: Slick stuff

Replace all the hydrogen atoms on ethylene with fluorine atoms, and you have tetrafluoroethylene. The tetrafluoroethylene can be polymerized to polytetrafluoroethylene, as shown in Figure 16-6.

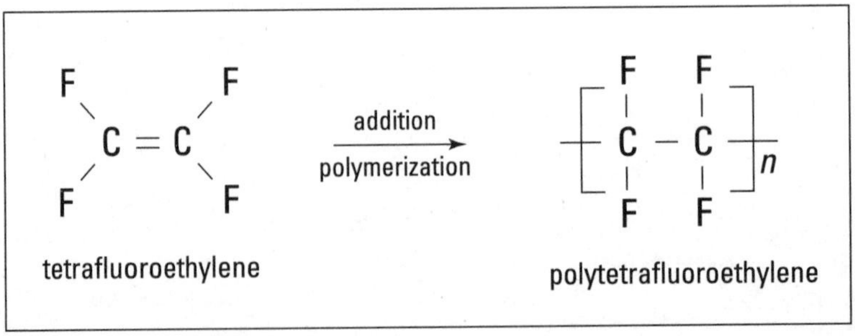

$$F_2C=CF_2 \xrightarrow[\text{polymerization}]{\text{addition}} \left[-CF_2-CF_2- \right]_n$$

tetrafluoroethylene | polytetrafluoroethylene

Figure 16-6: Tetrafluoroethylene and polytetrafluoroethylene.

Polytetrafluoroethylene is a material that's hard, heat resistant, and extremely slick. This material is used as bearings, valve seats, and (most importantly to me) nonstick coating for pots and pans.

You can find some other addition polymers in Table 16-1.

Table 16-1 Other Addition Polymers

Monomer	*Polymer*	*Uses*
$H_2C{=}CH{-}CN$ Acrylonitrile	$[-CH_2-CH(CN)-]_n$ Polyacrylonitrile	Wigs, rugs, yarn
$H_2C{=}CH{-}O{-}C({=}O){-}CH_3$ Vinyl acetate	$[-CH_2-CH(O{-}C({=}O){-}CH_3)-]_n$ Polyvinyl acetate	Adhesives, latex, paint, chewing gum resin, textile coatings
$H_2C{=}C(CH_3){-}C({=}O){-}O{-}CH_3$ Methyl methacrylate	$[-CH_2-C(CH_3)(C({=}O){-}O{-}CH_3)-]_n$ Polymethyl methacrylate	Contact lenses, glass substitute, bowling balls
$H_2C{=}CCl_2$ Vindylidine chloride	$[-CH_2-CCl_2-]_n$ Polyvindylidine chloride	food wrap

Let's get rid of something: Condensation polymerization

A reaction in which two chemical species combine with each other by eliminating a small molecule is called *condensation polymerization*. Polymers formed in this fashion are known as *condensation polymers*. Unlike in addition polymerization, no double bond is needed in this type of reaction.

A small molecule — normally water — is eliminated. Commonly, one molecule is an organic acid, and the other is an alcohol. These two molecules react, splitting off water and forming an organic compound called an ester. If a polymer chain grows, it forms a polyester.

Following are some examples of condensation polymers. These examples involve techno-speak specific to functional groups in organic chemistry, and they involve a lot of complexly named organic compounds. If you aren't familiar with functional groups or how organic compounds are named, just flip to Chapter 14 for the details.

Polyester: Leisure suits and soft drink bottles

If you take ethylene glycol, with its alcohol functional groups on both carbons, and react it with terephthalic acid, with its two organic acid functional groups, you can eliminate water and form the condensation polymer polyethylene terephthalate (PET), a polyester. Figure 16-7 shows the synthesis of PET.

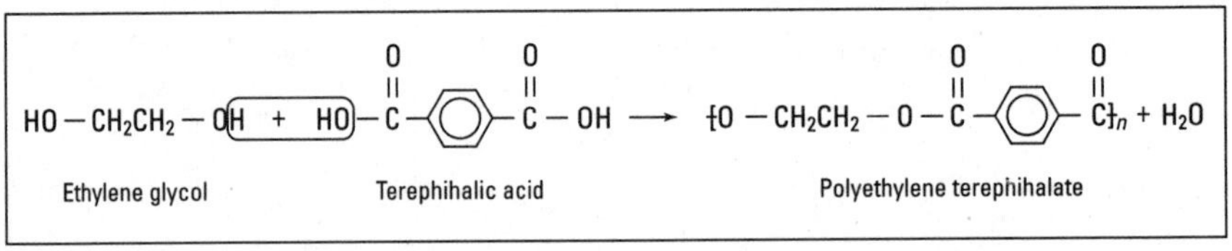

$$HO-CH_2CH_2-OH + HO-\overset{O}{\overset{\|}{C}}-C_6H_4-\overset{O}{\overset{\|}{C}}-OH \longrightarrow \left[O-CH_2CH_2-O-\overset{O}{\overset{\|}{C}}-C_6H_4-\overset{O}{\overset{\|}{C}}\right]_n + H_2O$$

Ethylene glycol   Terephihalic acid   Polyethylene terephihalate

Figure 16-7: Synthesis of PET.

This is the polyester you find in clothing (boy, did I ever love that baby-blue leisure suit I had in the 70s!), artificial automotive tire cord, substitute blood vessels, film, and soft drink bottles.

Polyamides: Sheer enough for a woman, strong enough for a (police)man

If you react an organic acid with an amine, you split off water and form an amide. If you use an organic acid that contains two acid ends and an amine that has two amine ends (a diamine), then you can polymerize a polyamide. The polyamide is commonly referred to as *nylon*. Figure 16-8 shows the reaction between 1,6-hexanediamine and adipic acid to form Nylon 66. (The 66 indicates that there are 6 carbon atoms in both the amine and the organic acid.)

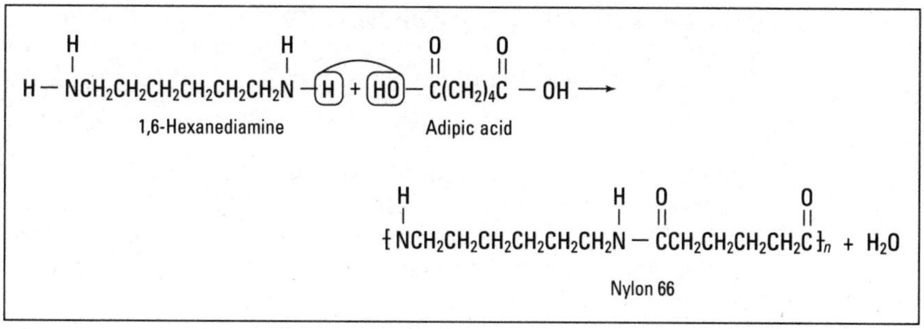

$$H-NHCH_2CH_2CH_2CH_2CH_2CH_2NH-H + HO-C(=O)(CH_2)_4C(=O)-OH \longrightarrow$$

1,6-Hexanediamine Adipic acid

$$\left[NHCH_2CH_2CH_2CH_2CH_2CH_2NH-C(=O)CH_2CH_2CH_2CH_2C(=O) \right]_n + H_2O$$

Nylon 66

Figure 16-8: Synthesis of Nylon 66.

The synthesis of nylon in 1935 had a major impact on the textile industry. Nylon stockings first went on sale in 1939, and nylon was used in parachutes extensively during World War II. Make a slight substitution in one of the carbon backbones, and you have a material strong enough for a bullet-proof vest.

Silicones: Bigger and better

Because silicon is in the same family as carbon, chemists can produce a class of polymers that contains silicon in its structure. These polymers are known as silicones. Figure 16-9 shows the synthesis of a typical silicone.

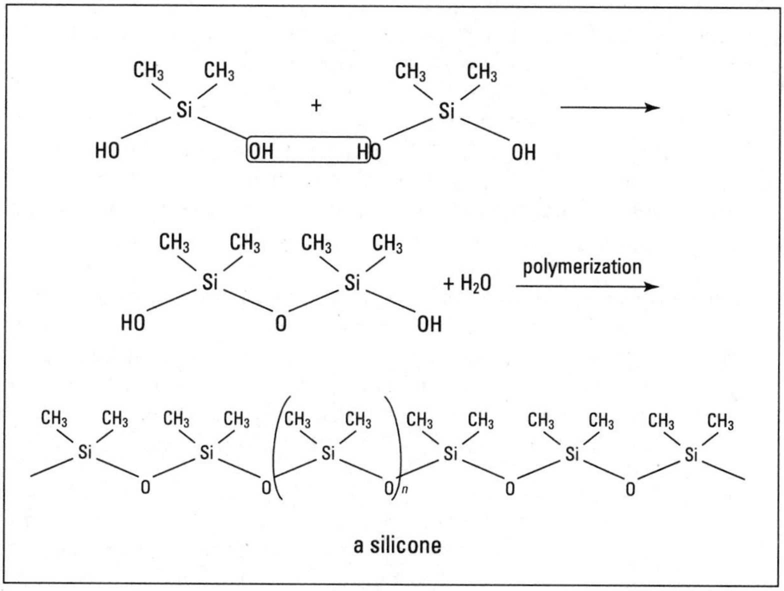

$$HO-Si(CH_3)_2-OH + HO-Si(CH_3)_2-OH \longrightarrow HO-Si(CH_3)_2-O-Si(CH_3)_2-OH + H_2O \xrightarrow{\text{polymerization}}$$

$$-Si(CH_3)_2-O-Si(CH_3)_2-O\left(-Si(CH_3)_2-O-\right)_n Si(CH_3)_2-O-Si(CH_3)_2-O-Si(CH_3)_2-$$

a silicone

Figure 16-9: Synthesis of a silicone.

The silicone polymers are held together by the strong silicon-oxygen bond, and they can have molecular weights in the millions. They're used as gaskets and seals, and they're found in waxes, polishes, and surgical implants. The press has given the most attention to their use as surgical implants.

Silicone-based implants and prostheses have been used for years. They've been used as shunts, ear prostheses, finger joints, and, of course, breast implants. The implants themselves are filled with silicone oil. Occasionally, an implant leaks, and the silicone oil escapes into the body. In 1992, some evidence was found that silicone oil may trigger an autoimmune response. Although studies have not established a cause-and-effect relationship, many implants have been removed, and silicone oil is no longer used in the United States.

Polymers have reshaped our society as well as our figures. They're useful in a very wide variety of ways, relatively inexpensive, and durable. Because they're so durable, figuring out how to dispose of them is a major problem.

Reduce, Reuse, Recycle — Plastics

Plastics have basically an infinite lifetime. Nothing in nature does a good job at degrading them. If you bury that plastic plate, Styrofoam cup, or disposable diaper in a landfill and then dig it up ten years later, there will be no change. You could even dig it up a hundred years later and get the same results. Waste containing plastics will be with us for a long time.

Some plastics can be burned as fuels. They have a high heat content, but they often produce gases that are toxic or corrosive. Society can reduce its reliance on plastics to a certain degree. Using cardboard hamburger boxes and cellulose shipping packing instead of Styrofoam helps, but our best answer so far has been in the area of recycling.

Thermoplastic polymers can be melted down and reformed. But in order to do this, the plastics must be separated into their various components. Most plastic containers contain a symbol on the bottom that indicates what type of plastic the containers are made of. Recyclers can use these symbols to separate the plastics into various categories to make recycling easier. Figure 16-10 shows the recycling symbols for plastics and indicates what type of plastic each symbol represents.

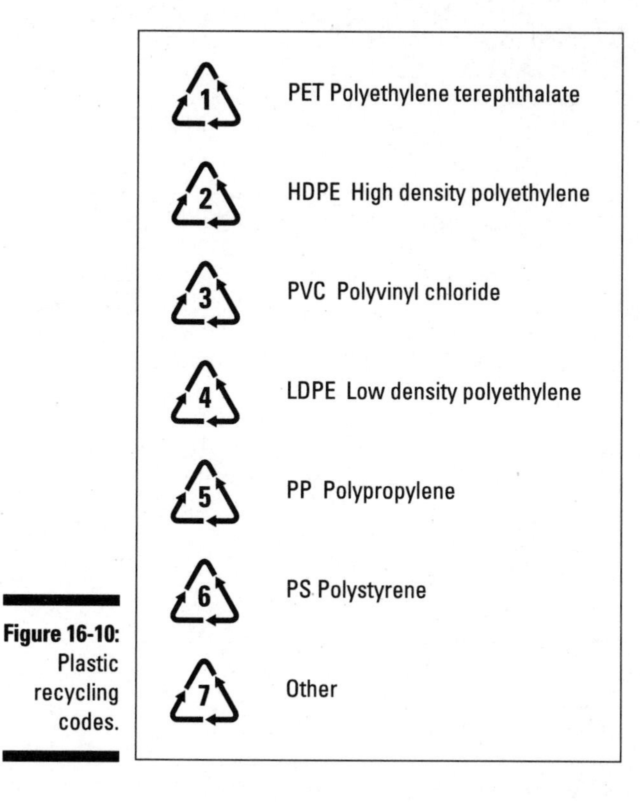

Figure 16-10: Plastic recycling codes.

PET bottles and HDPE milk jugs are probably the plastics recycled the most. But the major problem isn't the chemistry involved in the recycling process: The major problems are encouraging individuals, families, and businesses to recycle and developing an easy means to collect and sort the plastics for recycling. These polymers are too valuable a resource to simply be buried in some landfill.

Chapter 17

Chemistry in the Home

In This Chapter

- Discovering the chemistry behind cleaners and detergents
- Finding out about the chemistry of cosmetics
- Looking at the chemistry of drugs and medicines

You'll probably come into direct contact with more chemicals and chemistry in your own home than anyplace else. The kitchen is filled with cleaners, soaps, and detergents, most of which are contained in plastic bottles. The bathroom is filled with medicines, soaps, toothpaste, and cosmetics. My wife is glad to have her own private chemist handy, especially when it's time to clean the silver or find a solvent to remove an adhesive. And all that doesn't even cover the myriad chemical reactions that take place while cooking. No wonder consumer chemistry is sometimes called "kitchen chemistry."

In this chapter, I cover a few topics from the chemistry of consumer products. I show you the chemistry behind soaps, detergents, and cleaners. I talk a little bit about medicines and drugs, and I show you some things about personal care products, permanents, tanning products, and perfumes. I hope that you'll gain an appreciation for chemistry and what it has done to make your life better and easier. (Note that lots of common chemicals in the home are acids and bases. Chapter 12's main thing is acids and bases, which makes it nice complementary reading to this chapter.)

Chemistry in the Laundry Room

Have you ever become distracted and forgotten to put the laundry detergent in the washer? Or have you ever been suckered into trying one of those miracle solid ceramic laundry detergent discs? I doubt that the clothes came out very clean. You may have gotten some surface dirt off, but the grease and oil stayed right where it was. The grease and oil stayed on the clothes because "like dissolves like." Grease and oils are nonpolar materials, and water is a polar substance, so water isn't going to dissolve the grease and oil. (Chapter 7 gets into detail about this whole polar/nonpolar business, if you're interested.) I guess you could dump some gasoline (a nonpolar material) into the washer, but I don't think that's a good solution to the problem. Wouldn't it be wonderful if something existed that could bridge the gap between the nonpolar grease and oil and the polar water? Something does. It's called a *surfactant.*

Surfactants, which are also called surface active agents, reduce the surface tension of water, allowing it to "wet" nonpolar substances such as grease and oil. Surfactants are able to do this because they have both a nonpolar end and a polar end.

The nonpolar end is called the *hydrophobic* (water-fearing) end. This end is normally composed of a long hydrocarbon chain. (If you're feeling ambitious, Chapter 14 covers more than you'll probably ever want to know about hydrocarbons.) The nonpolar end dissolves in the nonpolar grease and oil.

The other part of the surfactant molecule, the polar end, is called the *hydrophilic* (water-loving) end. This end is normally an ionic end with a negative charge *(anionic)*, a positive charge *(cationic)*, or both *(amphoteric)*. There are even some surfactants that have no charge *(nonionic)*. (Ions, anions, cations — Chapter 6 explains 'em all.)

A vast majority of the surfactants on the market are anionic surfactants, because they're cheaper to produce. Figure 17-1 shows a typical anionic surfactant.

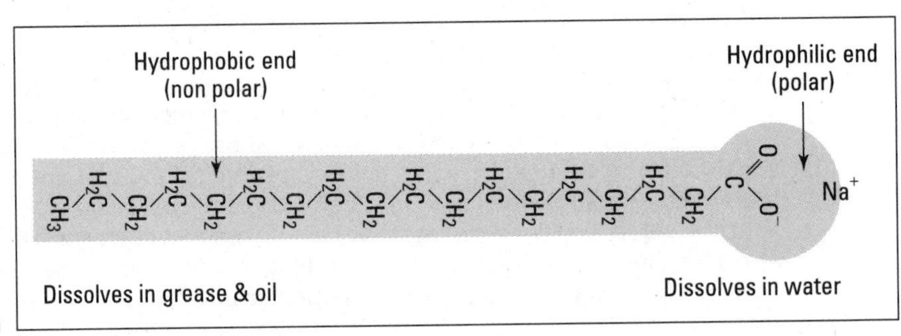

Figure 17-1: A typical anionic surfactant.

When a surfactant is added to water, the hydrophobic end dissolves in the oil and grease, while the hydrophilic end becomes attracted to the polar water molecules. The grease and oil are broken into very tiny droplets called *micelles,* with the hydrophobic (hydrocarbon) end of the surfactant sticking into the droplet and the hydrophilic end sticking out into the water. This gives the droplet a charge (a negative charge in the case of an anionic surfactant). These charged droplets repel each other and keep the oil and grease droplets from joining together. These micelles remain dispersed and eventually go down the drain with the used wash water.

The two general types of surfactants that are used in the cleaning of clothes are soaps and detergents.

Keep it clean: Soap

Soaps are certainly the oldest and most well-known surfactant for cleaning. The use of soap dates back almost 5,000 years. The specific type of organic reaction involved in the production of soap is a hydrolysis reaction of fats or oils in a basic solution. This reaction is commonly called *saponification.* The products of this reaction are glycerol and the salt of the fatty acid. Figure 17-2 shows the hydrolysis of tristearin to sodium stearate, a soap. (This is the same soap, or surfactant, shown in Figure 17-1.)

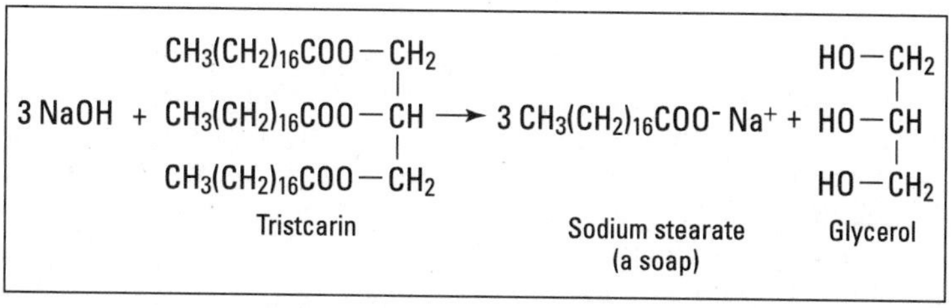

Figure 17-2: Production of a soap by saponification.

$$3\ NaOH + \begin{array}{l} CH_3(CH_2)_{16}COO-CH_2 \\ \qquad\qquad\qquad\quad\ | \\ CH_3(CH_2)_{16}COO-CH \\ \qquad\qquad\qquad\quad\ | \\ CH_3(CH_2)_{16}COO-CH_2 \end{array} \longrightarrow 3\ CH_3(CH_2)_{16}COO^-\ Na^+ + \begin{array}{l} HO-CH_2 \\ \qquad\ \ | \\ HO-CH \\ \qquad\ \ | \\ HO-CH_2 \end{array}$$

Tristcarin — Sodium stearate (a soap) — Glycerol

Grandma made her soap by taking animal fat, adding it to water and lye (sodium hydroxide, NaOH), and boiling it in a huge iron kettle. The lye came from wood ashes. After cooking for hours, the soap rose to the top. It was then skimmed off and pressed into bars. However, Grandma didn't know much about reaction stoichiometry. She usually had an excess of lye, so her soap was very alkaline.

Today, soap is made a little differently. The hydrolysis is generally accomplished without the use of lye. Coconut oil, palm oil, and cottonseed oil are used in addition to animal tallow. For bar soaps, an abrasive, such as pumice, is occasionally added to aid in the removal of tough grease and oil from your skin. In addition, perfumes may be added, and air may be mixed with the soap to get it to float.

Soap, however, has a couple of big disadvantages. If soap is used with acidic water, the soap is converted to fatty acids and loses its cleaning ability. And if soap is used with hard water (water containing calcium, magnesium, or iron ions), a greasy insoluble precipitate (solid) forms. This greasy deposit is commonly called bathtub ring. And it's a bummer. Not only does the deposit form in your bathtub, but it also appears on your clothes, dishes, and so on. A couple of ways are available to avoid this deposit. You can use a whole-house water softener (see "Make it soft: Water softeners," later in this chapter), or you can buy a synthetic soap that doesn't precipitate with hard-water ions. These synthetic soaps are called *detergents.*

Get rid of that bathtub ring: Detergents

Detergents have the same basic structure as the soap in Figure 17-1. Their hydrophobic end — composed of a long nonpolar hydrocarbon chain that dissolves in the grease and oil — is the same, but their hydrophilic (ionic) end is different. Instead of having a carboxylate ($-COO^-$), the hydrophilic end may have a sulfate ($-O-SO_3^-$), a hydroxyl (OH^-), or some other polar group that doesn't precipitate with hard water.

Laundry detergents contain a number of other compounds in addition to the detergent surfactant. The compounds in laundry detergents are

- **Builders:** These compounds increase the surfactant's efficiency by softening the water (removing the hard water ions) and making it alkaline. The builder that was used in early laundry detergents was sodium tripolyphosphate. It was cheap and safe. However, it was also an excellent nutrient for water plants and caused an increase in the growth of algae in lakes and streams, choking out fish and other aquatic life. States began banning the use of phosphates in detergents in order to control this problem. Sodium carbonate and zeolites (complex aluminosilicates — compounds of aluminum, oxygen, and silicon) have been used as replacements for the polyphosphates, but both are less than ideal. There really hasn't been an effective, cheap, and nontoxic replacement for the polyphosphate builders. This is an area of research that's still quite active.
- **Fillers:** Compounds such as sodium sulfate (Na_2SO_4) are added to give the detergent bulk and to keep it free-flowing.
- **Enzymes:** These biological catalysts are sometimes added to help remove protein-based stains such as blood and grass.
- **Sodium perborate:** $NaBO_3$ is sometimes added as a solid bleach to help remove stains. It works by generating hydrogen peroxide in water. It's much more gentle on textiles than chlorine bleach. However, it's most effective in hot water, which can present a problem for those of us who like to wash in cold water.
- **Suspension agents:** These compounds are added to help keep the dirt in solution in the wash water so that it doesn't redeposit itself on another portion of the clothes.
- **Corrosion inhibitors:** These compounds coat washer parts to help prevent rust.
- **Optical brighteners:** These compounds are used to make white clothes appear extra clean and bright. These very complex organic compounds deposit themselves as a thin coating on the clothes. They absorb ultraviolet light and re-emit it as a blue light in the visible part of the spectrum. This process is shown in Figure 17-3.

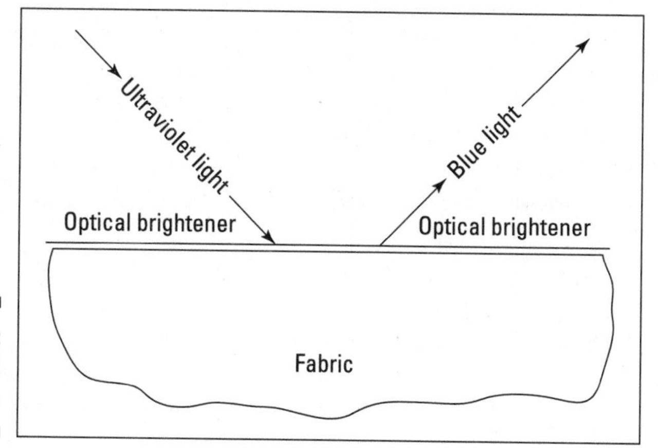

Figure 17-3: Optical brighteners.

Coloring agents and perfumes are added to laundry detergent, as well. I bet you didn't know that washing clothes was so complex.

Make it soft: Water softeners

Using synthetic detergents is one way to combat the problem of hard water and bathtub ring. Another way is to simply remove the cations responsible for the hard water before they reach the house. You can accomplish this feat through a home water softener (see Figure 17-4).

A water softener consists of a large tank containing an ion-exchange resin. The resin is charged when a concentrated sodium chloride solution runs through it. The sodium ions are held to the polymer material of the resin. The hard water passes through the polymer, and the calcium, magnesium, and iron ions are exchanged for the sodium ions on the resin (that's where the term *ion-exchange* resin comes from). The softened water contains sodium ions, but the hard-water ions remain in the resin. After a while, the resin must be recharged with more sodium chloride from the reservoir. The wastewater that contains the Ca^{2+}, Mg^{2+}, and Fe^{2+} is drained from the resin tank.

What's that froth in the lake?

The original synthetic detergents weren't capable of being broken down by bacteria and other natural forces. In other words, they weren't *biodegradable.* These detergents accumulated in lakes and streams and caused a thick coating of suds. They were quickly reformulated to solve the problem.

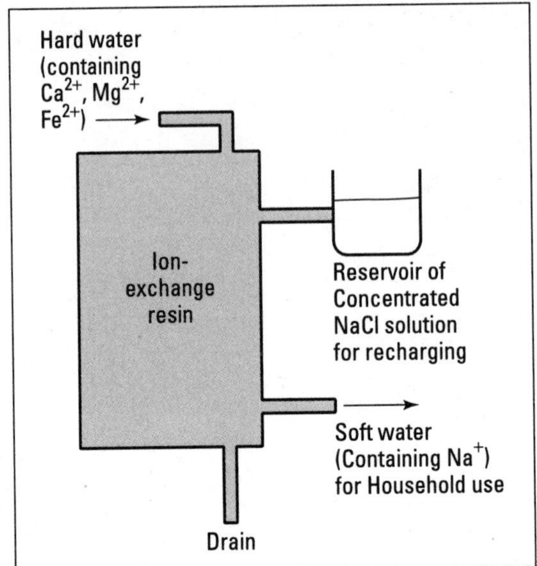

Figure 17-4: A whole-house water softener.

If you limit your sodium intake because of high blood pressure, you should avoid drinking softened water because it has a high sodium ion concentration.

Make it whiter: Bleach

Bleaches use redox reactions to remove color from material (see Chapter 9 for a discussion of redox reactions). Most bleaches are oxidizing agents. The most common bleach used in the home is a 5 percent solution of sodium hypochlorite. This type of bleach is produced by bubbling chlorine gas through a sodium hydroxide solution:

$$2\ NaOH(aq) + Cl_2(g) \rightarrow NaOCl(aq) + NaCl(aq) + H_2O(l)$$

The chlorine released by hypochlorite bleaches can damage fabrics. Also, these types of bleaches don't work very well on polyester fabrics.

Bleaches containing sodium perborate have been introduced to the market, and they're gentler on fabrics. This type of bleach generates hydrogen peroxide, which in turn decomposes with oxygen gas as one of the products:

$$2\ H_2O_2 \rightarrow 2\ H_2O(l) + O_2(g)$$

Chemistry in the Kitchen

You can take a peek under the kitchen sink and see countless products that are made with chemicals (and stored in plastic bottles that are made through chemistry).

Clean it all: Multipurpose cleaners

Most multipurpose cleaners are composed of some surfactant and disinfectant. Ammonia is commonly used because of its ability to react with grease and because it leaves no residue. Pine oil, a solution of compounds called terpenes, is used for its pleasant odor, its ability to dissolve grease, and its antibacterial nature.

Be careful when mixing household cleaning products — especially bleach with ammonia or muratic acid (HCl). This solution generates toxic gases that can be quite dangerous.

Wash those pots: Dishwashing products

Dishwashing detergent is much simpler than laundry detergent. It has some surfactant (normally a nonionic one), a little colorant, and something to make your hands feel soft.

Dishwashing detergent is not nearly as alkaline as laundry detergent. However, automatic dishwasher detergents are highly alkaline and contain only a little surfactant. They use the high pH to saponify the fats (like the process used to make soap) and a high water temperature as well as agitation to clean the dishes. They're composed mostly of sodium metasilicate (Na_2SiO_3), for its alkalinity; sodium tripolyphosphate ($Na_5P_3O_{10}$), which acts as a detergent; and a little chlorine bleach.

Chemistry in the Bathroom

A lot of chemistry goes on in the bathroom. There are all those skin and hair care products, as well as products to make you look good and smell good and even taste good.

Detergent for the mouth: Toothpaste

Walk down any toothpaste aisle, and you'll see a wide variety of toothpastes with different colors, flavors, and so on. Although they may look different, they all contain the same basic ingredients. The two primary ingredients are surfactant (detergent) and abrasive. The abrasive is for scraping the film off the teeth without damaging the teeth themselves. Common abrasives are chalk ($CaCO_3$), titanium dioxide (TiO_2), and calcium hydrogen phosphate ($CaHPO_4$). Other ingredients are added to give the toothpaste color, flavoring, and so on. Table 17-1 gives the general formula for toothpaste. The percentages and specific chemical compounds may vary from toothpaste to toothpaste.

Table 17-1 Typical Formulation for Toothpaste

Function	*Possible Ingredient*	*Percentage*
Solvent and filler	Water	30–40%
Detergent	Sodium lauryl sulfate, soap	4%
Abrasive	Calcium carbonate, calcium hydrogen phosphate, titanium dioxide, sodium metaphosphate, silicia, aluminia	30–50%
Sweetener	Glycerine, saccharin, sorbitol	15–20%
Thickener	Gum cellulose, carrageenan	1%
Fluoride	Stannous or sodium fluoride	1%
Flavoring	Oil of wintergreen, peppermint, strawberry, lime, and so on	1%

The addition of stannous or sodium fluoride is effective in the prevention of dental cavities, because the fluoride ion actually becomes part of the tooth enamel, making the enamel stronger and more resistant to the attack of acids.

Phew! Deodorants and antiperspirants

Sweating helps your body regulate its internal temperature. Sweat contains amines, low molecular weight fatty acids, and proteins, in addition to sodium chloride and other inorganic compounds. Some of these organic compounds have a disagreeable odor. Bacterial action can certainly make the odor worse. Deodorants and antiperspirants can be used to control the socially unacceptable odor. (Quite a professional way to discuss stinky B.O., eh?)

Deodorants contain fragrances to cover up the odor and an antibacterial agent to destroy the odor-causing bacteria. They may also contain substances such as zinc peroxide that oxidize the amines and fatty acids to less odorous compounds.

Antiperspirants inhibit or stop perspiration. They act as an *astringent,* constricting the sweat gland ducts. The most commonly used antiperspirants are compounds of aluminum — aluminum chlorohydrates ($Al_2(OH)_5Cl$, $Al_2(OH)_4Cl_2$, and so on), hydrated aluminum chloride ($AlCl_36H_2O$), and others.

Skin care chemistry: Keeping it soft and pretty

Beeswax. Whale wax. Borax. You may be surprised at what's in some of that stuff you put on your skin.

Creams and lotions

The skin is a complex organ composed primarily of protein and naturally occurring macromolecules (polymers — see Chapter 16). Healthy skin contains about 10 percent moisture. Creams and lotions work to soften and moisturize the skin.

Emollients are skin softeners. Petroleum jelly (mixture of alkanes, with 20-plus carbons, isolated from crude oil), lanolin (mixture of esters isolated from sheep wool fat), and coco butter (mixture of esters isolated from the cacao bean) are excellent skin softeners.

Skin creams are normally made of oil-in-water or water-in-oil emulsions. An *emulsion* is a colloidal dispersion of one liquid in another (see Chapter 11 for a discussion of colloids). It tries to soften and moisturize the skin at the same time. Cold creams are used in the removal of makeup and as moisturizers, while vanishing creams make the skin appear younger by filling in wrinkles. Typical formulations for cold cream and vanishing cream are

Cold Cream Formulation	***Vanishing Cream Formulation***
20–50% water	70% water
30–60% mineral oil	10% glycerin
12–15% beeswax	20% stearic acid/sodium stearate
	5–15% lanolin or whale wax
	1% borax
	trace of perfume

Body and face powders

Body and face powders are used to dry and smooth the skin. The main ingredient in both types of powder is talc ($Mg_3(Si_2O_5)_2(OH)_2$), a mineral that absorbs both oil and water. Astringents are added to reduce sweating, and binders are added to help the powders stick to the skin better. Face powders often contain dyes to give color to the skin. Table 17-2 shows a typical formulation for body powder, and Table 17-3 shows a typical formulation for face powder.

Table 17-2 Typical Formulation for Body Powder

Ingredient	*Function*	*Percentage*
Talc	Absorbent, bulk	50–60%
Chalk ($CaCO_3$)	Absorbent	10–15%
Zinc oxide (ZnO)	Astringent	15–25%
Zinc stearate	Binder	5–10%
Perfume, dye	Odor, color	trace

Table 17-3 Typical Formulation for Face Powder

Ingredient	*Function*	*Percentage*
Talc	Absorbent, bulk	60–70%
Zinc oxide	Astringent	10–15%
Kaolin (Al_2SiO_5)	Absorbent	10–15%
Magnesium and zinc stearates	Texture	5–15%
Cetyl alcohol	Binder	1%
Mineral oil	Emollient	2%
Lanolin, perfume, dyes	Softening, odor, color	2%

Making up those eyes

Eye shadow and mascara are composed primarily of emollients, lanolin, beeswax, and colorants. Mascara darkens the eyelashes, making them appear longer. Typical formulations for eye shadow and mascara are

Eye Shadow

55–60% petroleum jelly

5–15% fats and waxes

5–10% lanolin

15–25% zinc oxide

1–5% dyes

Mascara

45–50% soap

35–40% wax and paraffin

5–10% lanolin

1–5% dyes

Kissable lips: Lipstick

Lipstick keeps the lips soft and protects them from drying out, while adding a desirable color. It's composed mostly of wax and oil. These ingredients must be balanced carefully so that the lipstick goes on easily without running and comes off easily, but not too easily, when the wearer is ready to remove it. The color normally comes from a precipitate (solid) of some metal ion with an organic dye. This is commonly called a *lake.* The metal ion tends to intensify the color of the dye. A typical lipstick formulation is shown in Table 17-4.

Table 17-4 A Typical Lipstick Formulation

Ingredient	*Function*	*Percentage*
Castor oil, mineral oil, fats	Dye solvent	40–50%
Lanolin	Emollient	20–30%
Carnauba wax or beeswax	Stiffener	15–25%
Dye	Color	5–10%
Perfume and flavoring	Odor and taste	trace

Beautiful nails: Nail polish

Nail polish is a synthetic lacquer that owes its flexibility to a polymer and a plasticizer (a liquid mixed with plastics to soften them). The polymer is normally nitrocellulose. The solvents used in the polish are acetone and ethyl acetate, the same substances used for nail polish removers.

Smelling GOOD! Perfumes, colognes, and aftershaves

The major difference between perfume, cologne, and aftershave is the amount of fragrance used. Perfumes are commonly composed of 10 to 25 percent fragrance, while colognes use 1 to 3 percent, and aftershaves use less than 1 percent. These fragrances are usually organic esters, alcohols, ketones, and aldehydes. Perfumes also contain *fixatives,* compounds that help keep the fragrances from evaporating too rapidly.

Interestingly, several fixatives have disagreeable odors or histories themselves: Civetone comes from the glands of the skunk-like civet cat, ambergris is sperm-whale vomit, and indole is isolated from feces. I don't think I'll comment on this.

Perfumes are usually mixtures of *notes,* fragrances with similar aromas but different volatilities (the ease with which a substance is converted into a gas). The most volatile is called the *top note.* It's what you initially smell. The *middle note* is the most noticeable smell, while the *end note* fragrances are responsible for the lingering odor of the perfume. Figure 17-5 shows the chemical structure of several of the fragrances commonly used in perfumes.

Don't you think it's neat being able to see an odor?

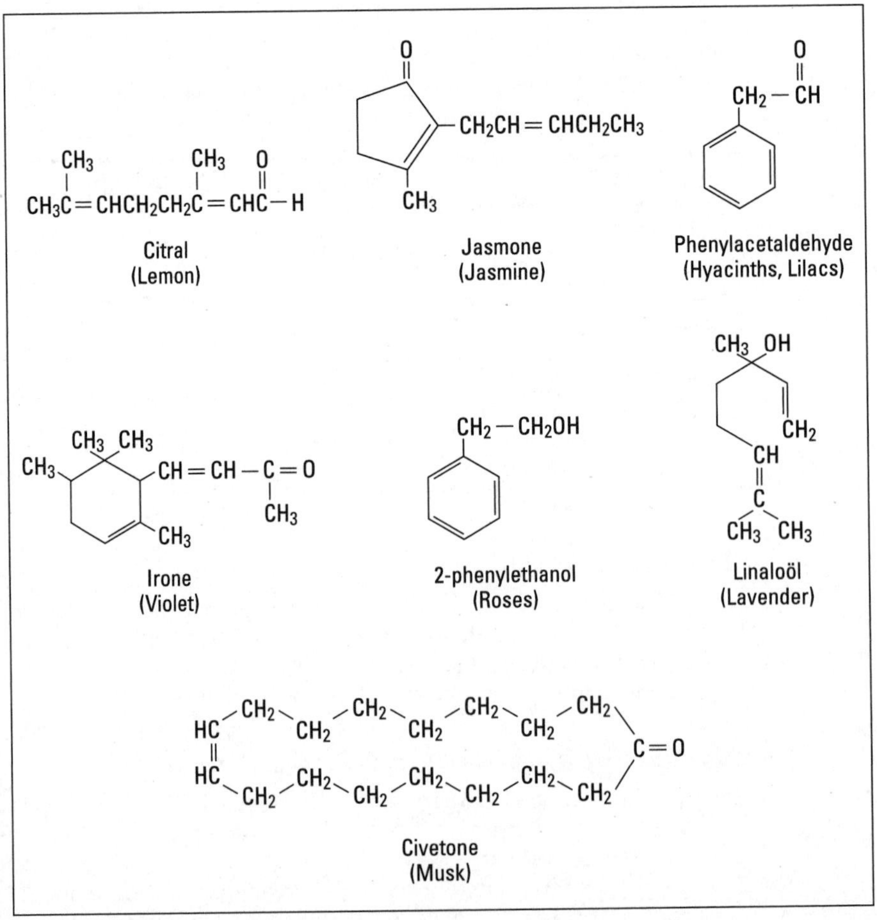

$CH_3C{=}CHCH_2CH_2C(CH_3){=}CHC({=}O){-}H$ (with CH_3 on the first C)

Citral
(Lemon)

$-CH_2CH{=}CHCH_2CH_3$, O, CH_3

Jasmone
(Jasmine)

CH_2-CH, O

Phenylacetaldehyde
(Hyacinths, Lilacs)

CH_3 CH_3, CH_3, $CH{=}CH-C{=}O$, CH_3, CH_3

Irone
(Violet)

CH_2-CH_2OH

2-phenylethanol
(Roses)

CH_3 OH, CH_2, CH, C, CH_3 CH_3

Linaloöl
(Lavender)

HC, HC, CH_2 ..., $C{=}O$

Civetone
(Musk)

Figure 17-5: Perfume fragrances.

Suntan lotion and sunscreen: Brown is beautiful

A suntan is nature's way of protecting our bodies against the harmful UV rays of the sun. The UV spectrum is composed of two regions: the UV-A region and the UV-B region. The UV-A region has slightly longer wavelengths and tends to produce a tan rather than a burn. UV-B radiation is what's responsible for those quick sunburns most of us are familiar with. Repeated exposure to both of these harmful UV rays, especially UV-B rays, is related to an increase in the occurrences of skin cancers, such as melanoma.

Suntan lotions and sunscreens protect the skin by partially or totally blocking the sun's radiation in the UV range, allowing you to be exposed to the sun for longer periods of time without burning. Some suntan lotions and sunscreens block both the UV-A and UV-B regions. Other types selectively block the UV-B regions, allowing the UV-A rays through, which gives the body a chance to produce *melanin,* a dark pigment that acts as a natural shield against the sun's UV rays, *and* that desirable brown skin tone.

These products are given a Sun Protection Factor (SPF) rating. The *SPF value* is a ratio of the amount of time required to tan (or burn) with versus without the product. An SPF value of 10, for example, indicates that when using the product, you can be exposed to the sun 10 times as long without burning.

There's some debate about whether SPF values above 15 are any more effective than the value 15, because few tanning products effectively block the UV-A radiation. The FDA is currently examining tanning products carefully.

A number of chemical substances are effective at blocking UV radiation. An opaque cream of zinc oxide and titanium dioxide is the most effective type of sunscreen. In addition, para-aminobenzoic acid (PABA), benzophenone, and cinnamates are commonly used to block UV radiation. Recently, there has been a move away from the use of PABA, though. It's somewhat toxic, and a significant number of individuals are allergic to it.

Figure 17-6 shows the structures of several compounds used in suntan and sunscreen products. The dihydroxyacetone shown in the figure produces a tan without exposure to the sun. It reacts with the skin to produce a brown pigment.

Clean it, color it, curl it: Hair care chemistry

Hair is composed of a protein called keratin. The protein chains in the hair strand are connected to each other by what's called a *disulfide bond,* a sulfur-to-sulfur bond from the cystine (an amino acid component of hair) on one protein chain to another cystine on another protein chain.

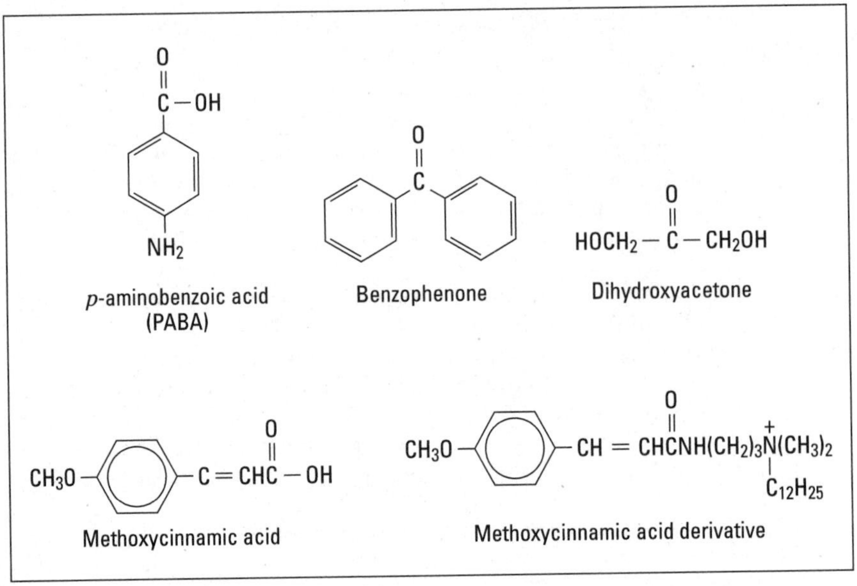

$HOCH_2-C(=O)-CH_2OH$

p-aminobenzoic acid (PABA)

Benzophenone

Dihydroxyacetone

$CH_3O-C_6H_4-C=CHC(=O)-OH$

Methoxycinnamic acid

$CH_3O-C_6H_4-CH=CHC(=O)NH(CH_2)_3N^+(CH_3)_2(C_{12}H_{25})$

Methoxycinnamic acid derivative

Figure 17-6: Tanning and sunscreen products.

Figure 17-7 shows a portion of hair and disulfide crosslinks joining two protein chains. These crosslinks give hair its strength. (I say more about this disulfide bond in "Permanents — that aren't," a little later in this section.)

Shampoos: Detergents for the hair

Modern shampoos are simple surfactants, such as sodium lauryl sulfate and sodium dodecyl sulfate. Shampoo contains other ingredients, however, that react with the metal ions in hard water to help prevent the soap from precipitating with these metal ions (in other words, to help prevent insoluble precipitates — solids, deposits, bathtub ring — from forming in your hair).

Other ingredients give a pleasant odor, replace some of the natural lubricants in the hair (conditioners), and adjust the hair's pH. (Hair and skin are slightly acidic. A very alkaline, or basic, shampoo will damage hair, so the pH is commonly adjusted to the 5 to 8 pH range. A higher pH may also make the scales on the hair cuticles fan out and reflect the light poorly, making the hair look flat and dull.) A protein is sometimes added to the shampoo to help glue damaged split-ends together. Colorants and preservatives are also commonly added, too.

Color that hair!

Hair contains two pigments — melanin and phaeomelanin. Melanin has a dark-brown color, and phaeomelanin has a reddish-brown color. The natural color of the hair is determined by the relative amounts of these pigments. Red heads have much less melanin; brunettes have much more. Blondes have very little of either.

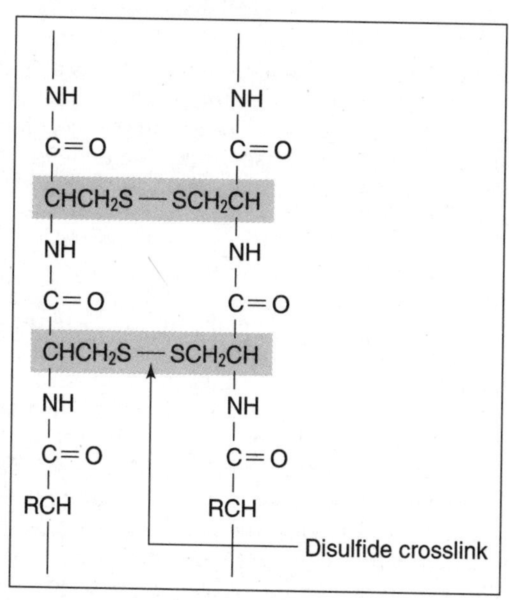

Figure 17-7: Disulfide bond in hair.

You can *bleach* hair by using hydrogen peroxide to oxidize these colored pigments to their colorless forms. However, bleached hair becomes weaker and more brittle, because the hair protein is broken down into lower molecular weight compounds. Perborate compounds, which tend to be more expensive than bleach, and chlorine-based bleaches are also sometimes used to bleach hair.

You can change the color of your hair temporarily by using dyes that simply coat the hair strands. These compounds are composed of complex organic molecules. They're too large to penetrate the hair strand, so they simply accumulate on the surface. You can add semi-permanent color by using dyes with smaller molecules that can penetrate into the hair. These dyes frequently contain complexes of chromium or cobalt. The dyes withstand repeated washing, but because the molecules contained in the dye were small enough to penetrate the hair initially, they eventually migrate out.

Permanent dyes are actually formed inside the hair. Small molecules are forced into the hair and then oxidized, normally by hydrogen peroxide, into colored complexes that are too large to migrate out of the hair. The color then becomes permanent on the portion of the hair that was treated. To maintain the color, you have to repeat the process as new hair grows out. This maintenance program keeps hairdressers in business.

Another type of hair coloring is made to change color gradually, over a period of weeks, so that the change goes unnoticed (fat chance!). A solution of lead acetate is applied to the hair. The lead ions react with the sulfur atoms in the hair protein, forming lead (II) sulfide (PbS), which is black and very insoluble. Instead of losing its color in the sunlight like other dyes, PbS-treated hair actually darkens.

Take it off, take it all off! Depilatories

Depilatories remove hair by chemical reaction. They contain a substance, usually sodium sulfide, calcium sulfide, or calcium thioglycolate, that disrupts the disulfide linkages in the hair and dissolves it. The formulations commonly contain a base such as calcium hydroxide to raise the pH and enhance the action of the depilatory. A detergent and a skin conditioner such as mineral oil are also generally added to depilatories.

Permanents — that aren't

Disulfide bonds are responsible for the shape of your hair, whether it's straight or curly. In order to affect the shape of hair, those disulfide bonds must be broken and reformed into a new orientation. Suppose, for example, that you want to make your straight hair curly, so you go to the beauty parlor for a permanent. The hairdresser initially treats your hair with some reducing agent that breaks the disulfide bonds; thioglycolic acid ($HS\text{-}CH_2\text{-}COOH$) is commonly used. Then the hairdresser changes the orientation of the protein chains of the hair by using curlers. Finally, the hairdresser treats your hair with an oxidizing agent such as hydrogen peroxide to reform the disulfide bonds in their new locations. Water-soluble polymers are used to thicken the solutions, ammonia is used to adjust the pH to a basic level, and a conditioner is used to complete the formulation. Figure 17-8 shows this process.

Hair is straightened in exactly the same fashion, except it's stretched straight instead of curled. Obviously, as new hair grows in, you need to repeat the process.

I guess *permanent* refers to the fact that trips to the beauty parlor become a permanent part of your life.

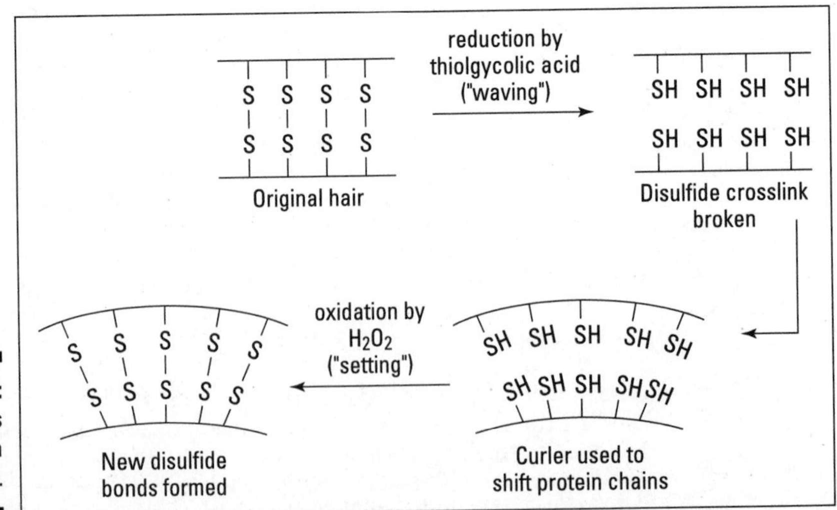

Figure 17-8: The process of getting a perm.

Chemistry in the Medicine Cabinet

Okay, take a quick peek in the medicine cabinet. There are a lot of drugs and medicines inside there. I could spend pages and pages talking about the chemistry of their reactions and interactions, but I'm just going to say a few brief words about a couple of them.

The aspirin story

As early as the fifth century B.C., it was known that chewing willow bark could relieve pain. But it wasn't until 1860 that the chemical compound responsible for the analgesic effect, salicylic acid, was isolated. It had a very sour taste and caused irritation of the stomach. In 1875, chemists created sodium salicylate. It caused stomach irritation, but it was less bitter than the salicylic acid. Finally, in 1899, the German Bayer Company began marketing acetylsalicylic acid, made by reacting salicylic acid with acetic anhydride, under the trade name of *aspirin.* Figure 17-9 traces the history of aspirin.

Aspirin is the most widely used drug in the world. More than 55 billion aspirin tablets are sold annually in the United States.

Minoxidil and Viagra

Science proceeds by hard work, training, intuition, hunches, and luck. That luck is sometimes called *serendipity,* which is another name for an accidental discovery. Or, as I like to say, "finding something you didn't know you were looking for." Chapter 20 tells the stories of ten serendipitous discoveries. But because I'm in the medicine cabinet anyway, I may as well mention a couple of serendipitous discoveries right now.

Male pattern baldness affects many millions of men and women in the world. Minoxidil, the current over-the-counter treatment for baldness, was discovered quite by accident. It was being used as an oral treatment for high blood pressure, when patients reported hair growth. Now it's usually applied topically instead of orally.

The much-publicized properties of Viagra were discovered in much the same fashion. It was also being used as a treatment for high blood pressure, as well as angina (heart pain), when its side effect was reported. In fact, it has been said that male patients refused to return the unused portion of their medications during clinical trials.

Both serendipitous discoveries have spawned multimillion-dollar industries and made countless men and women very happy. They're *growing* industries, for sure.

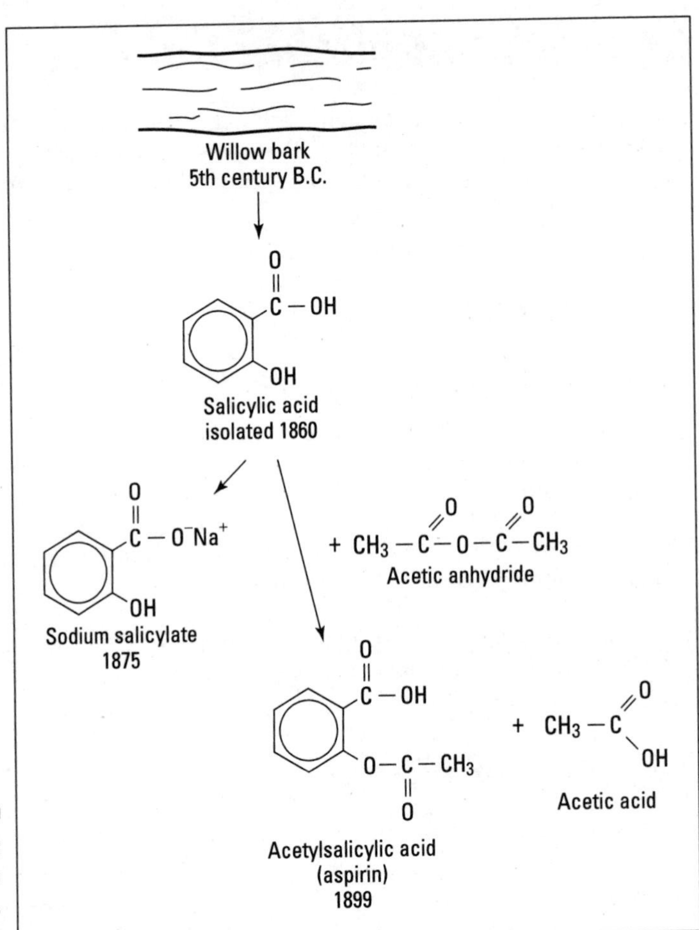

Figure 17-9: The aspirin story.

Chapter 18

Cough! Cough! Hack! Hack! Air Pollution

In This Chapter

- Finding out about the parts of the atmosphere involved in air pollution
- Looking at ozone depletion and the Greenhouse Effect
- Examining the causes of photochemical smog and acid rain

This chapter looks at the global problem of air pollution. (I consider the perfume department of a large department store at Christmas time the ultimate in air pollution, but I won't discuss that here.) I show you the chemical problems involved with air pollution, and I explain how air pollution is linked to modern society and its demand for energy and personal transportation.

Civilization's Effect on the Atmosphere (Or Where This Mess Began)

The air that surrounds the earth — our *atmosphere* — is absolutely necessary for life. The atmosphere provides oxygen (O_2) for respiration and carbon dioxide (CO_2) for *photosynthesis,* the process by which organisms (mainly plants) convert light energy into chemical energy; it moderates the temperature of the earth and plays an active part in many of the cycles that sustain life. The atmosphere is affected by many chemical reactions that take place or exist on earth.

When few humans were on earth, mankind's effect on the atmosphere was negligible. But as the world's population grew, the effect of civilization on the atmosphere became increasingly significant. The Industrial Revolution, which gave rise to the construction of large, concentrated industrial sites, added to man's effect on the atmosphere. As humans burned more *fossil fuels* — organic substances, such as coal, that are found in underground deposits and used for energy — the amount of carbon dioxide (CO_2) and *particulates* (small, solid particles suspended in the air) in the atmosphere increased significantly. During the Industrial Revolution, humans also began to use more items that released chemical pollutants into the atmosphere, including hairsprays and air conditioners.

The increase in CO_2 and particulates, combined with the increase in pollutants, has disrupted delicate balances in the atmosphere. High concentrations of these atmospheric pollutants have led to a multitude of problems such as *acid deposition,* acidic rain that damages living things, buildings, and statues, and *photochemical smog,* the brown, irritating haze that often sits over Los Angeles and other cities.

To Breathe or Not to Breathe: Our Atmosphere

The earth's atmosphere is divided into several layers: the troposphere, the stratosphere, the mesosphere, and the thermosphere. I want to focus on the two layers closest to the earth — the troposphere and stratosphere — because they're the layers affected the most by humans. They're also the layers that have the greatest direct effect on human life.

- The *troposphere* lies next to the earth and contains the gases we breathe and depend on for survival.
- The *stratosphere* contains the ozone layer, which protects us from ultraviolet radiation.

The troposphere: What humans affect most

The troposphere is composed of about 78.1 percent nitrogen (N_2), 20.9 percent oxygen (O_2), 0.9 percent argon (Ar), 0.03 percent carbon dioxide (CO_2), and smaller amounts of various other gases. The troposphere also contains varying amounts of water vapor. These gases are held tight to the earth by the force of gravity. If a balloonist were to rise high into the troposphere, he or she would find the atmospheric gases much thinner due to the decreased pull of gravity on the gases. This effect tells us that the dense layer of gases held tight to the earth is more at risk from the effects of pollution.

The troposphere is the layer where our weather occurs. It's also the layer that takes the brunt of both natural and man-made pollution because of its proximity to the earth.

Nature pollutes the atmosphere to a certain extent — with noxious hydrogen sulfide (H_2S) and particulate matter from volcanoes, and the release of organic compounds from plants such as pine trees. But these pollutants have a minor effect on the troposphere. Humankind, on the other hand, pollutes the troposphere with a large amount of chemicals from automobiles, power plants, and industries. Acid rain and photochemical smog are some of the results of man-made pollutants.

The stratosphere: Protecting humans with the ozone layer

Above the troposphere is the stratosphere, which is where jets and high-altitude balloons fly. The atmosphere is much thinner in this layer because of the decreasing pull of gravity. Few of the heavier pollutants are able to make it to the stratosphere, because gravity holds them tight and close to the surface of the earth. The protective *ozone layer* resides in the stratosphere; this protective barrier absorbs a large amount of harmful ultraviolet (UV) radiation from the sun and keeps it from reaching the earth.

Even though heavier pollutants don't make their way to the stratosphere, this layer isn't immune to the effects of mankind. Some lighter manmade gases do make it into the stratosphere, where they attack the protective ozone layer and destroy it. This destruction can have far-reaching effects on humans because UV radiation is a major cause of skin cancer.

A chemical substance can be both a good guy and a bad guy. The only difference is where, and in what concentration, it's found. For example, a person can overdose on water if he or she drinks enough of it. The same goes with the ozone in the stratosphere. On one hand, it shields us from harmful UV radiation. But on the other, it can be an irritant and destroy rubber products (see "Brown Air? (Photochemical Smog)" for details).

Leave My Ozone Alone: Hair Spray, CFCs, and Ozone Depletion

The ozone layer absorbs almost 99 percent of the ultraviolet radiation that reaches the earth from the sun. It protects us from the effects of too much ultraviolet radiation, including sunburns, skin cancers, cataracts, and premature aging of the skin. Because of the ozone layer, most of us can enjoy the outdoors without head-to-toe protection.

How is ozone (O_3) formed? Well, oxygen in the *mesosphere* — the part of the earth's atmosphere between the stratosphere and the *thermosphere* (the layer that extends to outer space) — is broken apart by ultraviolet radiation into highly reactive oxygen atoms. These oxygen atoms combine with oxygen molecules in the stratosphere to form ozone.

$O_2(g)$ + ultraviolet radiation → 2 $O(g)$

$O_2(g) + O(g) \rightarrow O_3(g)$

As a society, humans release many gaseous chemicals into the atmosphere. Many of the gaseous chemicals rapidly decompose through reaction with each other, or they react with the water vapor in the atmosphere to form compounds such as acids that fall to earth in the rain (see "'I'm Meltingggggg!' — Acid Rain," later in this chapter). Besides forming acid rain, some of these chemicals also form photochemical smog (see "Brown Air? (Photochemical Smog)," later in this chapter).

But these reactions occur rather quickly, and we can deal with them in a variety of ways, many of which are related to breaking the series of reactions that produce the pollutant by stopping the release of a critical chemical into the air.

Some classes of gaseous chemical compounds are rather *inert* (inactive and unreactive), so they remain with us for quite a while. Because these inert compounds stick around, they have a negative effect on the atmosphere. One such troublesome class of compounds are the *chlorofluorocarbons,* gaseous compounds composed of chlorine, fluorine, and carbon. These compounds are commonly called *CFCs.*

Because CFCs are relatively unreactive, they were extensively used in the past as refrigerants for such items as refrigerators and automobile air conditioners (Freon-12), foaming agents for plastics such as Styrofoam, and propellants for the aerosol cans of such consumer goods as hair spray and deodorants. As a result, they were released into the atmosphere in great quantities. Over the years, the CFCs have diffused into the stratosphere, and they're now doing damage to it.

How do CFCs hurt the ozone layer?

Although CFCs don't react much when they're close to earth — they're pretty inert — most scientists believe that they react with the ozone in the atmosphere and then harm the ozone layer in the stratosphere.

The reaction occurs in the following way:

1. A typical chlorofluorocarbon, CF_2Cl_2, reacts with ultraviolet radiation, and a highly reactive chlorine atom is formed.

   $CF_2Cl_2(g)$ + UV light → $CF_2Cl(g) + Cl(g)$

2. The reactive chlorine atom reacts with ozone in the stratosphere to produce oxygen gas molecules and chlorine oxide (ClO).

$Cl(g) + O_3(g) \rightarrow O_2(g) + ClO(g)$

This is the reaction that destroys the ozone layer. If things stopped here, the problems would actually be minimal.

3. The chlorine oxide (ClO) can then react with another oxygen atom in the stratosphere to produce an oxygen molecule and a chlorine atom; the newly created oxygen molecule and chlorine atom are now available to start the whole ozone-destroying process all over again.

   $CLO(g) + O(g) \rightarrow O_2(g) + Cl(g)$

   So one CFC molecule can initiate a process that can destroy many molecules of ozone.

Because they're harmful, are CFCs still produced?

The problem of ozone depletion was identified in the 1970s. As a result, the governments of many industrialized nations began to require the reduction of the amount of CFCs and halons released into the atmosphere. (*Halons,* which contain bromine in addition to fluorine and chlorine, were commonly used as fire-extinguishing agents, especially in fire extinguishers used around computers.)

CFCs were banned for use as propellants in aerosol cans in many countries, and the CFCs used in the production of plastics and foams were recovered instead of released into the air. Laws were enacted to ensure that the CFCs and halons used as refrigerants were recovered during the recharging and repair of units. In 1991, Du Pont started producing refrigerants that weren't harmful to the ozone layer. And in 1996, the United States, along with 140 other countries, stopped producing chlorofluorocarbons altogether.

Unfortunately, though, these compounds are extremely stable. They'll remain in our atmosphere for many years. If the damage man has done to the ozone layer isn't too great, it may replenish itself (like new skin grows to replace sunburned skin). But it may well be several years before the ozone layer returns to its former composition.

Is It Hot in Here to You? (The Greenhouse Effect)

When most people think about air pollutants, they think of such chemicals as carbon monoxide, chlorofluorocarbons, or hydrocarbons. Yet carbon dioxide, the product of animal respiration and the compound used by plants in the process of photosynthesis, can also be considered a pollutant if present in abnormally high amounts.

In the late 1970s and early 1980s, scientists realized that the average temperature of the earth was increasing. They determined that an increase in carbon dioxide (CO_2) and a few other gases, such as chlorofluorocarbons (CFCs), methane (CH_4, a hydrocarbon), and water vapor (H_2O), were responsible for the slight increase in temperature through a process called the *greenhouse effect* (named so because the gases serve pretty much the same purpose as the glass walls and roof of a greenhouse — the gases themselves are called *greenhouse gases*).

Here's how the greenhouse effect works: As radiation from the sun travels through the earth's atmosphere, it strikes the earth, heating the land and water. Some of this solar energy is sent back (reflected) into the atmosphere as heat (infrared radiation), which is then absorbed by certain gases (CO_2, CH_4, H_2O, and CFCs) in the atmosphere. These gases, in turn, warm the atmosphere. This process helps to keep the temperature of the earth and atmosphere moderate and relatively constant, and as a result, we don't experience dramatic day-to-day temperature fluctuations. So, in general, the greenhouse effect is a good thing, not a bad thing.

But if there's an excess of carbon dioxide and other greenhouse gases, too much heat gets trapped in the atmosphere. The atmosphere heats up, leading to the disruption of many of the delicate cycles of the earth. This process is commonly called *global warming*, and it's currently happening with the earth's atmosphere.

We depend on burning fossil fuels (coal, natural gas, or petroleum) for energy. We burn coal and natural gas to produce electricity, we burn gasoline in the internal combustion engine, and we burn natural gas, oil, wood, and coal to heat homes. In addition, industrial processes burn fuel to produce heat. As a result of all this burning of fossil fuels, the carbon dioxide level in the atmosphere has risen from 318 parts-per-million (ppm) in 1960 to 362 parts-per-million (ppm) in 1998. (For a discussion of the concentration unit *ppm*, see Chapter 11.) The excess carbon dioxide has led to an increase of about a half-degree in the average temperature of the atmosphere.

A half-degree increase in the average temperature of the atmosphere may not sound like much, but this global warming trend may have serious effects on several of the ecological systems of the world:

- The rising atmospheric temperature may melt ice masses and cause sea levels around the world to rise. Rising sea levels may result in the loss of coastal land (Houston might become a coastal town) and make many more people vulnerable to *storm surges* (those extremely damaging rushes of seawater that occur during very bad storms).

- The increased temperature may affect the growth patterns of plants.
- The tropical regions of the world may increase and lead to the spread of tropical diseases.

Brown Air? (Photochemical Smog)

Smog is a generic word people use to describe the combination of smoke and fog that's often irritating to breathe. There are two major types of smog:

- London smog
- Photochemical smog

London smog

London smog is a gaseous atmospheric mixture of fog, soot, ash, sulfuric acid (H_2SO_4 — battery acid), and sulfur dioxide (SO_2). The name comes from the air pollution that plagued London in the early part of the twentieth century. The burning of coal for heat in the highly populated city caused this smog. The dangerous mixture of gases and soot from the coal stoves and furnaces killed more than 8,000 people in London in 1952.

Electrostatic precipitators and scrubbers (see "Charge them up and drop them out: Electrostatic precipitators" and "Washing water: Scrubbers," later in this chapter), combined with filters, have been effective in reducing the release of soot, ash, and sulfur dioxide into the atmosphere and have reduced the occurrence of London smog.

Photochemical smog

Photochemical smog is produced after sunlight initiates certain chemical reactions involving unburned hydrocarbons and oxides of nitrogen (commonly shown as NO_x — which stands for a mixture of NO and NO_2). The common automobile engine produces both of these compounds when it's running.

Photochemical smog is the brown haze that sometimes makes it difficult to see in such cities as Los Angeles, Salt Lake City, Denver, and Phoenix. (This smog is sometimes called *Los Angeles smog* — sometimes sunny California isn't so sunny.) These cities are especially vulnerable to photochemical smog; they have a large number of automobiles, which emit the chemicals that react to produce the smog, and they're surrounded by mountain ranges. The mountain ranges and the westward winds create an ideal condition for thermal inversions, which trap the pollutants close to the cities. (In a *thermal inversion,* a layer of warmer air moves in over a layer of cooler air. The warm air traps the cooler air and its pollutants close to the ground. The process can be compared to sheets trapping certain noxious gases in a bed. The gaseous pollutants are trapped and can't move higher in the atmosphere. They stay close to us humans, causing all kinds of problems.)

The chemistry of photochemical smog is still not crystal clear (pun intended), but scientists do know the basics that go into creating the smog. Nitrogen from the atmosphere is oxidized to nitric oxide in internal combustion engines and then released into the atmosphere through the engines' exhaust systems:

$$N_2(g) + O_2(g) \rightarrow 2\ NO(g)$$

The nitric oxide is oxidized to nitrogen dioxide by atmospheric oxygen:

$$2\ NO(g) + O_2(g) \rightarrow 2\ NO_2(g)$$

Nitrogen dioxide is a brownish gas. It's irritating to the eyes and lungs. It absorbs sunlight and then produces nitric oxide and highly reactive oxygen atoms:

$$NO_2(g) + \text{sunlight} \rightarrow NO(g) + O(g)$$

These reactive oxygen atoms quickly react with diatomic (two-atom) oxygen gas molecules in the air to produce ozone (O_3):

$$O(g) + O_2(g) \rightarrow O_3(g)$$

This is the same ozone that acts as a shield against ultraviolet radiation in the stratosphere. But when it's down closer to the earth, it acts as a powerful irritant to the eyes and lungs. It attacks rubber, causing it to harden, and thus shortens the life of automobile tires and weather stripping. It also affects crops such as tomatoes and tobacco.

The unburned hydrocarbons from auto exhaust also react with the oxygen atoms and ozone to produce a variety of organic aldehydes that are also irritants. These hydrocarbons can react with diatomic oxygen and nitrogen dioxide to produce peroxyacetylnitrates (PANs):

$$\text{Hydrocarbons}(g) + O(g) + NO_2(g) \rightarrow \text{PANs}$$

These PANs are also eye and lung irritants; they tend to be very reactive, causing damage to living organisms.

The combination of the brown nitrogen dioxide, the ozone, and the PANs is photochemical smog. It reduces visibility and is a major cause of respiratory problems. And, unfortunately, controlling it has been difficult.

Auto emissions have been closely monitored, and strict controls have been put into place to minimize the amount of unburned hydrocarbons released into the atmosphere. The Clean Air Act of 1990 was passed to help reduce hydrocarbon emissions from automobiles. The catalytic converter was developed to help react the unburned hydrocarbon and produce a less dangerous emission of carbon dioxide and water. (As a side benefit, lead had to be eliminated from gasoline because it "poisoned" the catalyst and made the catalytic converter useless. The big campaign to "get the lead out" removed a major source of the deadly heavy metal from the environment.)

Although such measures as catalytic converters and activated carbon canisters, which are used to help reduce gasoline fumes, have been somewhat effective, photochemical smog still presents a problem. Until mankind develops an acceptable substitute for the internal combustion engine or requires mass transit, photochemical smog will remain with us for years to come.

"I'm Meltingggggg!" — Acid Rain

The wicked witch in *The Wizard of Oz* dissolved in water. Sometimes buildings do the same because of the action of acid rain on the limestone and marble.

Rainwater is naturally acidic (with a pH less than 7) as a result of the dissolving of carbon dioxide in the moisture of the atmosphere and the forming of carbonic acid. (See Chapter 12 for more about carbonic acid as well as the pH scale.) This interaction results in rainwater having a pH of around 5.6. The term *acid rain,* or *acid deposition,* is used to describe a situation in which rainfall has a much lower (more acidic) pH than can be explained by the simple dissolving of carbon dioxide. Specifically, acid rain is formed when certain pollutants in the atmosphere, primarily oxides of nitrogen and sulfur, dissolve in the moisture of the atmosphere and fall to earth as rain with a low pH value.

Oxides of nitrogen (NO, NO_2, and so on) are produced naturally during lightning discharges in the atmosphere. This is one way that nature "fixes" nitrogen, or puts it in a form that can be used by plants. However, man adds tremendously to the local amount of atmospheric nitrogen oxides through the use of automobiles. The internal combustion engine reacts the gasoline hydrocarbons with the oxygen in the air, producing carbon dioxide (and carbon monoxide) and water. But the nitrogen that's present in the air (about 78 percent of the air is nitrogen) may also react with the oxygen at the high temperatures present in the engine. This can produce nitric oxide (NO), which is then released into the atmosphere:

$$N_2(g) + O_2(g) \rightarrow 2\,NO(g)$$

As the NO enters the atmosphere, it reacts with additional oxygen gas to produce nitrogen dioxide (NO_2):

$$2\ NO(g) + O_2\ (g) \rightarrow 2\ NO_2(g)$$

This nitrogen dioxide can then react with the water vapor in the atmosphere to form nitric and nitrous acids:

$$2\ NO_2(g) + H_2O(g) \rightarrow HNO_3(aq) + HNO_2(aq)$$

These dilute acid solutions fall to earth as rain with a low pH value — generally in the 4.0 to 4.5 range (although rains with a pH as low as 1.5 have been reported).

A significant amount of acid rain in the eastern part of the United States is caused by oxides of nitrogen, but the acid rain of the Midwest and West is caused by mostly oxides of sulfur, which are primarily generated by power plants and the burning of coal and oil. Sulfur-containing compounds are found as impurities in coal and oil, sometimes as high as 4 percent by weight. These compounds, when burned, produce a sulfur dioxide gas (SO_2). Many millions of tons of SO_2 are released into the atmosphere each year from power-generating plants. The SO_2 reacts with the water vapor in the atmosphere to produce sulfurous acid (H_2SO_3), and with the oxygen in the atmosphere to produce sulfur trioxide (SO_3):

$$SO_2(g) + H_2O(g) \rightarrow H_2SO_3(aq)$$

$$2\ SO_2(g) + O_2(g) \rightarrow 2\ SO_3(g)$$

This sulfur trioxide then reacts with the moisture in the atmosphere to produce sulfuric acid (H_2SO_4), which is the same acid found in your automobile battery:

$$SO_3(g) + H_2O(g) \rightarrow H_2SO_4(aq)$$

So the sulfurous and sulfuric acids that are dissolved in the rainwater form the acid rain that falls to the earth. Anyone for a bath in battery acid?

The acids formed in the atmosphere can travel many hundreds of miles before falling to earth as acid rain and leaving their mark on both nonliving and living things. The acids of the rain react with the iron in buildings and automobiles, causing them to corrode. The acids also destroy the details of fine works of art when they react with marble statues and limestone buildings to form soluble compounds that wash away. (Want to see this in action? Put a drop of vinegar, an acid, on a piece of marble and then watch the bubbles form as the acid dissolves the marble. Careful, though. Don't try this on anything too valuable — maybe that ugly marble cheese slicer that Aunt Gertrude gave you last Christmas.)

It's not surprising that acid rain has a bad effect on vegetation. Acid rain has been identified as the major cause of death of many trees and even whole forests. Even if trees aren't killed immediately by acid rain, forests sometimes grow slower because of its effects. The growth may be hindered by the release of aluminum from the soil, which interferes with the absorption of nutrients, or it may be slowed by the bacteria found in the soil.

In addition, acid rain has altered the ecosystems of many lakes in Canada and the United States. Fish kills have been reported, and entire species of fish have vanished from certain lakes. In fact, the ecosystems of entire lakes have been destroyed by acid rain, rendering the lakes lifeless.

Steps have been taken to reduce acid rain and its effects. Increasing fuel efficiency and the use of pollution control devices on automobiles have helped reduce the amount of nitrogen oxide released into the atmosphere. But fossil fuel power plants produce the most tonnage of acid-causing pollutants. A number of controls have been adopted to decrease the amount of sulfur-containing gases released into the atmosphere, including electrostatic precipitators and scrubbers, which are discussed in the following two sections. But although they've been effective in reducing the amount of acid-causing material released into the atmosphere, much more still needs to be done before the problem of acid rain is reduced to a manageable level.

Charge them up and drop them out: Electrostatic precipitators

When you were a child, did you ever run a comb through your hair on a cold winter morning and then use it to pick up little scraps of paper? An electrostatic precipitator does much the same thing.

Electrostatic precipitators give a negative electrical charge to pollutant particles. The sides of the precipitator have a positive charge, so the negative particles are then pulled to the positively charged walls. They stick to the walls and accumulate there. Then they can be removed (it's like sweeping out those dust bunnies from under the bed).

In one type of electrostatic precipitation system, the SO_2 produced by the burning of fossil fuels is reacted with lime (CaO) to produce solid calcium sulfite ($CaSO_3$):

$$SO_2(g) + CaO(s) \rightarrow CaSO_3(s)$$

The finely divided calcium sulfite is electrostatically precipitated and collected. It can then be disposed of properly in a chemical landfill.

Washing water: Scrubbers

Scrubbers are thingies that remove impurities from pollutant gases by using a fine spray of water to trap the gases as an aqueous solution or force them through a reacting mixture. The process is similar to using a water spray to settle dust in arid regions.

You can use a scrubber as an especially efficient system for removing sulfur dioxide by forcing the SO_2 through a slurry of magnesium hydroxide and converting it to magnesium sulfite, which can then easily be collected:

$$SO_2(g) + Mg(OH)_2(aq) \rightarrow MgSO_3(s) + H_2O\ (l)$$

Is the quality of air getting better?

The quality of air in cities such as Los Angeles has improved over the last 15 years. Pollution controls have reduced the oxides of nitrogen and unburned hydrocarbons released by automobiles, and the levels of photochemical smog have been significantly reduced.

Pollution controls have also reduced the levels of sulfur dioxides released from power plants, which has helped lower the occurrence of acid rain. In addition, the ban on CFCs released into the atmosphere should eventually have an effect on ozone depletion. So in many respects, yes, the quality of our air is improving.

But humans are still releasing tremendous amounts of carbon dioxide into the atmosphere and using up large amounts of the earth's valuable plant and animal material (biomass). It is this biomass that really would tend to use up this excess carbon dioxide.

The effect on the environment is debated on a daily basis. All agree that the effect is negative. The question is simply a matter of degree. If mankind can reduce its dependence on fossil fuels for electricity and heat by using solar, nuclear, or perhaps even fusion power, then maybe we'll be able to make advances in reducing the amount of carbon dioxide released into the atmosphere. This strategy, combined with limits on the destruction of forests, may bring the problem of global warming under control.

Chapter 19

Brown, Chunky Water? Water Pollution

In This Chapter

- Understanding where our water supply comes from
- Clarifying how the structure of water makes it vulnerable to pollution
- Taking a look at several types of water pollutants
- Finding out about water treatment

Water is absolutely necessary to our survival. After all, the human body is about 70 percent water. Most of the water on earth, however, is found as seawater. Only about 2 percent of the water on the earth is fresh water, and a little over three-quarters of that is in the form of ice and glaciers. But it's that very small amount of fresh water suitable for drinking (*potable* water) that most people are concerned about.

I'm sure you're quite aware of the water you drink and the water you use for bathing, cooking, and watering your lawn. But unless you live in a rural agricultural area, I doubt that you think much about the water used to grow the plants and animals we depend on for food.

In addition, water is used to carry waste products from our homes and to generate electricity. It's also used in chemical reactions and cooling towers. And then there's recreation — boating, swimming, and fishing. All these things depend on an adequate supply of good, pure water.

But where does water come from? How does it get contaminated, and how does it get cleaned? These are some of the questions I discuss in this chapter. So sit back, grab your glass of water, and dive in.

Where Does Our Water Come From, and Where Is It Going?

The actual amount of water on earth is relatively constant, but its location and purity may vary. Water moves throughout the environment by what is called the *water cycle,* or the *hydrologic cycle.* Figure 19-1 diagrams this cycle.

Evaporate, condense, repeat

Water *evaporates* (goes from a liquid to a gas when heated) from lakes, streams, oceans, trees, and even humans. As water evaporates, it leaves behind any contaminates that it may have accumulated. (That's where the salt comes from on your sweatband and cap.) This process of evaporation is one of nature's ways of purifying water.

The water vapor may then travel many miles, or it may stay relatively local, depending on the prevailing winds. Sooner or later, the vapor *condenses* (goes from a gas to a liquid when cooled) and falls back to the earth as rain, snow, or sleet.

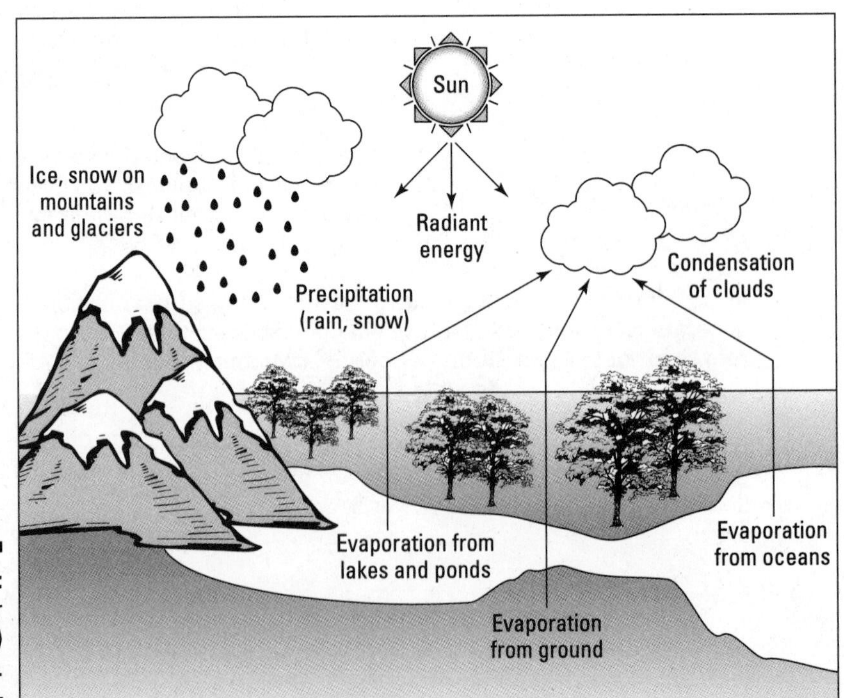

Figure 19-1: The water (hydrologic) cycle.

Where the water goes

Water may fall to earth and collect in a lake or stream. If it does, it eventually finds its way back to the sea. If it falls onto the land, it can form *runoff* and eventually enter a lake or stream, or it can soak into the ground and become *groundwater.* The porous layer of soil and rock that holds the groundwater forms a zone called an *aquifer.* This zone provides us with a good source of groundwater. We tap into these aquifers by using wells.

Human activities can affect this water cycle. Cutting vegetation can increase the rate of runoff, causing less water to become absorbed into the soil. Man-made dams and reservoirs increase the surface area available for water evaporation. Using more groundwater than can be replenished may deplete the aquifers and lead to water shortages. And society can contaminate the water in a wide variety of ways that I discuss in this chapter.

Water: A Most Unusual Substance

Water is a polar molecule. Chapter 7 covers polar molecules in detail, but here's a quickie version relating to water: The oxygen in water (H_20) has a higher *electronegativity* (attraction for a bonding pair of electrons) than the hydrogen atoms, so the bonding electrons are pulled in closer to the oxygen. The oxygen end of the water molecule then acquires a partial negative charge, and the hydrogen atoms take on a partial positive charge. When the partially positively charged hydrogen (where's my editor on *that* clunker of a phrase?) of one water molecule is attracted to the partially negatively charged oxygen of another water molecule, there can be a rather strong interaction between the water molecules. This interaction is called a hydrogen bond (H-bond). This is not to be confused with a hydrogen *bomb.* Two very different things. Figure 19-2 shows the hydrogen bonds that occur in water.

Hydrogen bonds, caused by the polar covalent bonds of water molecules, give water some very unusual properties:

- **Water has a very high surface tension.** The water molecules at the surface of the water are only attracted downward into the body of the liquid. The molecules in the body of the liquid, on the other hand, are attracted into all different directions. Bugs and small lizards can walk across water because they don't exert enough force to break the surface tension. The high surface tension of water also means that evaporation rates are really less than you'd expect.
- **Water becomes a liquid at temperatures commonly found on earth.** The boiling point of a liquid is normally related to its molecular weight. Substances that have molecular weights close to the molecular weight of water (18 g/mol) boil at far lower temperatures; these substances become gases at normal room temperature.

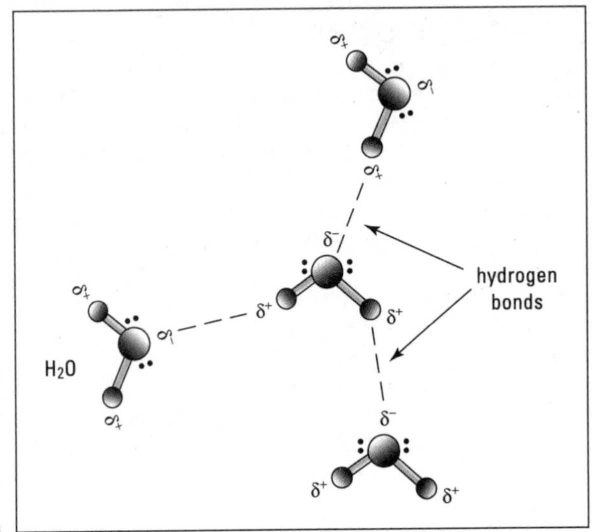

Figure 19-2: Hydrogen bonding in water.

- **Ice, the solid state of water, floats when placed in water.** Normally, you may think of a solid as having a higher density than its corresponding liquid, because the particles are closer together in the solid. When water freezes, however, it's locked into a crystal lattice that has large holes incorporated into it by its hydrogen bonds. So the density of ice is less than that of water (see Figure 19-3).

  The floating property of ice is one of the reasons that life, in all its diversity and magnitude, is able to exist on earth. If ice were denser than water in the winter, the water at the top of lakes would freeze and sink. Then more water would freeze and sink, and so on. Pretty soon, the lake would be frozen solid, destroying most of the life — such as plants and fish — in the lake. Instead, ice floats and forms an insulating layer over the water, which allows life to exist, even in the winter.

- **Water has a relatively high heat capacity.** The *heat capacity* of a substance is the amount of heat a substance can absorb or release in order to change its temperature 1 degree Celsius. Water's heat capacity is almost 10 times greater than iron and 5 times greater than aluminum. This means that lakes and oceans can absorb and release large amounts of heat without a dramatic change in temperature, which moderates the temperature on earth. Lakes absorb heat during the day and release it at night. Without water's high heat capacity, the earth would undergo dramatic swings in temperature during its day/night cycle.

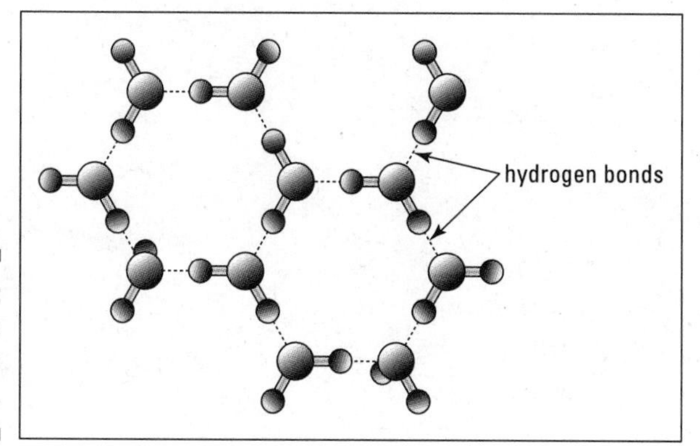

Figure 19-3: The structure of ice.

- **Water has a high heat of vaporization.** The *heat of vaporization* of a liquid is the amount of energy needed to convert a gram of the liquid to a gas. Water has a heat of vaporization of 54 calories per gram (see Chapter 2 for more about the calorie, a metric unit of heat). This high heat of vaporization allows us to rid our bodies of a great deal of heat when sweat is evaporated from the skin. This property also helps to keep the climate on earth relatively moderate without extreme short-term swings.
- **Water is an excellent solvent for a large number of substances.** In fact, water is sometimes called the universal solvent, because it dissolves so many things. Water is a polar molecule, so it acts as a solvent for polar solutes. It dissolves ionic substances easily; the negative ends of the water molecules surround the cations (positively charged ions), while the positive ends of the water molecules surround the anions (negatively charged ions). (Turn to Chapter 6 for specifics on ions, cations, and anions.) With the same process, water can dissolve many polar covalent compounds, such as alcohols and sugars (see Chapter 7 for more on these types of compounds). This is a desirable property, but it also means that water dissolves many substances that are not desirable to us or that make water unusable. We lump all those substances together under the terms *pollutants* or *contaminants.*

Yuck! Some Common Water Pollutants

Because water is such an excellent solvent, it easily picks up unwanted substances from a variety of sources. Figure 19-4 shows some sources of water contamination.

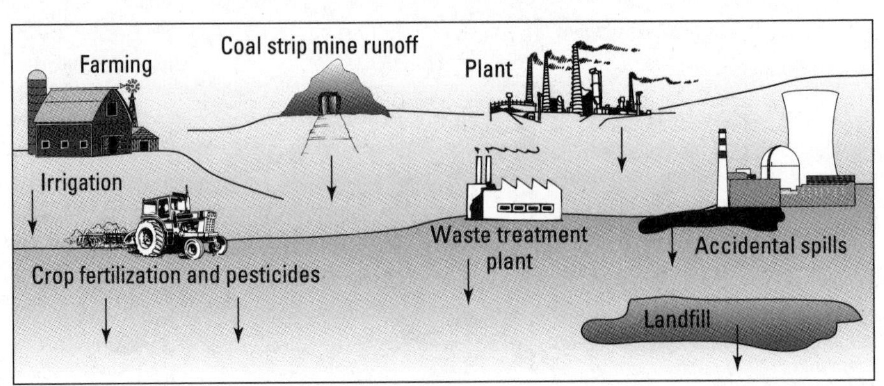

Figure 19-4: Some sources of water pollution.

I call Figure 19-4 Pollution Place, because it shows so many pollution sources in the same place. Naturally, you won't find this many pollution sites this close together in too many places in the United States.

Pollution sources are normally classified as point sources or non-point sources:

- *Point sources* are pollution sites that have a definite identifiable source. Discharges from a chemical industry or raw sewage from wastewater treatment plants are common examples of point-source pollution. Point sources are easy to identify, control, and regulate. The Environmental Protection Agency (EPA) is the governmental agency that regulates point sources.
- *Non-point sources* are pollution sources that are rather diffuse in nature. Good examples of this type of pollution are water contamination caused by agricultural runoff or acid rain. Controlling and regulating this type of pollution is much more difficult because you can't identify a particular company or individual as the polluter. In recent years, federal and state agencies have attempted to address non-point source pollution. The Clean Water Action Plan of 1998 was one such attempt that focused on watersheds and runoffs.

We really didn't get the lead out: Heavy metal contamination

Water supplies are closely monitored for heavy metals, because they tend to be very toxic. Major sources of heavy metal contamination include landfills, industries, agriculture, mining, and old water distribution systems.

Lead is one type of heavy metal pollutant that has received a lot of press in recent years. Large amounts of lead entered the environment from the use of leaded gasoline: The tetraethyl lead that was used to boost the gasoline's octane was oxidized in the combustion process, and a large amount of lead was emitted through exhaust systems. Rain runoff carried the lead into streams where it was deposited. Another source of lead was old pipes in municipal buildings and homes. These pipes were commonly joined with a lead solder that then leached lead into the drinking water.

Mercury is released into the aquatic environment from mercury compounds used to treat seeds from fungus and rot. Runoff from fields washes the mercury compounds into the surface water and sometimes into the groundwater supply.

The automobile is also an indirect source of another heavy metal contaminant, chromium. Chromium compounds (such as CrO_4^{2-}) are used in chrome plating for bumpers and grills. This plating also requires the use of the cyanide ion (CN^-), another major pollutant. These contaminants used to be discharged directly into streams, but now they're either pretreated to reduce to a less-toxic form or precipitated (formed into a solid) and disposed of in landfills.

Mining also adds to the heavy metal pollution problem. As the earth is mined, deposits of minerals, which contain metals, are exposed. If the chemicals used in extracting ore or coal deposits are acidic, then the metals in the minerals are dissolved, and they may make their way into the surface water and sometimes the groundwater. This problem is sometimes controlled with a process that isolates mine drainage and then treats it to remove the metal ions.

Biological concentration is a problem that occurs when industries release heavy metal ions into the waterways. As metal ions move through the ecosystem, they become more and more concentrated. (The same thing happens with radioisotopes — see Chapter 5 for details.) The ions may be released at a very low concentration level, but by the time they move up the food chain to us, the concentration may be at the toxic level. This situation happened in Minamata Bay, Japan. An industry was dumping mercury metal into the bay. As the metal moved through the ecosystem, it was eventually converted to the extremely toxic methylmercury compound. People died as a result of the toxins, and others became permanently affected.

Acid rain

Oxides of nitrogen and sulfur can combine with the moisture in the atmosphere to form rain that can be highly acidic — acid rain. This rain can affect the pH of lakes and streams and has been known to seriously affect aquatic life. In fact, it's made some lakes devoid of life altogether.

Acid rain is a good example of a non-point source of pollution. It's difficult to pinpoint a single entity as the cause. Air pollution controls have decreased the amount of acid rain produced, but it's still a major problem. (If you'd like more info on acid rain, flip to Chapter 18.)

Infectious agents

This category of contamination includes fecal coliform bacteria from human wastes and the wastes of birds and other animals. Fecal coliform bacteria was once a major problem in the United States and most parts of the world. Epidemics of typhoid, cholera, and dysentery were common. Treatment of wastewater has minimized this problem in industrialized nations, but it's still a definite problem in underdeveloped nations.

Many experts think that more than three-quarters of the sicknesses in the world are related to biological water contaminates. And even now in the United States, beaches and lakes are still closed at times because of biological contamination.

Stricter controls on municipal water treatment, septic tanks, and runoff from feedlots will help decrease the biological contamination of our water.

Landfills and LUST

Landfills — both the public and hazardous chemical kind — are a major source of groundwater contamination. The landfills that are constructed today require special liners to prevent hazardous materials from leaching into the groundwater. Monitoring equipment is also required to confirm that the hazardous materials don't leak from the landfills. However, very few landfills in the United States have liners and monitoring systems.

Many landfills contain *VOCs* (volatile organic chemicals). This group of chemicals includes benzene and toluene (both carcinogens), chlorinated hydrocarbons, such as carbon tetrachloride, and trichloroethylene, which previously was used as a dry-cleaning solvent. Even though these compounds are not very soluble in water, they do accumulate at the parts-per-million level. Their long-term effect on human health is unknown at this time.

Most people think of toxic wastes in terms of an industrial dump, but the municipal landfill is becoming a more popular site for the disposal of hazardous household wastes. Every year, tons of the following toxic materials are placed in commercial landfills:

- Batteries containing heavy metals like mercury
- Oil-based paints containing organic solvents
- Motor oil containing metals and organic compounds

- Gasoline containing organic solvents
- Automobile batteries containing sulfuric acid and lead
- Antifreeze containing organic solvents
- Household insecticides containing organic solvents and pesticides
- Fire and smoke detectors containing radioactive isotopes
- Nail polish remover containing organic solvents

Some cities and states are trying to reduce the amount of toxic substances released into the environment by providing special collection sites for materials such as used motor oil. But much more needs to be done.

LUST (leaking underground storage tanks) is another source of VOCs. The major culprit? Old, rusted gasoline storage tanks — especially from filling stations that have long been out of business. It takes less than a gallon of leaked gasoline to contaminate the water supply of a mid-sized town. Recent federal regulations have required the identification and replacement of leaking tanks, but abandoned service stations have become a major problem. It's estimated that as many as 200,000 tanks still need to be replaced.

The problem of hazardous materials in landfills and the contamination of our water supply prompted Congress to pass the Superfund program. This program was designed to identify and clean up potentially harmful landfills and dumps. Some progress has been made, but there may be thousands of dump sites that need to be cleaned at a monumental cost to the taxpayer.

The alternatives to landfills are recycling and incineration. Some of the material that commonly goes into a public landfill can be recycled — paper, glass, aluminum, and some plastics, for example — but more needs to be done. Incineration of some materials can be accomplished with the generation of electricity. Modern incineration produces very little air pollution.

Agricultural water pollution

Many types of water pollution are associated with the agricultural industry. For example, the excessive use of fertilizers, which contain nitrate and phosphate compounds, have caused a dramatic increase in the growth of algae and plants in lakes and streams. This increased growth may interfere with the normal cycles that occur in these aquatic systems, causing them to age prematurely — a process known as *eutrophication.*

In addition, pesticides used in treating crops may be released into the waterways. These pesticides, especially the organo-phosphorus ones, may undergo biological concentration (see "We really didn't get the lead out: Heavy metal contamination," earlier in this chapter). Many of us remember the reported impact of DDT on fish and birds. Because of the effects of DDT, the United States has banned its use, but it's still manufactured and sold overseas.

The release of soil and silt into the waterways is another form of pollution associated with the agricultural industry. The soil builds up in the water and interferes with the normal cycles of lakes and streams. It also carries agricultural chemicals into the waterways.

Polluting with heat: Thermal pollution

People usually think of things like lead, mercury, toxic organic compounds, and bacteria as being major pollutants. However, heat can also be a major pollutant. The solubility of a gas in a liquid *decreases* as the temperature increases (see Chapter 13 for more about the solubility of gases). This means that warm water doesn't contain as much dissolved oxygen as cool water. And how is this related to pollution? The amount of oxygen in water has a direct impact on aquatic life. The reduction of the dissolved oxygen content of water caused by heat is called *thermal pollution.*

Industries, especially those that generate electric power, use a tremendous amount of water to cool steam and condense it back to water. This water is normally taken from a lake or stream, used in the cooling process, and then returned to the same body of water. If the heated water is returned directly to the lake or stream, the increase in temperature may cause the oxygen levels to decrease below those required for the survival of certain types of fish. The increased temperature may also trigger or repress natural cycles of aquatic life, such as spawning.

Federal regulations prohibit the release of heated water back into lakes or streams. Industries cool the water by allowing it to remain in pools or running it over the outside of cooling towers. The cooling towers help the water release its heat to the atmosphere. Both of these methods, however, lose a lot of water to evaporation. (And believe me, there are some places in the United States that certainly don't need the increased humidity.)

Using up oxygen — BOD

If organic material (such as raw sewage, organic chemicals, or a dead cow) finds its way into the water, it decays. The decaying process is basically the oxidation of the organic compounds by *aerobic bacteria,* or oxygen-consuming bacteria, into simpler molecules such as carbon dioxide and water.

The process requires dissolved oxygen (DO) from the water. The amount of oxygen needed to oxidize the organic material is called the *biological oxygen demand (BOD),* and it's normally measured in parts per million (ppm) of oxygen needed. If the BOD is too high, too much dissolved oxygen is used, and there's not enough oxygen remaining for the fish. Fish kills occur, leading to an even higher BOD.

In extreme cases, there's not enough oxygen for the aerobic bacteria to survive, so another group of bacteria, *anaerobic bacteria,* assumes the job of decomposing the organic material. Anaerobic bacteria doesn't use oxygen in the water; instead, it uses oxygen that's in the organic compounds. Anaerobic bacteria reduces the waste instead of oxidizing it. (See Chapter 9 for a discussion of oxidation and reduction.) The bad news is that anaerobic bacteria decomposes organic matter into foul-smelling compounds such as hydrogen sulfide (H_2S), ammonia, and amines.

In order to stop overloading the BOD of the waterways, most chemical industries pretreat (normally with oxidization) their waste chemicals before releasing them into the water. Cities and towns do the same with their wastewater treatment plants.

Wastewater Treatment

The days when towns and cities in the United States could dump raw, untreated sewage into the waterways are largely over. Every once in a while, a treatment plant malfunctions or becomes overloaded due to some natural disaster and has to dump raw sewage, but those situations are few and far between.

It's not like that in the rest of the world. In South Asia and most of Africa, for example, very little of the sewage gets treated. But in the United States, sewage gets at least primary treatment; it often gets secondary and tertiary treatment, too. Figure 19-5 diagrams both the primary and secondary treatment of sewage.

Primary sewage treatment

In *primary sewage treatment,* raw sewage basically undergoes settling and filtration. The sewage first goes through a grate and screen system to remove large items. (I don't even want to talk about what those items are.) It then moves through a grit chamber where more material is filtered. Finally, it goes to a primary sedimentation tank, where the material is treated with solutions of aluminum sulfate and calcium hydroxide. The two solutions form aluminum hydroxide, a gelatinous precipitate (solid) that accumulates dirt and bacteria as it settles. This primary treatment removes about 50 to 75 percent of the solids, but it only reduces the biological oxygen demand (BOD) by about 30 percent.

If primary treatment is all that the wastewater undergoes, then sometimes chlorine is added to kill a majority of the bacteria before the wastewater is returned to the waterways. It still contains a high BOD, though. So if the waterway is a lake or slow-moving stream, then the high BOD causes problems — especially if a number of towns use the same type of sewage treatment. The problems can be prevented with secondary sewage treatment.

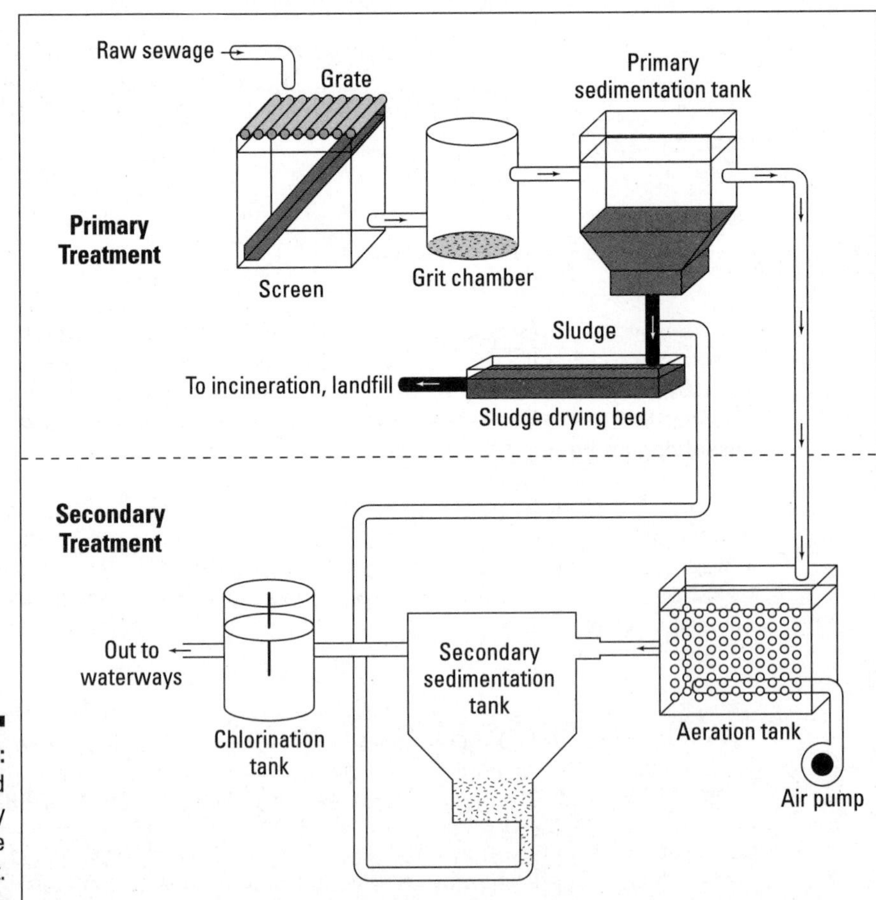

Figure 19-5: Primary and secondary sewage treatment.

Secondary sewage treatment

In *secondary sewage treatment,* bacteria and other microorganisms are given the opportunity to decompose the organic compounds in the wastewater. Because aerobic bacteria (oxygen-consuming bacteria) produces products that are less noxious than those produced by anaerobic bacteria (bacteria that uses oxygen in the organic compounds instead of oxygen in the water), the sewage is commonly aerated in order to provide the needed oxygen.

Both the primary and secondary processes produce a material called *sludge,* which is a mixture of particulate matter and living and dead microorganisms. The sludge is dried and disposed of by incineration or in a landfill. It can even be spread on certain types of cropland, where it acts as a fertilizer.

But even secondary treatment can't remove some substances that are potentially harmful to the environment. These substances include certain organic compounds, certain metals such as aluminum, and fertilizers such as phosphates and nitrates. Tertiary sewage treatment can be used to remove these substances.

Tertiary sewage treatment

Tertiary sewage treatment is essentially a chemical treatment that removes the fine particles, nitrates, and phosphates in wastewater. The basic procedure is adjusted for the specific substance to be removed. Activated charcoal filtration, for example, is used to remove most of the dissolved organic compounds. And alum ($Al_2(SO_4)_3$) is used to precipitate phosphate ions: by dissolving and freeing the aluminum cation.

$$Al^{3+} + PO_4^{3-} \rightarrow AlPO_4(s)$$

Ion exchange (Chapter 9), reverse osmosis (Chapter 11), and distillation (Chapter 15) are also occasionally used with this type of treatment. All these procedures are relatively expensive, though, so tertiary treatment isn't done unless really necessary.

Even after tertiary treatment is completed, the wastewater must still be disinfected before it's released back into the waterways. It's commonly disinfected by bubbling chlorine gas (Cl_2) into the water. Chlorine gas is an extremely powerful oxidizing agent, and it's very effective at killing the organisms responsible for cholera, dysentery, and typhoid. But the use of chlorine has come under question lately. If residual organic compounds are in the wastewater, they can be converted into chlorinated hydrocarbons. Several of these compounds have been shown to be carcinogenic. The levels of these compounds are being closely monitored during testing of wastewater.

Ozone (O_3) can also be used to disinfect wastewater. It's effective at killing viruses that chlorine can't kill. It's more expensive, however, and doesn't provide the residual protection against bacteria.

Drinking Water Treatment

One of the things we tend to take for granted is the availability of good drinking water. Most people of the world aren't as fortunate as we are.

The water is brought in from a lake, stream, or reservoir and initially filtered to remove sticks, leaves, dead fish, and such. The turbidity (haziness) that's commonly present in river or lake water is removed through treatment with a mixture of alum (aluminum sulfate) and lime (calcium hydroxide), which forms gelatinous aluminum hydroxide and traps the suspended solids. This is basically the same treatment used in wastewater treatment plants (see "Primary sewage treatment").

Then the water is filtered again to remove the solid mass of fine particles (called a *flocculate* or *floc*) leftover from the initial filtering treatment. Chlorine is added to kill any bacteria in the water. Then it's run through an activated charcoal filter that absorbs (collects on its surface) and removes substances responsible for taste, odor, and color. Fluoride may be added at this time to help prevent tooth decay. Finally, the purified water is collected in a holding tank, ready for your use.

Part V
The Part of Tens

In this part . . .

Chemistry thrives on discovery. And sometimes the discoveries are accidental. The first chapter in this part shows my top-ten favorite accidental discoveries. I also present ten great chemistry nerds and ten useful chemistry Web sites that you can use to expand your knowledge.

Chapter 20

Ten Serendipitous Discoveries in Chemistry

In This Chapter

- Discovering how some discoveries are made
- Looking at some famous people of science

This chapter presents ten stories of good scientists — individuals who discovered something they didn't know they were looking for.

Archimedes: Streaking Around

Archimedes was a Greek mathematician who lived in the third century B.C. I know this is supposed to be about scientists and not mathematicians, but back then, Archimedes was as close to a scientist as you could get.

Hero, the king of Syracuse, gave Archimedes the task of determining whether Hero's new gold crown was composed of pure gold, which it was supposed to be, or whether the jeweler had substituted an alloy and pocketed the extra gold. Now Archimedes knew about density, and he knew the density of pure gold. He figured that if he could measure the density of the crown and compare it to that of pure gold, he'd know whether the jeweler had been dishonest. But although he knew how to measure the weight of the crown, he couldn't figure out how to measure its volume in order to get the density.

Needing some relaxation, he decided to bathe at the public baths. As he stepped into the full tub and saw the water overflow, he realized that the volume of his body that was submerged was equal to the volume of water that overflowed. He had his answer for measuring the volume of the crown. He got so excited that he ran home naked through the streets, yelling "Eureka, eureka!" (I've found it!) And this method of determining the volume of an irregular solid is still used today. (By the way, the crown was an alloy, and the dishonest jeweler received swift justice.)

Vulcanization of Rubber

Rubber, in the form of latex, was discovered in the early 16th century in South America, but it gained little acceptance because it became sticky and lost its shape in the heat.

Charles Goodyear was trying to find a way to make the rubber stable when he accidentally spilled a batch of rubber mixed with sulfur on a hot stove. He noticed that the resulting compound didn't lose its shape in the heat. Goodyear went on to patent the *vulcanization process,* which is the chemical process used to treat crude or synthetic rubber or plastics to give them useful properties such as elasticity, strength, and stability.

Right- and Left-Handed Molecules

In 1884, the French wine industry hired Louis Pasteur to study a compound left on wine casks during fermentation — racemic acid. Pasteur knew that racemic acid was identical to tartaric acid, which was known to be *optically active* — that is, it rotated polarized light in one direction or another.

When Pasteur examined the salt of racemic acid under a microscope, he noticed that two types of crystals were present and that they were mirror images of each other. Using a pair of tweezers, Pasteur laboriously separated the two types of crystals and determined that they were both optically active, rotating polarized light the same amount but in different directions. This discovery opened up a new area of chemistry and showed how important molecular geometry is to the properties of molecules.

William Perkin and a Mauve Dye

In 1856, William Perkin, a student at The Royal College of Chemistry in London, decided to stay home during the Easter break and work in his lab on the synthesis of quinine. (I guarantee you that working in the lab isn't what my students do during their Easter break!)

During the course of his experiments, Perkin created some black gunk. As he was cleaning the reaction flask with alcohol, he noticed that the gunk dissolved and turned the alcohol purple — mauve, actually. This was the synthesis of the first artificial dye. As luck would have it, mauve was "in" that year, and this dye quickly became in great demand. So Perkin quit school and, with the help of his wealthy parents, built a factory to produce the dye.

Now if this were the entire story, it would've had little effect on history. However, the Germans saw the potential in this chemical industry and invested a great deal of time and resources in it. They began building up and investigating great supplies of chemical compounds, and soon Germany led the world in chemical research and manufacturing.

Kekule: The Beautiful Dreamer

Friedrich Kekule, a German chemist, was working on the structural formula of benzene, C_6H_6, in the mid-1860s. Late one night he was sitting in his apartment in front of a fire. He began dozing off and, in the process, saw groups of atoms dancing in the flames like snakes. Then, suddenly, one of the snakes reached around and made a circle, or a ring. This vision startled Kekule to full consciousness, and he realized that benzene had a ring structure. He stayed up all night working out the consequences of his discovery. Kekule's model for benzene paved the way for the modern study of aromatic compounds.

Discovering Radioactivity

In 1856, Henri Becquerel was studying the *phosphorescence* (glowing) of certain minerals when exposed to light. In his experiments, he'd take a mineral sample, place it on top of a heavily wrapped photographic plate, and expose it to strong sunlight.

He was preparing to conduct one of these experiments when a cloudy spell hit Paris. Becquerel put a mineral sample on top of the plate and put it in a drawer for safekeeping. Days later, he went ahead and developed the photographic plate and, to his surprise, found the brilliant image of the crystal, even though it hadn't been exposed to light. The mineral sample contained uranium. Becquerel had discovered radioactivity.

Finding Really Slick Stuff: Teflon

Roy Plunkett, a Du Pont chemist, discovered Teflon in 1938. He was working on the synthesis of new refrigerants. He had a full tank of tetrafluoroethylene gas delivered to his lab, but when he opened the valve, nothing came out. He wondered what had happened, so he cut the tank open. He found a white substance that was very slick and unreactive. The gas had polymerized into the substance now called Teflon. It was used during World War II to make gaskets and valves for the atomic bomb processing plant. After the war, Teflon finally made its way into the kitchen as a nonstick coating for frying pans.

Stick 'Em Up!! Sticky Notes

In the mid-1970s, a chemist by the name of Art Frey was working for 3M in its adhesives division. Frey, who sang in a choir, used little scraps of paper to keep his place in his choir book, but they kept falling out. At one point, he remembered an adhesive that had been developed but rejected a couple years earlier because it didn't hold things together well. The next Monday, he smeared some of this "lousy" adhesive on a piece of paper and found that it worked very well as a bookmark — and it peeled right off without leaving a residue. Thus was born those little yellow sticky notes you now find posted everywhere.

Growing Hair

In the late 1970s, Minoxidil, patented by Upjohn, was used to control high blood pressure. In 1980, Dr. Anthony Zappacosta mentioned in a letter published in *The New England Journal of Medicine* that one of his patients using Minoxidil for high blood pressure was starting to grow hair on his nearly bald head.

Dermatologists took note, and one — Dr. Virginia Fiedler-Weiss — crushed up some of the tablets and made a solution that some of her patients applied topically. It worked in enough cases that you now see Minoxidil as an over-the-counter hair-growth medicine.

Sweeter Than Sugar

In 1879, a chemist by the name of Fahlberg was working on a synthesis problem in the lab. He accidentally spilled on his hand one of the new compounds he'd made, and he noticed that it tasted sweet. (Wouldn't the government's Occupational Safety and Health Administration (OSHA) have loved that!) He called this new substance *saccharin*.

James Schlatter discovered the sweetness of *aspartame* while working on a compound used in ulcer research. He accidentally got a bit of one of the esters he'd made on his fingers. He noticed its sweetness when he licked his fingers while picking up a piece of paper.

Chapter 21

Ten Great Chemistry Nerds

In This Chapter

- Finding out how some scientists have influenced the field of chemistry
- Discovering some great discoveries
- Accepting the role of individuals in science

Science is a human enterprise. Scientists draw on their knowledge, training, intuition, and hunches. (And as I show you in Chapter 20, serendipity and luck come into play, also.) In this chapter, I introduce you to ten scientists who made discoveries that have advanced the field of chemistry. There are literally hundreds of choices, but these are mine for the top ten.

Amedeo Avogadro

In 1811, the Italian lawyer-turned-scientist Avogadro was investigating the properties of gases when he derived his now-famous law: Equal volumes of any two gases at the same temperature and pressure contain the same number of particles. From this law, the number of particles in a mole of any substance was determined. It was named Avogadro's number. Every chemistry student and chemist has Avogadro's number. Do you? See Chapter 10 if you don't.

Niels Bohr

Niels Bohr, a Danish scientist, used the observation that elements, if heated, emit energy in a set of distinct lines called a *line spectrum* to develop the idea that electrons can exist only in certain distinct, discrete energy levels in the atom. Bohr reasoned that the spectral lines resulted from the transition between these energy levels.

Bohr's model of the atom was the first to incorporate the idea of energy levels, a concept that's now universally accepted. For his work, Bohr received the Nobel Prize in 1922.

Marie (Madame) Curie

Madame Curie was born in Poland, but she did most of her work in France. Her husband, Pierre, was a physicist, and both were involved in the initial studies of radioactivity. Marie discovered that the mineral pitchblende contained two elements more radioactive than uranium. These elements turned out to be polonium and radium. Madame Curie coined the term *radioactivity.* She and her husband shared the Nobel Prize with Henri Becquerel in 1903.

John Dalton

In 1803, John Dalton introduced the first modern atomic theory. He developed the relationship between elements and atoms and established that compounds were combinations of elements. He also introduced the concept of atomic mass.

Unlike many other scientists who had to wait many years to see their ideas accepted, Dalton watched the scientific community readily embrace his theories. His ideas explained several laws that had already been observed and laid the groundwork for the quantitative aspects of chemistry. Not too bad for an individual who started teaching at the age of 12!

Michael Faraday

Michael Faraday made a tremendous contribution to the area of electrochemistry. He coined the terms *electrolyte, anion, cation,* and *electrode*. He established the laws governing electrolysis, discovered that matter has magnetic properties, and discovered several organic compounds, including benzene. He also discovered the magnetic induction effect, laying the groundwork for the electric motor and transformer. Without Faraday's discoveries, I may have had to write this book with a quill pen by lamplight.

Antoine Lavoisier

Antoine Lavoisier was a careful scientist who made detailed observations and planned his experiments. These characteristics allowed him to relate the process of respiration to the process of combustion. He coined the term *oxygen* for the gas that had been isolated by Priestly. His studies led him to the Law of Conservation of Matter, which states that matter can neither be created nor destroyed. This law was instrumental in helping Dalton develop his atomic theory. Lavoisier is sometimes called the father of chemistry.

Dmitri Mendeleev

Mendeleev is regarded as the originator of the periodic table, a tool that's indispensable in chemistry. He discovered the similarities in the elements while preparing a textbook in 1869. He found that if he arranged the then-known elements in order of increasing atomic weight, a pattern of repeating properties emerged. He used this concept of *periodic,* or repeating, properties to develop the first periodic table.

Mendeleev even recognized that there were holes in his periodic table where unknown elements should be found. Based on the periodic properties, Mendeleev predicted the properties of these elements. Later, when gallium and germanium were discovered, scientists found that these elements had properties that were very close to those predicted by Mendeleev.

Linus Pauling

If Lavoisier is the father of chemistry, then Linus Pauling is the father of the chemical bond. His investigations into the exact nature of how bonding occurs between elements were critical in the development of our modern understanding of bonding. His book, *The Nature of the Chemical Bond,* is a classic in the field of chemistry.

Pauling received a Nobel Prize in 1954 for his work in chemistry. He received another Nobel Prize, for peace, in 1963 for his work on limiting the testing of nuclear weapons. He's the only individual to receive two unshared Nobel Prizes. (He's also well known for his advocacy of using megadoses of Vitamin C to cure the common cold.)

Ernest Rutherford

Although Rutherford is perhaps better classified as a physicist, his work on the development of the modern model of the atom allows him to be placed with chemists.

He did some pioneer work in the field of radioactivity, discovering and characterizing alpha and beta particles — and received a Nobel Prize in chemistry for this work. But he's perhaps better known for his scattering experiments in which he realized that the atom was mostly empty space and that there had to be a dense, positive core at the center of the atom, which is now known as the nucleus. Inspired by Rutherford, many of his former students went on to receive their own Nobel Prizes.

Glenn Seaborg

Glenn Seaborg, while working on the Manhattan Project (that's the atomic bomb project), became involved in the discovery of several of the *transuranium elements* — elements with an atomic number greater than 92. Seaborg came up with the idea that the elements Th, Pa, and U were misplaced on the periodic table and should be the first three members of a new rare earth series under the lanthanides.

After World War II, he published his idea, which was met with strong opposition. He was told that he would ruin his scientific reputation if he continued to express his theory. But, as he said, he had no scientific reputation at that point. He persevered and was proven correct. And he received the Nobel Prize in 1951.

That Third-Grade Girl Experimenting with Vinegar and Baking Soda

This third-grade girl represents all those children out there who, each and every day, are making great discoveries. They explore the world around them with magnifying glasses. They pry open owl pellets and see what animals the owl ate. They experiment with magnets. They watch while baby animals are being born. They build vinegar-and-baking-soda volcanoes. They discover that science is fun.

They listen when they're told that scientists must keep on trying and that they must not give up. Their parents and teachers encourage them. They aren't told that they can't do science. If they are, they don't believe it.

They ask questions, lots of questions. They love the diversity of science, and they appreciate the beauty of science. They may never become professional scientists themselves, but they'll sit at the dining room table someday, laughing and joking with their kids as they help them build vinegar-and-baking-soda volcanoes.

Chapter 22

Ten Useful Chemistry Web Sites

In This Chapter

- Looking for Web sites related to chemistry
- Surfing the sites for the chemical information you want
- Navigating through additional links that you find

The Web is a gold mine of useful information, with a lot of fool's gold thrown in. In this chapter, I provide you with some good starting places to find cool chemical information. Because Web sites come and go, I'm not promising that all, or even any, of these sites will be there when you start looking for them, but I tried to choose ones that have a good chance of being around. (Even though I know people who wish the EPA would go away, it's quite likely that it'll still be there.) Use the additional links that you find at each site to branch out and fulfill your interest in chemistry.

American Chemical Society

`www.acs.org/portal/Chemistry`

The American Chemical Society (ACS) is the largest scientific organization in the world devoted to a single scientific discipline. Its Web site offers a wealth of information and links to other sites. A chemical search engine is available, along with a molecule of the week. It has links to chemistry-related news stories, an online store (if you simply *must* have that coffee cup made out of a beaker — I have two!), and links to various divisions within the American Chemical Society. You can even join the ACS online.

Material Safety Data Sheets

http://siri.uvm.edu/msds/

A Material Safety Data Sheet (MSDS) provides a wealth of information concerning the safe handling, spill control, health hazards, and so on, of a chemical. Most places are required by law to maintain an MSDS for every chemical in stock. At this site, you can search by name, product name, or CAS (Chemical Abstracts Service) registry number for an MSDS on a particular chemical and then print it out. This site also provides lots of information about chemicals, as well as links to other Internet MSDS sites and hazardous chemical reference sites.

U.S. Environmental Protection Agency

www.epa.gov/

This is the official EPA site. It has links to lots of information concerning the environment, hazardous chemicals, and such. You can browse through environmental laws and regulations, read the latest news articles concerning environmental issues, and check the status on toxic substances such as lead. The site even features a For Kids section that explains environmental issues in a way that's appropriate for children. You can order EPA publications online and obtain educational materials, too. It's your tax dollar — get the most out of it.

Chemistry.About.Com

www.chemistry.about.com/

Chemistry.about.com is a commercial Web site geared to a wide range of ages. It has a section on homework for those in high school and college, a description of scientific toys, links to companies that sell scientific equipment, and links to the specific areas of chemistry: organic, physical, analytical, and so on. One of the most useful sets of links is to chemistry clip art. Beware, though: This site can be frustrating because of the advertising windows that keep opening. If you can deal with that aspect, this is a good general site.

Webelements.com

`www.webelements.com/`

This great British Web site is set up as a periodic table. Want to know something about the element tantalum? What about osmium? Need the melting point of zinc? Just click on an element, and you get all its pertinent physical properties and common compounds — and in most cases, you even get a photograph of it. This Web site also keeps you up to date on the discovery of new elements. You can even print a copy of the periodic table. This site definitely belongs on your favorites list.

Plastics.com

`www.plastics.com/`

`Plastics.com` is primarily for those who want to know a little more about plastics or are in the plastics industry. This site has numerous news articles about current events in the industry, and you can get information about many different types of plastics. This is a great Web site for someone with an upcoming job interview at a plastics-related company.

Webbook

`http://webbook.nist.gov/`

This site from the National Institute of Standards and Technology is a great source of data on thousands of chemical compounds. You can access *thermochemical data* (data dealing with the relationship of heat in chemical reactions) on more than 6,000 organic and inorganic compounds, and you can get infrared, *mass spectrum* (the spectrum of a stream of gaseous ions separated according to their mass and charge — it's a way to identify the chemical constitution of a substance), and ultra violet and visible (*UV/Vis)* spectra (another way to determine the structure of molecules using energy) for numerous compounds. You can search the database by name, formula, CAS number, molecular weight, or numerous other properties.

ChemClub.com

www.chemclub.com/

ChemClub is a commercial site that provides access to a broad range of information about chemistry in general. It has numerous links to search engines, current events related to chemistry, and more. This well-developed site is useful to chemistry professionals or members of the general public who want an overall view of industrial chemistry.

Institute of Chemical Education

http://ice.chem.wisc.edu/

The Institute of Chemical Education (ICE) is associated with the University of Wisconsin. Its main emphasis is training in-service teachers. The institute's Web site has links to other chemical-education sites. Information concerning the institute's workshops and other presentations is available, too.

The Exploratorium

www.exploratorium.edu/

The Exploratorium, the "museum of science, art, and human perception," in San Francisco, California, is one of the foremost science museums in the country. This must-see Web site is geared toward kids and families. It's updated daily with news articles and current events. You can learn the science behind baseball, hockey, and other sports in its *Sports! Science* feature. The site presents a lot of activities for kids and adults in all the areas of science. The Exploratorium publications have been among my favorites for years.

Appendix A

Scientific Units: The Metric System

Much of the work chemists do involves measuring — things like the mass, volume, or length of a substance.

Because chemists must be able to communicate their measurements to other chemists all over the world, they need to speak the same measurement language. This language is the SI system of measurement (from the French *Systeme International*), commonly referred to as the *metric system.* There are actually minor differences between the SI and metric systems, but for the most part, they're interchangeable.

The SI system is a decimal system. There are base units for mass, length, volume, and so on, and there are prefixes that modify the base units. For example, *kilo-* means 1,000; a kilogram is 1,000 grams, and a kilometer is 1,000 meters.

This appendix lists the SI prefixes, base units for physical quantities in the SI system, and some useful SI-English conversions.

SI Prefixes

Use Table A-1 as a handy reference for the abbreviations and meanings of various SI prefixes.

Table A-1 SI (Metric) Prefixes

Prefix	*Abbreviation*	*Meaning*
Tera-	T	1,000,000,000,000 or 10^{12}
Giga-	G	1,000,000,000 or 10^{9}
Mega-	M	1,000,000 or 10^{6}

(continued)

Table A-1 (continued)

Prefix	*Abbreviation*	*Meaning*
Kilo-	K	1,000 or 10^3
Hecto-	H	100 or 10^2
Deka-	Da	10 or 10^1
Deci-	D	0.1 or 10^{-1}
Centi-	C	0.01 or 10^{-2}
Milli-	M	0.001 or 10^{-3}
Micro-	µ	0.000001 or 10^{-6}
Nano-	N	0.000000001 or 10^{-9}
Pico-	P	0.000000000001 or 10^{-12}

Length

The base unit for length in the SI system is the *meter.* The exact definition of meter has changed over the years, but it's now defined as the distance that light travels in a vacuum in $\frac{1}{299,792,458}$ of a second. Here are some SI units of length:

1 millimeter (mm) = 1,000 micrometers (µm)

1 centimeter (cm) = 10 millimeters (mm)

1 meter (m) = 100 centimeters (cm)

1 kilometer (km) = 1,000 meters (m)

Some common English to SI system length conversions are

1 mile (mi) = 1.61 kilometers (km)

1 yard (yd) = 0.914 meters (m)

1 inch (in) = 2.54 centimeters (cm)

Mass

The base unit for mass in the SI system is the *kilogram.* It's the weight of the standard platinum-iridium bar found at the International Bureau of Weights and Measures. Here are some SI units of mass:

1 milligram (mg) = 1,000 micrograms (μg)

1 gram (g) = 1,000 milligrams (mg)

1 kilogram (kg) = 1,000 grams (g)

Some common English to SI system mass conversions are

1 pound (lb) = 454 grams (g)

1 ounce (oz) = 28.4 grams (g)

1 pound (lb) = 0.454 kilograms (kg)

1 grain (gr) = 0.0648 grams (g)

1 carat (car) = 200 milligrams (mg)

Volume

The base unit for volume in the SI system is the *cubic meter.* But chemists normally use the *liter.* A liter is 0.001 m^3. Here are some SI units of volume:

1 milliliter (mL) = 1 cubic centimeter (cm^3)

1 milliliter (mL) = 1,000 microliters (μL)

1 liter (L) = 1,000 milliliters (mL)

Some common English to SI system volume conversions are

1 quart (qt) = 0.946 liters (L)

1 pint (pt) = 0.473 liter (L)

1 fluid ounce (fl oz) = 29.6 milliliters (mL)

1 gallon (gal) = 3.78 liters (L)

Temperature

The base unit for temperature in the SI system is *Kelvin.* Here are the three major temperature conversion formulas:

Celsius to Fahrenheit: $°F = (9/5)°C + 32$

Fahrenheit to Celsius: $°C = (5/9)(°F–32)$

Celsius to Kelvin: $°K = °C + 273$

Pressure

The SI unit for pressure is the *pascal*, where 1 pascal equals 1 newton per square meter. But pressure can also be expressed in a number of different ways, so here are some common pressure conversions:

1 millimeter of mercury (mm Hg) = 1 torr

1 atmosphere (atm) = 760 millimeters of mercury (mm Hg) = 760 torr

1 atmosphere (atm) = 29.9 inches of mercury (in Hg)

1 atmosphere (atm) = 14.7 pounds per square inch (psi)

1 atmosphere (atm) = 101 kilopascals (kPa)

Energy

The SI unit for energy (heat being one form) is the *joule,* but most folks still use the metric unit of heat, the *calorie*. Here are some common energy conversions:

1 calorie (cal) = 4.184 joules (J)

1 food Calorie (Cal) = 1 kilocalorie (kcal) = 4,184 joules (J)

1 British thermal unit (BTU) = 252 calories (cal) = 1,053 joules (J)

Appendix B

How to Handle Really Big or Really Small Numbers

Those who work in chemistry become quite comfortable working with very large and very small numbers. For example, when chemists talk about the number of sucrose molecules in a gram of table sugar, they're talking about a very large number. But when they talk about how much a single sucrose molecule weighs in grams, they're talking about a very small number. Chemists can use regular longhand expressions, but they become very bulky. It's far easier and quicker to use exponential or scientific notation.

Exponential Notation

In *exponential notation,* a number is represented as a value raised to a power of ten. The decimal point can be located anywhere within the number as long as the power of ten is correct. In *scientific notation,* the decimal point is always located between the first and second digit — and the first digit must be a number other than zero.

Suppose, for example, that you have an object that's 0.00125 meters in length. You can express that number in a variety of exponential forms:

0.00125 m = 0.0125×10^{-1} m, or 0.125×10^{-2} m, or 1.25×10^{-3} m, or 12.5×10^{-4} m, and so on.

All these forms are mathematically correct as numbers expressed in exponential notation. In scientific notation, the decimal point is placed so that there's one digit other than zero to the left of the decimal point. In the preceding example, the number expressed in scientific notation is 1.25×10^{-3} m. Most scientists automatically express numbers in scientific notation.

Here are some positive and negative powers of ten and the numbers they represent:

$1 \times 10^0 = 1$

$1 \times 10^1 = 10$

$1 \times 10^2 = 1 \times 10 \times 10 = 100$

$1 \times 10^3 = 1 \times 10 \times 10 \times 10 = 1{,}000$

$1 \times 10^4 = 1 \times 10 \times 10 \times 10 \times 10 = 10{,}000$

$1 \times 10^5 = 1 \times 10 \times 10 \times 10 \times 10 \times 10 = 100{,}000$

$1 \times 10^{10} = 1 \times 10 \times 10 \times 10 \times 10 \times 10 \times 10 \times 10 \times 10 \times 10 \times 10 = 10{,}000{,}000{,}000$

$1 \times 10^{-1} = 1/10 = 0.1$

$1 \times 10^{-2} = 1/100 = 0.01$

$1 \times 10^{-3} = 1/1000 = 0.001$

$1 \times 10^{-10} = 1/10{,}000{,}000{,}000 = 0.0000000001$

Addition and Subtraction

To add or subtract numbers in exponential or scientific notation, both numbers must have the same power of ten. If they don't, you must convert them to the same power. Here's an addition example:

$$(1.5 \times 10^3 \text{ g}) + (2.3 \times 10^2 \text{ g}) = (15 \times 10^2 \text{ g}) + (2.3 \times 10^2 \text{ g}) = 17.3 \times 10^2 \text{ g (exponential notation)} = 1.73 \times 10^3 \text{ g (scientific notation)}$$

Subtraction is done exactly the same way.

Multiplication and Division

To multiply numbers expressed in exponential notation, multiply the coefficients (the numbers) and add the exponents (powers of ten):

$$(9.25 \times 10^{-2} \text{ m}) \times (1.37 \times 10^{-5} \text{ m}) = (9.25 \times 1.37) \times 10^{(-2+-5)} = 12.7 \times 10^{-7} = 1.27 \times 10^{-6}$$

To divide numbers expressed in exponential notation, divide the coefficients and subtract the exponent of the denominator from the exponent of the numerator:

$$(8.27 \times 10^5 \text{ g}) \div (3.25 \times 10^3 \text{ mL}) = (8.27 \div 3.25) \times 10^{5-3} \text{ g/mL} = 2.54 \times 10^2 \text{ g/mL}$$

Raising a Number to a Power

To raise a number in exponential notation to a certain power, raise the coefficient to the power and then multiply the exponent by the power:

$$(4.33 \times 10^{-5}\text{cm})^3 = (4.33)^3 \times 10^{-5\times3} \text{ cm}^3 = 81.2 \times 10^{-15} \text{ cm}^3 = 8.12 \times 10^{-14} \text{ cm}^3$$

Using a Calculator

Scientific calculators take a lot of drudgery out of doing calculations. They enable you to spend more time thinking about the problem itself.

You can use a calculator to add and subtract numbers in exponential notation without first converting them to the same power of ten. The only thing you need to be careful about is entering the exponential number correctly. I'm going to show you how to do that right now:

I assume that your calculator has a key labeled *EXP*. The EXP stands for × *10*. After you press the EXP key, you enter the power. For example, to enter the number 6.25×10^3, you type 6.25, press the EXP key, and then type 3.

What about a negative exponent? If you want to enter the number 6.05×10^{-12}, you type 6.05, press the EXP key, type 12, and then press the +/- key.

When using a scientific calculator, *don't* enter the × *10* part of your exponential number. Press the EXP key to enter this part of the number.

Appendix C

Unit Conversion Method

You'll find that it's often unclear how to actually set up chemistry problems to solve them. A scientific calculator will handle the math, but it won't tell you what you need to multiply or what you need to divide.

That's why you need to know about the *unit conversion method,* which is sometimes called the *factor label method.* It will help you set up chemistry problems and calculate them correctly. Two basic rules are associated with the unit conversion method:

- **Rule 1:** Always write the unit and the number associated with the unit. Rarely in chemistry will you have a number without a unit. Pi is the major exception that comes to mind.
- **Rule 2:** Carry out mathematical operations *with* the units, canceling them until you end up with the unit you want in the final answer. In every step, you must have a correct mathematical statement.

How about an example so you can see those rules in action? Suppose that you have an object traveling at 75 miles per hour, and you want to calculate its speed in kilometers per second. The first thing you do is write down what you start with:

$$\frac{75\,\text{mi}}{1\,\text{hr}}$$

Note that per Rule #1, the equation shows the unit and the number associated with it.

Now convert miles to feet, canceling the unit of miles per Rule #2:

$$\frac{75\,\cancel{\text{mi}}}{1\,\text{hr}} \times \frac{5{,}280\,\text{ft}}{1\,\cancel{\text{mi}}}$$

Next, convert feet to inches:

$$\frac{75\,\cancel{\text{mi}}}{1\,\text{hr}} \times \frac{5{,}280\,\cancel{\text{ft}}}{1\,\cancel{\text{mi}}} \times \frac{12\,\text{in}}{1\,\cancel{\text{ft}}}$$

Convert inches to centimeters:

$$\frac{75\,\text{mi}}{1\,\text{hr}} \times \frac{5{,}280\,\text{ft}}{1\,\text{mi}} \times \frac{12\,\text{in}}{1\,\text{ft}} \times \frac{2.54\,\text{cm}}{1\,\text{in}}$$

Convert centimeters to meters:

$$\frac{75\,\text{mi}}{1\,\text{hr}} \times \frac{5{,}280\,\text{ft}}{1\,\text{mi}} \times \frac{12\,\text{in}}{1\,\text{ft}} \times \frac{2.54\,\text{cm}}{1\,\text{in}} \times \frac{1\,\text{m}}{100\,\text{cm}}$$

And convert meters to kilometers:

$$\frac{75\,\text{mi}}{1\,\text{hr}} \times \frac{5{,}280\,\text{ft}}{1\,\text{mi}} \times \frac{12\,\text{in}}{1\,\text{ft}} \times \frac{2.54\,\text{cm}}{1\,\text{in}} \times \frac{1\,\text{m}}{100\,\text{cm}} \times \frac{1\,\text{km}}{1000\,\text{m}}$$

Stop and stretch. Now you can start working on the denominator of the original fraction by converting hours to minutes:

$$\frac{75\,\text{mi}}{1\,\text{hr}} \times \frac{5{,}280\,\text{ft}}{1\,\text{mi}} \times \frac{12\,\text{in}}{1\,\text{ft}} \times \frac{2.54\,\text{cm}}{1\,\text{in}} \times \frac{1\,\text{m}}{100\,\text{cm}} \times \frac{1\,\text{km}}{1000\,\text{m}} \times \frac{1\,\text{hr}}{60\,\text{min}}$$

Next, convert minutes to seconds:

$$\frac{75\,\text{mi}}{1\,\text{hr}} \times \frac{5{,}280\,\text{ft}}{1\,\text{mi}} \times \frac{12\,\text{in}}{1\,\text{ft}} \times \frac{2.54\,\text{cm}}{1\,\text{in}} \times \frac{1\,\text{m}}{100\,\text{cm}} \times \frac{1\,\text{km}}{1000\,\text{m}} \times \frac{1\,\text{hr}}{60\,\text{min}} \times \frac{1\,\text{min}}{60\,\text{s}}$$

Now that you have the units of kilometers per second (km/s), you can do the math to get the answer:

0.033528 km/s

Note that you can round off your answer to the correct number of significant figures. Appendix D gives you details on how to do so, if you're interested. The rounded-off answer to this problem is

0.034 km/s or 3.4×10^{-2} km/s

Note that although the setup of the preceding example is correct, it's certainly not the only correct setup. Depending on what conversion factors you know and use, there may be many correct ways to set up a problem and get the correct answer.

Now I want to show you one more example to illustrate an additional point. Suppose that you have an object with an area of 35 inches squared, and you want to figure out the area in meters squared. Again, the first step is to write down what you start with:

$$\frac{35.0\,\text{in}^2}{1}$$

Now convert from inches to centimeters, but remember that you have to cancel inches *squared.* You must square the inches in the new fraction, and if you square the unit, you have to square the number also. And if you square the denominator, you have to square the numerator, too:

$$\frac{35.0\,\cancel{\text{in}}^2}{1}\frac{(2.54\,\text{cm})^2}{(1\,\cancel{\text{in}})^2}$$

Now convert from centimeters squared to meters squared in the same way:

$$\frac{35.0\,\cancel{\text{in}}^2}{1}\frac{(2.54\,\cancel{\text{cm}})^2}{(1\,\cancel{\text{in}})^2}\times\frac{(1\,\text{m})^2}{(100\,\cancel{\text{cm}})^2}$$

Now that you have the units of meters squared (m^2), you can do the math to get your answer:

$$0.0225806\ \text{m}^2$$

And if you want to round off your answer to the correct number of significant figures (see Appendix D for details), you get

$$0.023\ \text{m}^2 \text{ or } 2.3\times10^{-2}\ \text{m}^2$$

With a little practice, you'll really like and appreciate the unit conversion method. It got me through my introductory physics course!

Appendix D

Significant Figures and Rounding Off

Significant figures (no, I'm not talking about some supermodel) are the number of digits that you report in the final answer of the mathematical problem you are calculating. If I told you that one student determined the density of an object to be 2.3 g/mL and another student figured the density of the same object to be 2.272589 g/mL, I bet that you would naturally believe that the second figure was the result of a more accurate experiment. You might be right, but then again, you might be wrong. You have no way of knowing whether the second student's experiment was more accurate unless both students obeyed the significant figure convention. The number of digits that a person reports in his or her final answer is going to give a reader some information about how accurately the measurements were made. The number of the significant figures is limited by the accuracy of the measurement. This appendix shows you how to determine the number of significant figures in a number, how to determine how many significant figures you need to report in your final answer, and how to round your answer off to the correct number of significant figures.

Numbers: Exact and Counted Versus Measured

If I ask you to count the number of automobiles that you and your family own, you can do it without any guesswork involved. Your answer might be 0, 1, 2, or 10, but you would know exactly how many autos you have. Those are what are called *counted numbers.* If I ask you how many inches there are in a foot, your answer will be 12. That is an *exact number.* Another exact number is the number of centimeters per inch — 2.54. This number is exact by definition. In both exact and counted numbers, there is no doubt what the answer is. When you work with these types of numbers, you don't have to worry about significant figures.

Now suppose that I ask you and four friends to individually measure the length of an object as accurately as you possibly can with a meter stick. You then report the results of your measurements: 2.67 meters, 2.65 meters, 2.68 meters, 2.61 meters, and 2.63 meters. Which of you is right? You are all within experimental error. These measurements are measured numbers, and measured values always have some error associated with them. You determine the number of significant figures in your answer by your least reliable *measured* number.

Determining the Number of Significant Figures in a Measured Number

Here are the rules you need to determine the number of significant figures, or *sig. figs.*, in a measured number.

- **Rule 1:** All nonzero digits are significant. All numbers, one through nine, are significant, so 676 contains three sig. figs., 5.3×10^5 contains two, and 0.2456 contains four. The zeroes are the only numbers that you have to worry about.
- **Rule 2:** All of the zeroes between nonzero digits are significant. For example, 303 contains 3 sig. figs., 425003704 contains nine, and 2.037×10^{-6} contains four.
- **Rule 3:** All zeros to the left of the first nonzero digit are *not* significant. For example, 0.0023 contains two sig. figs. and 0.0000050023 contains five (expressed in scientific notation it would be 5.0023×10^{-6}).
- **Rule 4:** Zeroes to the right of the last nonzero digit are significant if there is a decimal point present. For example, 3030.0 contains five sig. figs., 0.000230340 contains six, and 6.30300×10^7 also contains six sig. figs.
- **Rule 5:** Zeroes to the right of the last nonzero digit are *not* significant if there is not a decimal point present. (Actually, a more correct statement is that I really don't know about those zeroes if there is not a decimal point. I would have to know something about how the value was measured. But most scientists use the convention that if there is no decimal point present, the zeroes to the right of the last nonzero digit are not significant.) For example, 72000 would contain two sig. figs and 50500 would contain three.

Reporting the Correct Number of Significant Figures

In general, the number of significant figures that you will report in your calculation will be determined by the *least* precise measured value. What values qualify as the least precise measurement will vary depending on the mathematical operations involved.

Addition and subtraction

In addition and subtraction, your answer should be reported to the number of decimal places used in the number that has the fewest decimal places. For example, suppose you're adding the following amounts:

2.675 g + 3.25 g + 8.872 g + 4.5675 g

Your calculator will show 19.3645, but you are going to round off to the hundredths place based on the 3.25, because it has the fewest number of decimal places. You then round the figure off to 19.36.

Multiplication and division

In multiplication and division, you can report the answer to the same number of significant figures as the number that has the *least* significant figures. Remember that counted and exact numbers don't count in the consideration of significant numbers. For example, suppose that you are calculating the density in grams per liter of an object that weighs 25.3573 (6 sig. figs.) grams and has a volume of 10.50 milliliters (4 sig. figs.). The setup looks like this:

(25.3573 grams/10.50 mL) × 1000 mL/L

Your calculator will read 2414.981000. You have six significant figures in the first number and four in the second number (the 1000 mL/L does not count because it is a exact conversion). You should have four significant figures in your final answer, so round the answer off to 2415 g/L. Only round off your final answer. Do not round off any intermediate values.

Rounding Off Numbers

When rounding off numbers, use the following rules:

- **Rule 1:** Look at the first number to be dropped; if it is 5 or greater, drop it and all the numbers that follow it, and increase the last retained number by 1. For example, suppose that you want to round off 237.768 to four significant figures. You drop the 6 and the 8. The 6, the first dropped number, is greater than 5, so you increase the retained 7 to 8. Your final answer is 237.8.
- **Rule 2:** If the first number to be dropped is less than 5, drop it and all the numbers that follow it, and leave the last retained number unchanged. If you're rounding 2.35427 to three significant figures, you drop the 4, the 2, and the 7. The first number to be dropped is 4, which is less than 5. The 5, the last retained number, stays the same. So you report your answer as 2.35.

Index

• A •

• B •

• C •

• D •

• E •

• F •

• G •

• H •

• I •

• J •

• K •

• L •

• M •

• N •

• O •

• P •

• Q •

• R •

• S •

• T •

• Y •

• Z •

FOR DUMMIES®

A world of resources to help you grow

TRAVEL

0-7645-5453-0

0-7645-5438-7

0-7645-5444-1

Also available:

America's National Parks For Dummies (0-7645-6204-5)
Caribbean For Dummies (0-7645-5445-X)
Cruise Vacations For Dummies 2003 (0-7645-5459-X)
Europe For Dummies (0-7645-5456-5)
Ireland For Dummies (0-7645-6199-5)
France For Dummies (0-7645-6292-4)
Las Vegas For Dummies (0-7645-5448-4)
London For Dummies (0-7645-5416-6)
Mexico's Beach Resorts For Dummies (0-7645-6262-2)
Paris For Dummies (0-7645-5494-8)
RV Vacations For Dummies (0-7645-5443-3)

EDUCATION & TEST PREPARATION

0-7645-5194-9

0-7645-5325-9

0-7645-5249-X

Also available:

The ACT For Dummies (0-7645-5210-4)
Chemistry For Dummies (0-7645-5430-1)
English Grammar For Dummies (0-7645-5322-4)
French For Dummies (0-7645-5193-0)
GMAT For Dummies (0-7645-5251-1)
Inglés Para Dummies (0-7645-5427-1)
Italian For Dummies (0-7645-5196-5)
Research Papers For Dummies (0-7645-5426-3)
SAT I For Dummies (0-7645-5472-7)
U.S. History For Dummies (0-7645-5249-X)
World History For Dummies (0-7645-5242-2)

HEALTH, SELF-HELP & SPIRITUALITY

0-7645-5154-X

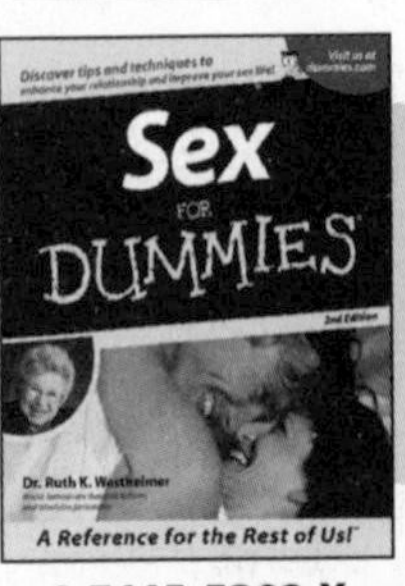

0-7645-5302-X

0-7645-5418-2

Also available:

The Bible For Dummies (0-7645-5296-1)
Controlling Cholesterol For Dummies (0-7645-5440-9)
Dating For Dummies (0-7645-5072-1)
Dieting For Dummies (0-7645-5126-4)
High Blood Pressure For Dummies (0-7645-5424-7)
Judaism For Dummies (0-7645-5299-6)
Menopause For Dummies (0-7645-5458-1)
Nutrition For Dummies (0-7645-5180-9)
Potty Training For Dummies (0-7645-5417-4)
Pregnancy For Dummies (0-7645-5074-8)
Rekindling Romance For Dummies (0-7645-5303-8)
Religion For Dummies (0-7645-5264-3)

Available wherever books are sold. Go to www.dummies.com or call 1-877-762-2974 to order direct